# PHYSIOLOGICAL CONTROL SYSTEMS

## IEEE PRESS SERIES IN BIOMEDICAL ENGINEERING

The focus of our series is to introduce current and emerging technologies to biomedical and electrical engineering practitioners, researchers, and students. This series seeks to foster interdisciplinary biomedical engineering education to satisfy the needs of the industrial and academic areas. This requires an innovative approach that overcomes the difficulties associated with the traditional textbook and edited collections.



**Books in the IEEE Press Series in Biomedical Engineering**

*TIME FREQUENCY AND WAVELETS IN BIOMEDICAL SIGNAL PROCESSING*
Edited by Metin Akay
1998 Hardcover 768 pp IEEE Order No. PC5619 ISBN 0-7803-1147-7

*PRINCIPLES OF MAGNETIC RESONANCE IMAGING: A SIGNAL PROCESSING PERSPECTIVE*
Zhi-Pei Liang and Paul C. Lauterbur
2000 Hardcover 416 pp IEEE Order No. PC5769 ISBN 0-7803-4723-4

*NEURAL NETWORKS AND ARTIFICIAL INTELLIGENCE FOR BIOMEDICAL ENGINEERING*
Donna L. Hudson and Maurice E. Cohen
2000 Hardcover 256 pp IEEE Order No. PC5675 ISBN 0-7803-3404-3

# PHYSIOLOGICAL CONTROL SYSTEMS

## *Analysis, Simulation, and Estimation*


**Michael C. K. Khoo**
*Biomedical Engineering Department*
*University of Southern California*



IEEE Engineering in Medicine and Biology Society, *Sponsor*

IEEE Press Series on Biomedical Engineering
Metin Akay, *Series Editor*

The Institute of Electrical and Electronics Engineers, Inc., New York

IEEE Press
445 Hoes Lane, P.O. Box 1331
Piscataway, NJ 08855-1331



IEEE Engineering in Medicine and Biology Society, *Sponsor*
EMB-S Liaison to IEEE Press, Metin Akay

Cover design: William T. Donnelly, *WT Design*

**Technical Reviewers**

Professor Giuseppe Baselli, *Polytechnic University Milano, Italy*
Dr. Susan M. Blanchard, *North Carolina State University*
Professor Sergio Cerutti, *Polytechnic University Milano, Italy*
Professor J. Andrew Daubenspeck, *Dartmouth Medical School*
Professor Alan V. Sahakian, *Northwestern University*
Dr. Hun Sun, *Drexel University*

To

Pam, Bryant, and Mason
and in memory of
John H. K. Khoo

This book and other books may be purchased at a discount
from the publisher when ordered in bulk quantities. Contact:

IEEE Press Marketing
Attn: Special Sales
445 Hoes Lane
Piscataway, NJ 08855-1331
Fax: (732) 981-9334

For more information about IEEE Press products, visit the
IEEE Press Home Page: http://www.ieee.org/organizations/pubs/press



Printed in the United States of America

10 9 8 7 6 5 4 3 2 1

**ISBN 0-7803-3408-6**
**IEEE Order Number PC5680**

**Library of Congress Cataloging-in-Publication Data**

Khoo, Michael C. K.
Physiological control systems : analysis, simulation, and estimation / Michael C. K. Khoo
p. cm. — (IEEE Press series on biomedical engineering)
Includes bibliographical references and index.
ISBN 0-7803-3408-6
1. Physiology–Mathematical models. 2. Biological control systems—Mathematical models, I. Title. II. Series.
QP33.6.M36K48 1999 99-27107
571.7′015118—dc 21 CIP

# Contents

# Preface

Control mechanisms provide the basis for the maintenance of homeostasis at all levels of organization in the hierarchy of living systems. As such, one's knowledge of the workings of a given biological system is incomplete unless one can arrive at some understanding of the regulatory processes that contribute to its natural operating characteristics. In order to attain this understanding, a conceptual model of the various interacting processes involved is necessary—but not sufficient. To determine whether one's model reflects the underlying reality, one has to make predictions with the model. However, more often than not, the factors in play are complex and dynamic, and the behavior of the model may depend strongly on the numerical values of certain key parameters. Under such circumstances, the rigorous framework provided by a quantitative approach becomes indispensable. Indeed, some of the most notable advances in the physiological sciences over the past several decades have been made through the application of quantitative models. Physiological control modeling also has been critical, directly or indirectly, for the development of many improved medical diagnostic techniques and new technological therapeutic innovations in recent times.

Because of its importance, the study of physiological control systems is generally incorporated, in one form or another, into the typical undergraduate biomedical engineering curriculum. Some programs offer courses that deal explicitly with physiological control systems, whereas in others, basic control theory may be incorporated into a course on quantitative physiology. Numerous high-quality research volumes on this subject have been published over the years, but there exist only a few books that most instructors would consider suitable for use as a comprehensive text in an upper-level undergraduate or first-year graduate course. Milsum's *Biological Control Systems Analysis* and Milhorn's *The Application of Control Theory to Physiological Systems* are two classic examples of possible texts, but these and the handful of other alternatives were published in the 1960s or early 1970s. The present book is aimed at filling this void. In addition to the classical methods that were covered in previous texts, this book also includes more contemporary topics and methodologies that continue to be employed in bioengineering research today.

The primary goals of this book are to highlight the basic techniques employed in control theory, systems analysis, and model identification, and to give the biomedical engineering student an appreciation of how these principles can be applied to better understand the processes involved in physiological regulation. The assumption made here is that the book would be used in a one-semester course on physiological control systems or physiological systems analysis taken by undergraduates in the junior or senior year. The book and its accompanying programs may also prove to be a useful resource for first-year biomedical engineering graduate students, as well as interested life science or clinical researchers who have had little formal training in systems or control theory. Throughout this book, I have emphasized the physiological applications of control engineering, focusing in particular on the analysis of feedback regulation. In contrast, the basic concepts and methods of control theory are introduced with little attention paid to mathematical derivations or proofs. For this reason, I would recommend the inclusion of a more traditional, engineering-oriented control theory course as a supplement to the material covered in this volume.

The book begins with a presentation of some historical perspectives, a discussion of the differences between technological and physiological control systems, and an introduction to the basic concepts of systems analysis and mathematical modeling. The subsequent five chapters cover classical control theory and its application to physiological systems. These begin in Chapter 2 with a tutorial on linear modeling. Here, we discuss generalized system properties, model analogs, lumped-parameter versus distributed-parameter models, and the utility of employing time-domain and frequency-domain descriptions of linear systems. In Chapter 3, we explore the techniques for steady-state analysis of physiological closed-loop systems. These problems traditionally have relied on graphical solution, as exemplified by the classic cardiac output–venous return analyses of Guyton and coworkers. Here, we also explore a decidedly more "modern" approach—that of employing computer analysis to solve the problems. Chapter 4 covers the transient response analyses of simple linear open-loop and closed-loop systems. We discuss the effect on system dynamics of "closing the loop," as well as changing the type of feedback from proportional to integral or derivative. In Chapter 5, we present the major methods for representing the frequency response of linear models, and also discuss the relationship between time-domain and frequency-domain approaches. Chapter 6 deals with the topic of stability, an issue of critical importance to physiological regulation. We discuss a range of techniques for assessing stability under conditions in which the assumption of linearity can be made. Chapter 7 addresses the problem of system identification, particularly in systems that operate under closed-loop conditions. Previous texts on physiological control have paid little attention to this important topic in spite of the fact that every bioengineering researcher has had to confront this problem at some point or other. In this chapter, we also discuss the related issues of parameter identifiability, sensitivity to noise, and input design. In Chapter 8, we move on to the application of "modern" control theory to physiological systems: these methods are based on the principle of optimization. We end this chapter with a brief exposition of how adaptive control theory may be applied in practice to regulate spontaneous fluctuations in a physiological signal. Chapter 9 presents a survey of some of the more common nonlinear analysis methodologies employed for investigating physiological systems. We recognize that this limited coverage, due to space constraints, does not do justice to the many other important nonlinear techniques and applications that have appeared in the research literature in the past two decades. Nevertheless, we believe we have included sufficient material to give the student a good "feel" of this area of study. Finally, we conclude the book in Chapter 10 with an examination of the potential mechanisms that could give rise to complex dynamic behavior in physiological control systems. These include spontaneous variability arising from structural nonlinearity in the system—the phenomenon of "chaos," interactions between different control systems, and nonstationarity in the system parameters. Throughout the book, I have attempted to include models of a wide range of physiological systems, although I cannot deny that there is somewhat of a bias towards my own favorite area of interest: cardiopulmonary control.

The ubiquity of personal computers among today's college students and the widespread use of MATLAB® and SIMULINK® (The Mathworks, Natick, MA) for systems analysis and simulation in the vast majority of engineering curricula have presented us with the opportunity to add a more "hands-on" flavor to the teaching of physiological control systems. As such, almost all chapters of the book include physiological applications that have accompanying MATLAB/SIMULINK simulation models. These, along with the computer exercises that accompany the end of each chapter, should aid the learning process by allowing the student the opportunity to explore "first-hand" the dynamics underlying the biological mechanisms being studied. This feature of the book is nonexistent in previous texts on physiological control systems. However, in incorporating this feature, we make the implicit assumption that the reader has some basic familiarity with MATLAB and/or SIMULINK. For the reader who has not used MATLAB or SIMULINK, it is fortunate that there are currently many "primers" on the subject that can be easily found in any academic bookstore. Appendix II lists and explains the MATLAB/SIMULINK functions used in the examples presented in this book, along with the names of the files that contain the model simulation/analysis programs discussed in the text. Details of how one can obtain these program files through the Internet are also given in the appendix.

Michael C. K. Khoo
*Biomedical Engineering Department*
*University of Southern California*

# Acknowledgments

The "birth" of this book took place some two and one-half years ago when Dr. Metin Akay, the editor in chief of this series of monographs, first encouraged me to submit a proposal to the IEEE Press. It has been a long, arduous journey since that moment—indeed, there were times when it felt as if the end would never be in sight. I thank Metin for his faith in me and for his continual support throughout the entire process. I am also thankful to John Griffin and Linda Matarazzo at IEEE Press for their assistance and cordiality. The anonymous reviewers who had to suffer through the drafts of various chapters in this book have my deepest gratitude for pointing out errors and for giving me valuable feedback. The collegial environment provided by my faculty colleagues, the staff, and students of the Biomedical Engineering Department at the University of Southern California (USC) has been an essential ingredient in the development of this book. I would be remiss if I did not mention my eternal indebtedness to the late Professor Fred Grodins, who was always my role model of what a true bioengineer should be like. Much of the intellectual stimulation that has led to the writing of this book is drawn from my own research work in the modeling of neurocardiorespiratory control, an activity sponsored by the National Institutes of Health through the Biomedical Simulations Resource (BMSR) at USC. I thank my colleagues and codirectors of the BMSR, Professors Vasilis Marmarelis and David D'Argenio, for their encouragement and support over the years. Finally, this endeavor certainly would not have been possible if I had not been blessed with the environment of love, joy, and understanding created so generously by those closest to me. My deepest feelings of appreciation go deservedly to my wife, Pam, and my sons, Bryant and Mason. It is to them that I dedicate this book.

Michael C. K. Khoo
*Biomedical Engineering Department*
*University of Southern California*

face of disturbances to the overall physiology of the organism. The maintenance of these relatively constant conditions was achieved by the organism itself. This observation so impressed him that he wrote: "It is the fixity of the '*millieu interieur*' which is the condition of free and independent life." He added further that "all the vital mechanisms, however varied they may be, have only one object, that of preserving constant the conditions of life in the internal environment." In the earlier half of this century, Harvard physiologist Walter Cannon refined Bernard's ideas further and demonstrated systematically these concepts in the workings of various physiological processes, such as the regulation of adequate water and food supply through thirst and hunger sensors, the role of the kidneys in regulating excess water and the maintenance of blood acid–base balance. He went on to coin the word *homeostasis* to describe the maintenance of relatively constant physiological conditions. However, he was careful to distinguish the second part of the term, i.e., "stasis", from the word "statics," since he was well aware that although the end result was a relatively unchanging condition, the coordinated physiological processes that produce this state are highly dynamic.

Armed with the tools of mathematics, Wiener in the 1940s explored the notion of feedback to a greater level of detail than had been done previously. Mindful that most physiological systems were nonlinear, he laid the foundation for modeling nonlinear dynamics from a Volterra series perspective. He looked into the problem of instability in neurological control systems and examined the connections between instability and physiological oscillations. He coined the word "cybernetics" to describe the application of control theory to physiology, but with the passage of time this term has come to take on a meaning more closely associated with robotics. The race to develop automatic airplane, radar, and other military control systems in the Second World War provided a tremendous boost to the development of control theory. In the postwar period, an added catalyst for even greater progress was the development of digital computers and the growing availability of facilities for the numerical solution of the complex control problems. Since then, research on physiological control systems has become a field of study in its own, with major contributions coming from a mix of physiologists, mathematicians, and engineers. Pioneering works in "modern" physiological control systems analysis include Grodins (1963), Clynes and Milsum (1970), Milsum (1966), Bayliss (1966), Guyton, Jones and Coleman (1973), Stark (1968) and Milhorn (1966), to name a few.

## 1.3 SYSTEMS ANALYSIS: FUNDAMENTAL CONCEPTS

Prior to analyzing or designing a control system, it is useful to define explicitly the major variables and structures involved in the problem. One common way of doing this is to construct a *block diagram*. The block diagram captures in schematic form the relationships among the variables and processes that comprise the control system in question. Figure 1.1 shows block diagrams that represent open-loop and closed-loop control systems in canonical form. Consider first the open-loop system (Figure 1.1a). Here, the *controller* component of the system translates the input ($r$) into a controller action ($u$), which affects the *controlled system* or "*plant*" thereby influencing the system output ($y$). At the same time, however, external disturbances ($x$) also affect plant behavior; thus, any changes in $y$ reflect contributions from both the controller and the external disturbances. If we consider this open-loop system in the context of our previous example of the heating system, the heater would be the controller and the room would represent the plant. Since the function of this control system is to regulate the temperature of the room, it is useful to define a *set-point*, which would

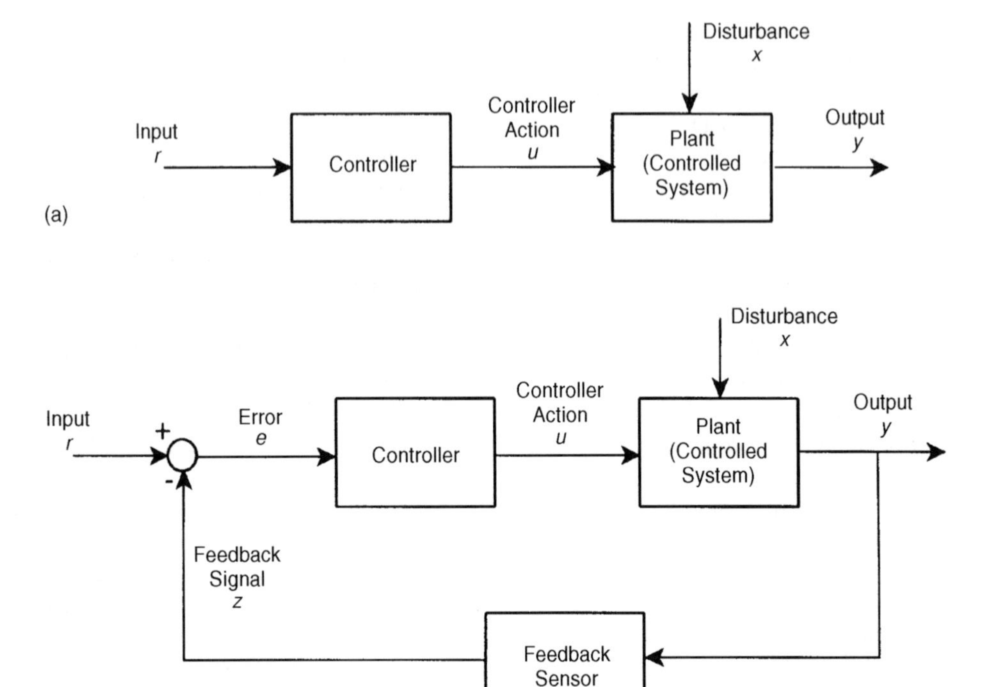


**Figure 1.1** Block diagrams of (a) an open-loop control system and a (b) closed-loop control system (bottom).

correspond to the desired room temperature. In the ideal situation of no fluctuations in external temperature (i.e., $x = 0$), a particular input voltage setting would place the room temperature exactly at the set-point. This input level may be referred to as the *reference input value*. In linear control systems analysis, it is useful (and often preferable from a computational viewpoint) to consider the system variables in terms of *changes* from these reference levels instead of their absolute values. Thus, in our example, the input ($r$) and controller action ($u$) would represent the deviation from the reference input value and the corresponding change in heat generated by the heater, respectively, while the output ($y$) would reflect the resulting change in room temperature. Due to the influence of changes in external temperature ($x$), $r$ must be adjusted continually to offset the effect of these disturbances on $y$.

As mentioned earlier, we can circumvent this limitation by "closing the loop." Figure 1.1b shows the closed-loop configuration. The change in room temperature ($y$) is now measured and transduced into the feedback signal ($z$) by means of a feedback sensor, i.e., the thermostat. The feedback signal is subsequently subtracted from the reference input and the error signal ($e$) is used to change the controller output. If room temperature falls below the set-point (i.e., $y$ becomes negative), the feedback signal ($z$) would also be negative. This feedback signal is subtracted from the reference input setting ($r = 0$) at the *mixing point* or *comparator* (shown as the circular object in Figure 1.1b), producing the error signal ($e$), which is used to adjust the heater setting. Since $z$ is negative, $e$ will be positive. Thus, the heater setting will be raised, increasing the flow of heat to the room and consequently raising the room temperature. Conversely, if room temperature becomes higher than its set-point, the feedback signal now becomes positive, leading to a negative error signal, which in turn lowers the heater output. This kind of closed-loop system is said to have *negative feedback*, since any changes in system output are compensated for by changes in controller action in the *opposite* direction.

Negative feedback is the key attribute that allows closed-loop control systems to act as regulators. What would happen if, rather than being subtracted, the feedback signal were to be *added* to the input? Going back to our example, if the room temperature were to rise and the feedback signal were to be added at the comparator, the error signal would become positive. The heater setting would be raised and the heat flow into the room would be increased, thereby increasing the room temperature further. This, in turn, would increase the feedback signal and the error signal, and thus produce even further increases in room temperature. This kind of situation represents the *runaway* effect that can result from *positive feedback*. In lay language, one would refer to this as a *vicious cycle* of events. Dangerous as it may seem, positive feedback is actually employed in many physiological processes. However, in these processes, there are constraints built in that limit the extent to which the system variables can change. Nevertheless, there are also many positive feedback processes (e.g., circulatory shock) that in extreme circumstances can lead to the shut-down of various system components, leading eventually to the demise of the organism.

## 1.4 PHYSIOLOGICAL CONTROL SYSTEMS ANALYSIS: A SIMPLE EXAMPLE

One of the simplest and most fundamental of all physiological control systems is the *muscle stretch reflex*. The most notable example of this kind of reflex is the *knee jerk*, which is used in routine medical examinations as an assessment of the state of the nervous system. A sharp tap to the patellar tendon in the knee leads to an abrupt stretching of the extensor muscle in the thigh to which the tendon is attached. This activates the muscle spindles, which are stretch receptors. Neural impulses, which encode information about the magnitude of the stretch, are sent along afferent nerve fibers to the spinal cord. Since each afferent nerve is synaptically connected with one motorneuron in the spinal cord, the motorneurons get activated and, in turn, send efferent neural impulses back to the same thigh muscle. These produce a contraction of the muscle, which acts to straighten the lower leg. Figure 1.2 shows the basic components of this reflex. A number of important features of this system should be highlighted. First, this and other stretch reflexes involve reflex arcs that are monosynaptic, i.e. only two neurons and one synapse are employed in the reflex. Other reflexes have at least one interneuron connecting the afferent and efferent pathways. Secondly, this closed-loop regulation of muscle length is accomplished in a completely involuntary fashion, as the name "reflex" suggests.

A third important feature of the muscle stretch reflex is that it provides a good example of negative feedback in physiological control systems. Consider the block diagram representation of this reflex, as shown in Figure 1.3. Comparing this configuration with the general closed-loop control system of Figure 1.1, one can see that the thigh muscle now corresponds to the plant or controlled system. The disturbance, $x$, is the amount of initial stretch produced by the tap to the knee. This produces a proportionate amount of stretch, $y$, in the muscle spindles, which act as the feedback sensor. The spindles translate this mechanical quantity into an increase in afferent neural traffic ($z$) sent back to the reflex center in the spinal cord, which corresponds to our controller. In turn, the controller action is an increase in efferent neural traffic ($u$) directed back to the thigh muscle, which subsequently contracts in order to

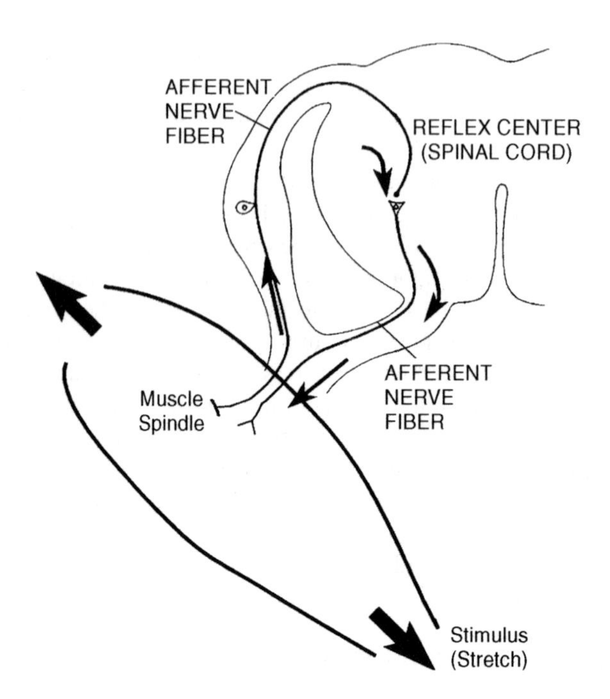


**Figure 1.2** Schematic illustration of the muscle stretch reflex. (Adapted from Vander *et al.*, 1997).

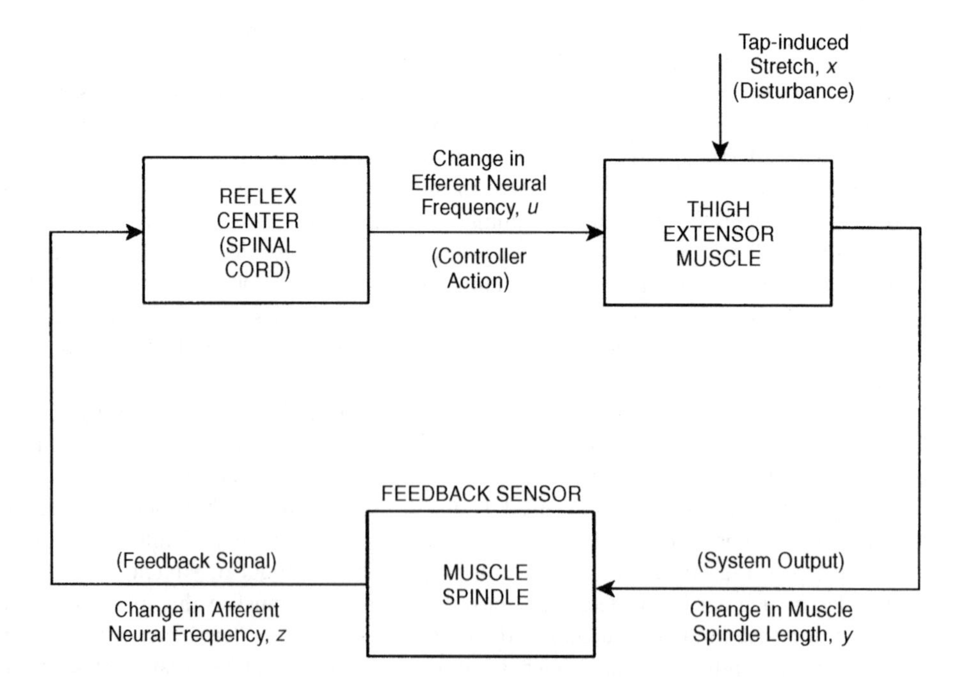


**Figure 1.3** Block diagram representation of the muscle stretch reflex.

offset the initial stretch. Although this closed-loop control system differs in some details from the canonical structure shown in Figure 1.1, it is indeed a *negative feedback* system, since the initial disturbance (tap-induced stretch) leads to a controller action that is aimed at *reducing* the effect of the disturbance.

## 1.5 DIFFERENCES BETWEEN ENGINEERING AND PHYSIOLOGICAL CONTROL SYSTEMS

While the methodology of systems analysis can be applied to both engineering and physiological control systems, it is important to recognize some key differences:

- An engineering control system is designed to accomplish a defined task, and frequently, the governing parameters have been fine-tuned extensively so that the system will perform its task in an "optimal" manner (at least, under the circumstances in which it is tested). In contrast, physiological control systems are built for versatility and may be capable of serving several different functions. For instance, although the primary purpose of the respiratory system is to provide gas exchange, a secondary but also important function is to facilitate the elimination of heat from the body. Indeed, some of the greatest advances in physiological research have been directed at discovering the functional significance of various biological processes.
- Since the engineering control system is synthesized by the designer, the characteristics of its various components are generally known. On the other hand, the physiological control system usually consists of components that are unknown and difficult to analyze. Thus, we are confronted with the need to apply system identification techniques to determine how these various subsystems behave before we are able to proceed to analyzing the overall control system.
- There is an extensive degree of *cross-coupling* or interaction among different physiological control systems. The proper functioning of the cardiovascular system, for instance, is to a large extent dependent on interactions with the respiratory, renal, endocrine, and other organ systems. In the example of the muscle stretch reflex considered earlier, the block diagram shown in Figure 1.3 oversimplifies the actual underlying physiology. There are other factors involved that we omitted and these are shown in the modified block diagram shown in Figure 1.4. First, some branches of the afferent nerves also synapse with the motorneurons that lead to other extensor muscles in the thigh that act synergistically with the primary muscle to straighten the lower leg. Secondly, other branches of the afferent nerves synapse with interneurons which, in turn, synapse with motorneurons that lead to the flexor or antagonist muscles. However, here, the interneurons introduce a polarity change in the signal, so that an increase in afferent neural frequency produces a decrease in the efferent neural traffic that is sent to the flexor muscles. This has the effect of relaxing the flexor muscles so that they do not counteract the activity of the extensor muscles.
- Physiological control systems, in general, are *adaptive*. This means that the system may be able to offset any change in output not only through feedback but also by allowing the controller or plant characteristics to change. As an example of this type of feature, consider again the operation of the muscle stretch reflex. While this reflex plays a protective role in regulating muscle stretch, it also can hinder the effects of

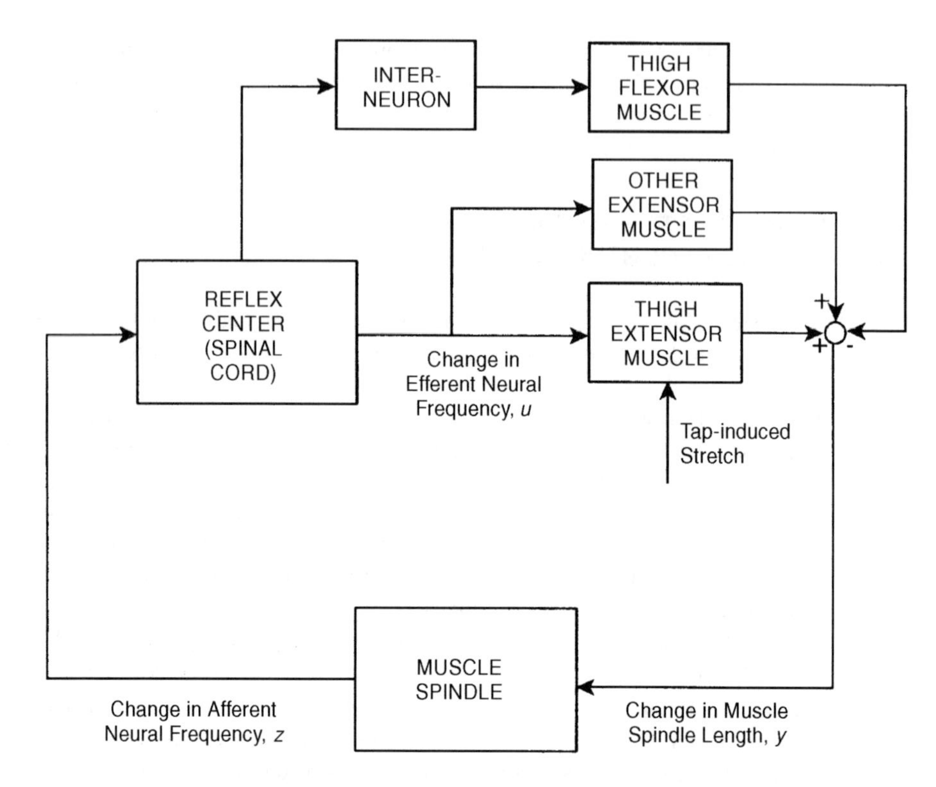


**Figure 1.4** Contributions of interrelated systems to the muscle stretch reflex.

voluntary control of the muscles involved. For instance, if one voluntarily flexed the knee, the stretch reflex, if kept unchanged, would come into play and this would produce effects that oppose the intended movement. Figure 1.5 illustrates the solution chosen by Nature to circumvent this problem. When the higher centers send signals down the alpha motorneurons to elicit the contraction of the flexor muscles and the relaxation of the extensor muscle, signals are sent simultaneously down the efferent gamma nerves that innervate the muscle spindles. These gamma signals produce in effect a resetting of the operating lengths of the muscle spindles, so that the voluntarily induced stretch in the extensor muscles is no longer detected by the spindles. Thus, by this clever, adaptive arrangement, the muscle stretch reflex is basically neutralized.

- At the end of Section 1.4, we alluded to another difference that may be found between physiological control systems and simpler forms of engineering control systems. In Figure 1.1, the feedback signal is explicitly subtracted from the reference input, demonstrating clearly the use of negative feedback. However, in the stretch reflex block diagram of Figure 1.3, the comparator is nowhere to be found. Furthermore, muscle stretch leads to an *increase* in both afferent and efferent neural traffic. So, how is negative feedback achieved? The answer is that negative feedback in this system is "built into" the plant characteristics: increased efferent neural input produces a *contraction* of the extensor muscle, thereby acting to counteract the initial stretch. This kind of *embedded* feedback is highly common in physiological systems.

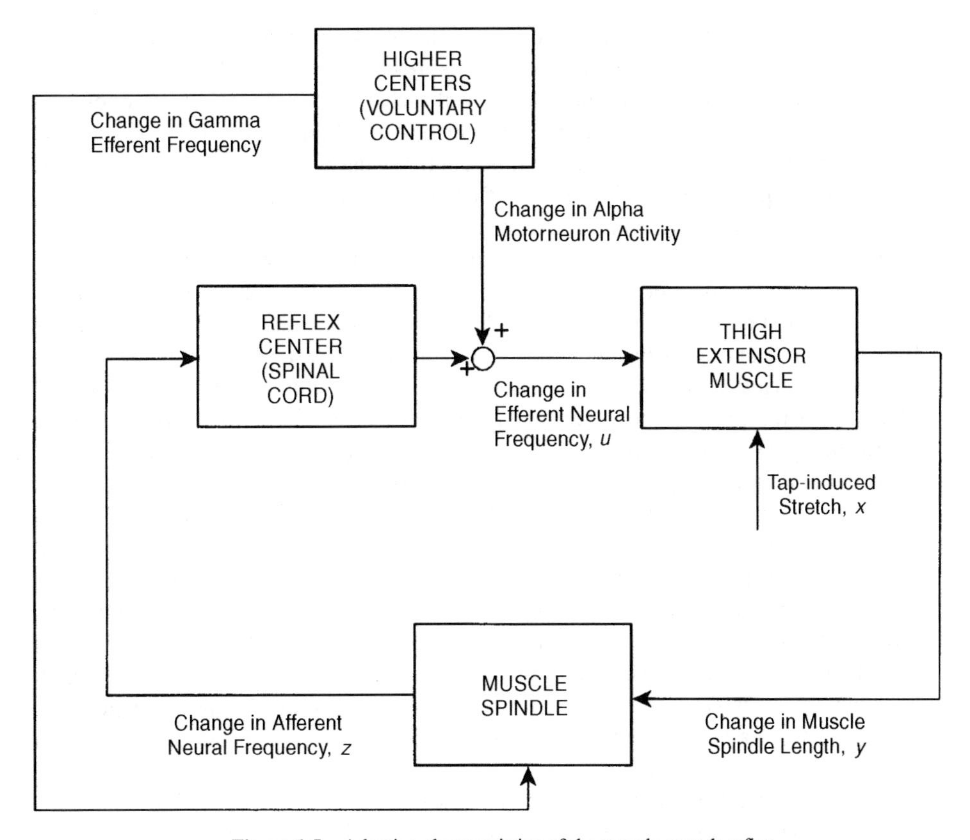


**Figure 1.5** Adaptive characteristics of the muscle stretch reflex.

- One final difference is that physiological systems are generally nonlinear, while engineering control systems can be linear or nonlinear. Frequently, the engineering designer prefers the use of linear system components since they have properties that are well-behaved and easy to predict. This issue will be revisited many times over in the chapters to follow.

## 1.6 THE SCIENCE (AND ART) OF MODELING

As we have shown, the construction of block diagrams is useful in helping us clarify in our own minds what key variables best represent the system under study. It is also helpful in allowing us to formalize our belief (which is usually based partly on other people's or our own observations and partly on intuition) about how the various processes involved are causally related. The block diagram that emerges from these considerations, therefore, represents a *conceptual model* of the physiological control system under study. However, such a model is limited in its abiliity to enhance our understanding or make predictions, since it only allows qualitative inferences to be made.

To advance the analysis to the next level involves the upgrading of the conceptual model into a *mathematical model*. The mathematical model allows us to make hypotheses about the contents in each of the "boxes" of the block diagram. For instance, in Figure 1.3, the box labeled "controller" will contain an expression of our belief about how the change in afferent neural frequency may be related to the change in efferent neural frequency. Is the relationship between afferent frequency and efferent frequency linear? If the changes in afferent frequency follow a particular time-course, what will the time-course of the response in efferent frequency be like? One way of answering these questions would be to isolate this part of the physiological control system and perform experiments that would allow us to measure this relationship. In this case, the relationship between the controller input and controller output is derived purely on the basis of observations, and therefore, it may take the form of a table or a curve best-fitted to the data. Alternatively, these data may already have been measured, and one can simply turn to the literature to establish the required input–output relationship. This kind of model assumes no internal structure and has been given a number of labels in the physiological control literature, such as *black-box*, *empirical*, or *nonparametric* model. Frequently, on the basis of previous knowledge, we also have some idea of what the underlying physical or chemical processes are likely to be. In such situations, we might propose an hypothesis that reflects this belief. On the basis of the particular physical or chemical laws involved, we would then proceed to derive an algebraic, differential, or integral equation that relates the "input" to the "output" of the system component we are studying. This type of model is said to possess an internal structure, i.e., it places some constraints on how the input may affect the output. As such, we might call this a *structural* or *gray-box* model. In spite of the constraints built into this kind of model, the range of input–output behavior that it is capable of characterizing can still be quite extensive, depending on the number of free parameters (or coefficients) it incorporates. For this reason, this type of model is frequently referred to as a *parametric* model.

Mathematical modeling may be seen as the use of a "language" to elaborate on the details of the conceptual model. However, unlike verbal languages, mathematics provides descriptions that are unambiguous, concise, and self-consistent. By being unambiguous, it enables different researchers to use and test the same model without being confused about the hypotheses built into the model. Since the equations employed in the model are based, at least in large part, on existing knowledge of the physiological processes in question, they also serve the useful purpose of archiving past knowledge and compressing all that information into a compact format. The inherent self-consistency of the model derives from the operational rules of mathematics, which provide a logical accounting system for dealing with the multiple system variables and their interactions with one another. On the other hand, the hypotheses embedded in some components of the model *are* only *hypotheses*, reflecting our best belief regarding the underlying process. More often than not, these are incorrect or oversimplistic. As a consequence, the behavior of the model may not reflect the corresponding reality. Yet, the power of the modeling process lies in its replication of the scientific method: the discrepancy between model prediction and physiological observation can be used as "feedback" to alert us to the inadequacies of one or more component hypotheses. This allows us to return to the model development stage once again in order to modify our previous assumptions. Subsequently, we would retest the revised model against experimental observations, and so the alternating process of induction and deduction continues until we are satisfied that the model can "explain" most of the observed behavior. Then, having arrived at a "good" model, we could venture to use this model to predict how the system might behave under experimental conditions that have not been employed previously. These predictions would serve as a guide for the planning and design of future experiments.

## BIBLIOGRAPHY

Adolph, E. Early concepts in physiological regulation. *Physiol. Rev.* **41**:737–770, 1961.

Bayliss, L.E. *Living Control Systems*. English University Press, London, 1966.

Cannon, W. *The Wisdom of the Body*. Norton, New York, 1939.

Clynes, M., and J.H. Milsum. *Biomedical Engineering Systems*. McGraw-Hill, New York, 1970.

Grodins, F.S. *Control Theory and Biological Systems*. Columbia University Press, New York, 1963.

Guyton, A.C., C.E. Jones, and T.G. Coleman. *Circulatory Physiology: Cardiac Output and Its Regulation*. Saunders, Philadelphia, 1973.

Jones, R.W. *Principles of Biological Regulation*. Academic Press, New York, 1973.

Milhorn, H.T. *The Application of Control Theory to Physiological Systems*. Saunders, Philadelphia, 1966.

Miller, J. *The Body in Question*. Random House, New York, 1978.

Milsum, J.H. *Biological Control Systems Analysis*. McGraw-Hill, New York, 1966.

Riggs, D.S. *Control Theory and Physiological Feedback Mechanisms*. Williams & Wilkins, Baltimore, 1970.

Stark, L. *Neurological Control Systems*. Plenum Press, New York, 1968.

Unschuld, P.U. *Medicine in China: A History of Ideas*. Berkeley, London, 1985.

Vander, A.J., J.H. Sherman, and D.S. Luciano. *Human Physiology: The Mechanisms of Body Function*, 7th ed. McGraw-Hill, New York, 1997.

Wiener, N. *Cybernetics: Control and Communication in the Animal and the Machine*. Wiley, New York, 1961.

## PROBLEMS

Based on the verbal descriptions of the following physiological reflex systems, construct block diagrams to represent the major control mechanisms involved. Clearly identify the physiological correlates of the controller, the plant, and the feedback element, as well as the controlling, controlled, and feedback variables. Describe how negative (or positive) feedback is achieved in each case.

**P1.1.** The Bainbridge reflex is a cardiac reflex that aids in the matching of cardiac output (the flow rate at which blood is pumped out of the heart) to venous return (the flow rate at which blood returns to the heart). Suppose there is a transient increase in the amount of venous blood returning to the right atrium. This increases blood pressure in the right atrium, stimulating the atrial stretch receptors. As a result, neural traffic in the vagal afferents to the medulla is increased. This, in turn, leads to an increase in efferent activity in the cardiac sympathetic nerves as well as a parallel decrease in efferent parasympathetic activity. Consequently, both heart rate and cardiac contractility are increased, raising cardiac output. In this way, the reflex acts like a servomechanism, adjusting cardiac output to track venous return.

**P1.2.** The pupillary light reflex is another classic example of a negative feedback control system. In response to a decrease in light intensity, receptors in the retina transmit neural impulses at a higher rate to the pretectal nuclei in the midbrain, and subsequently, to the Edinger–Westphal nuclei. From the Edinger–Westphal nuclei, a change in neural traffic down the efferent nerves back to the eyes leads to a relaxation of the sphincter muscles and contraction of the radial dilator muscles that together produce an increase in pupil area, which increases the total flux of light falling on the retina.

**P1.3.** The regulation of water balance in the body is intimately connected with the control of sodium excretion. One major mechanism of sodium reabsorption involves the renin–angiotensin–aldosterone system. Loss of water and sodium from the body, e.g., due to diarrhea, leads to a drop in plasma volume, which lowers mean systemic blood pressure. This stimulates the venous and arterial baroreflexes, which produce an increase in activity of the renal sympathetic nerves, in turn stimulating the release of renin by the kidneys into the circulation. The increase in plasma renin concentration leads to an increase in plasma angiotensin, which stimulates the release of aldosterone by the adrenal cortex. Subsequently, the increased plasma aldosterone stimulates the reabsorption of sodium by the distal tubules in the kidneys, thereby increasing plasma sodium levels.

**P1.4.** The control system that regulates water balance is intimately coupled with the control of sodium excretion. When sodium is reabsorbed by the distal tubules of the kidneys, water will also be reabsorbed if the permeability of the tubular epithelium is lowered. This is achieved in the following way. When there is a drop in plasma volume, mean systemic pressure decreases, leading to a change in stimulation of the left atrial pressure receptors. The latter send signals to a group of neurons in the hypothalamus, increasing its production of vasopressin or antidiuretic hormone (ADH). As a result, the ADH concentration in blood plasma increases, which leads to an increase in water permeability of the kidney distal tubules and collecting ducts.

**P1.5.** Arterial blood pressure is regulated by means of the baroreceptor reflex. Suppose arterial blood pressure falls. This reduces the stimulation of the baroreceptors located in the aortic arch and the carotid sinus, which lowers the rate at which neural impulses are sent along the glossopharyngeal and vagal afferents to the autonomic centers in the medulla. Consequently, sympathetic neural outflow is increased, leading to an increase in heart rate and cardiac contractility, as well as vasoconstriction of the peripheral vascular system. At the same time, a decreased parasympathetic outflow aids in the heart rate increase. These factors together act to raise arterial pressure.

**P1.6.** A prolonged reduction in blood pressure due to massive loss of blood can lead to "hemorrhagic shock" in which the decreased blood volume lowers mean systemic pressure, venous return and thus, cardiac output. Consequently, arterial blood pressure is also decreased, leading to decreased coronary blood flow, reduction in myocardial oxygenation, loss in the pumping ability of the heart, and therefore, further reduction in cardiac output. The decreased cardiac output also leads to decreased oxygenation of the peripheral tissues, which can increase capillary permeability, thereby allowing fluid to be lost from the blood to the extravascular spaces. This produces further loss of blood volume and mean systemic pressure, and therefore, further reduction in cardiac output.

# 2

# Mathematical Modeling

## 2.1 GENERALIZED SYSTEM PROPERTIES

In this chapter, we will review the basic concepts and methods employed in the development of "gray-box" models. Models of very different systems often contain properties that can be characterized using the same mathematical expression. The first of these is the *resistive* property. Everyone is familiar with the concept of electrical resistance ($R$), which is defined by Ohm's law as

$$V = RI \tag{2.1}$$

where $V$ is the voltage or driving potential across the resistor and $I$ represents the current that flows through it. Note that $V$ is an "across" variable and may be viewed as a measure of "effort," $I$, on the other hand, is a "through" variable and represents a measure of "flow." Thus, if we define the generalized "effort" variable, $\psi$, and the generalized "flow" variable, $\zeta$, Ohm's law becomes:

$$\psi = R\zeta \tag{2.2}$$

where $R$ now represents a generalized resistance. Figure 2.1 shows the application of this concept of generalized resistance to different kinds of systems. In the mechanical dashpot, when a force $F$ is applied to the plunger (and, of course, an equal and opposite force is applied to the dashpot casing), it will move with a velocity $v$ that is proportional to $F$. As illustrated in Figure 2.1a, this relationship takes on the same form as the generalized Ohm's law (Equation (2.2)), when $F$ and $v$ are made to correspond to $\psi$ and $\zeta$, respectively. The constant of proportionality, $R_m$, which is related to the viscosity of the fluid inside the dashpot, provides a measure of "mechanical resistance." In fact, $R_m$ determines the performance of the dashpot as a shock absorber and is more commonly known as the "damping coefficient." In fluid flow, the generalized Ohm's law assumes the form of Poiseuille's law, which states that the volumetric flow of fluid ($Q$) through a rigid tube is proportional to the pressure difference ($\Delta P$) across the two ends of the tube. This is illustrated in Figure 2.1b. Poiseuille further showed that the fluid resistance, $R_f$, is directly related to the viscosity of the fluid and the length of the tube, and inversely proportional to the square of the tube cross-sectional area. In Fourier's law of thermal transfer, the flow of heat, $Q$, conducted through a given material is directly proportional to the temperature difference that exists across the material (Figure 2.1c). Thermal resistance, $R_t$, can be shown to be inversely related to the thermal conductivity of the material. Finally, in chemical systems, the flux, $Q$, of a given chemical species across a permeable membrane separating two fluids with different species concentrations is proportional to the concentration difference, $\Delta\phi$ (Figure 2.1d). This is known as Fick's law of diffusion. The diffusion resistance, $R_c$, is inversely proportional to the more commonly used parameter, the membrane diffusivity.

The second generalized system property is that of *storage*. In electrical systems, this takes the form of capacitance, defined as the amount of electrical charge ($q$) stored in the capacitor per unit voltage ($V$) that exists across the capacitor:

$$C = \frac{q}{V} \tag{2.3}$$

F

$F = vR_m$

v

(a)

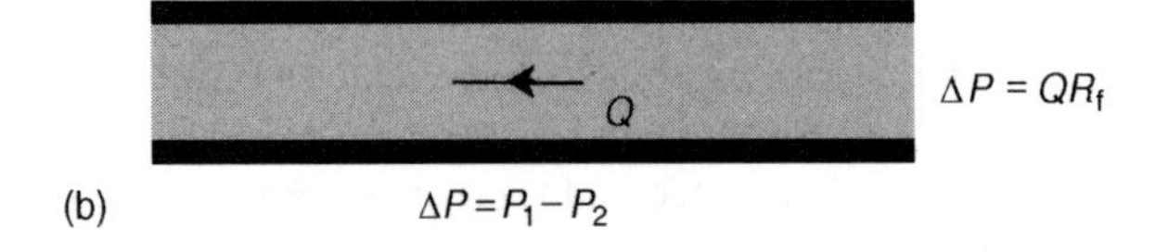


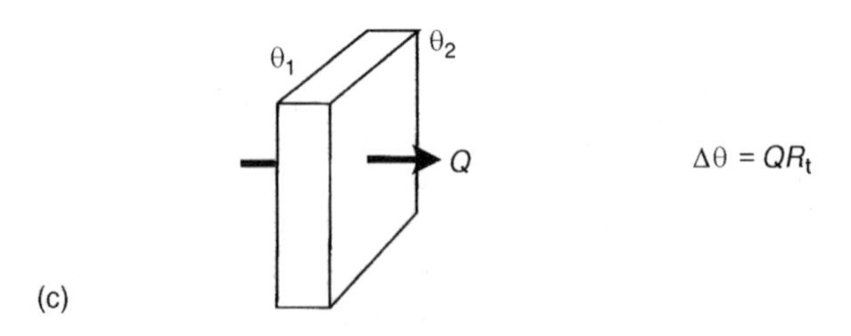


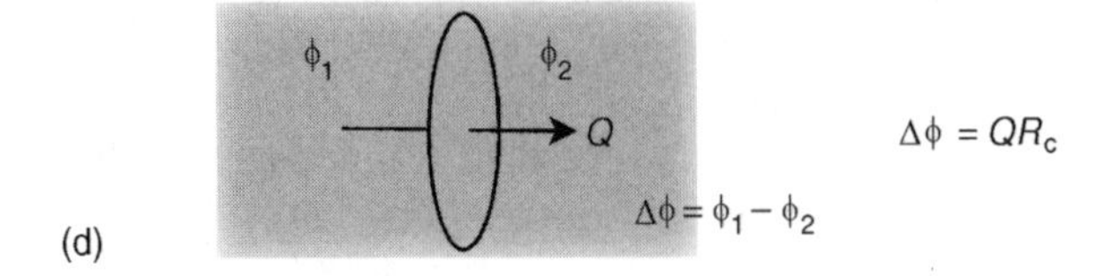


**Figure 2.1** "Resistance" in (a) mechanical, (b) fluidic, (c) thermal, and (d) chemical systems.

Note that $q$ represents the accumulation of all electric charge delivered via current flow to the capacitor, so that the following relationship exists between $q$ and $I$:

$$q = \int_0^t I \, dt \tag{2.4}$$

Thus, using Equation (2.4) in Equation (2.3) and rewriting the result in generalized form, we obtain the following expression:

$$\psi = \frac{1}{C} \int_0^t \zeta \, dt \tag{2.5}$$

For mechanical systems, this storage property takes the form of "compliance," as in the case of the elastic spring shown in Figure 2.2a. For a given applied force, the mechanical compliance determines the extent to which the spring will be extended or compressed. This property is also inversely related to the stiffness or elastic modulus of the spring: the more compliant the spring, the less stiff it will be, and the more it will extend for a given applied tension. Similarly, in fluidic systems, as represented in the example of the fluid-filled balloon in Figure 2.2b, compliance determines the volume by which the balloon will expand or contract per unit change in applied pressure. The compliance here is determined primarily by the elasticity of the balloon material: the stiffer the material, the less "compliant" the balloon. A much smaller contribution to the compliance arises from the compressibility of the fluid inside the balloon. In thermal systems, the "thermal mass" represents the amount of heat stored in a certain medium per unit difference in temperature that exists between the interior and exterior (Figure 2.2c). This thermal capacitance depends on the dimensions and specific heat of the medium in question. Finally, in chemical systems, the storage property is

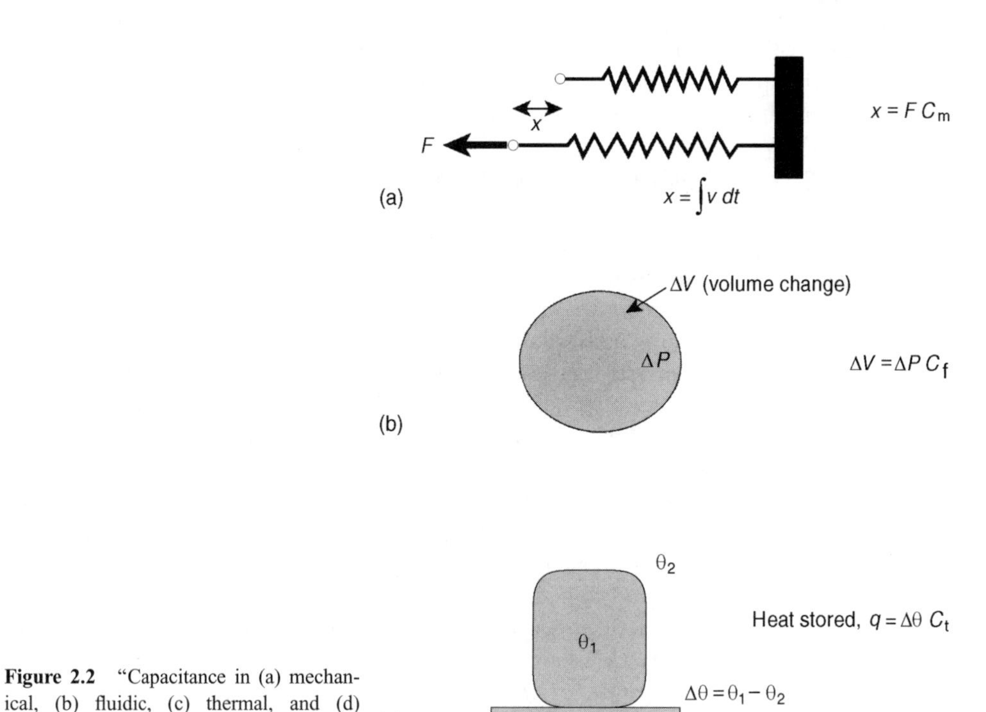


**Figure 2.2** "Capacitance in (a) mechanical, (b) fluidic, (c) thermal, and (d) chemical systems.

represented by the total volume of the fluid in which the chemical species exists (Figure 2.2d), i.e., for given volume, the total mass of the chemical species present is proportional to its concentration. This property, in fact, is used in the *definition* of concentration.

The resistive and storage properties also represent elements through which energy is dissipated or stored, respectively. In the context of electrical resistance, note that the product of voltage and current yields power. In mechanical systems, force multiplied by velocity also yields power. Similarly, in fluidic systems, power is defined as the product of pressure and flow rate. Thus, in any general system, energy is *dissipated* when effort is applied to produce flow through the resistive element. On the other hand, the storage element allows the accumulation of "static" or *potential energy.* For instance, in the mechanical spring, application of force $F$ which produces an extension of $x$ length units will lead to the storage of $Fx$ units of potential energy. Similarly, in the balloon system, potential energy is stored when the balloon is expanded with the application of internal pressure.

The final generalized system property, *inertance*, allows the storage of *kinetic energy* in electrical systems. This property is also known as *inductance* ($L$), which is defined as the voltage required to produce a given rate of change of electrical current:

$$V = L \frac{dI}{dt} \tag{2.6}$$

Replacing $V$ and $I$ by the corresponding generalized system variables, $\psi$ and $\zeta$, we have:

$$\psi = L \frac{d\zeta}{dt} \tag{2.7}$$

Note that in the context of mechanical systems, $\psi$ becomes $F$ (force) and $\zeta$ becomes velocity, so that Equation (2.7) becomes Newton's Second Law of Motion: force equals mass times acceleration. Thus, in this case, the inertance is simply the mass of the system. Inertance is also present in fluidic systems, so that fluid acceleration is proportional to the pressure differential applied. On the other hand, there is no element that represents inertance in thermal and chemical systems; kinetic energy storage does not exist in these systems.

## 2.2 MODELS WITH COMBINATIONS OF SYSTEM ELEMENTS

An assumption implicit in the previous section is that the system properties are time-invariant and independent of the values of the generalized system variables. In other words, the three basic model elements are *linear.* In reality, this will not be the case. An electrical resistor will heat up as the current that passes through it increases, thereby raising its resistance. Similarly, fluid resistance remains relatively constant only under conditions of steady, laminar flow; as flow increases and becomes turbulent, the resistance becomes a function of flow itself. Thus, as the effect of nonlinearities increases, the similarities in behavior among these different systems will be progressively reduced. Nevertheless, in scientific exploration, we always have to start somewhere—and past scientific history has shown that it is wise to begin with the simplest model. Therefore, linear analysis plays an important role in physiological systems modeling by allowing us to obtain a first approximation to the underlying reality.

We will proceed with our discussion of linear analysis by considering how we can derive the overall model equations from various combinations of the three basic types of system elements. Figure 2.3 shows a simple "circuit" linking three generalized system elements in series and parallel. The node labeled "0" represents the reference level of the "effort" variable $\psi$ to which all other nodes are compared; this is set equal to zero. In electrical systems, this node would be called "electrical ground." $\psi_a$ and $\psi_c$ represent the values of the *across-variables* at nodes a and b, respectively, relative to node 0. $\zeta_1$, $\zeta_2$ and $\zeta_3$ represent the values of the *through-variables* that pass through the elements E1, E2, and E3, respectively. The mathematical relationships among these variables can be derived by applying two fundamental physical principles:

1. The algebraic sum of the "across-variable" values around any closed circuit must equal zero. Thus, in the loop a–b–0–a in Figure 2.3, we have

$$(\psi_a - \psi_b) + (\psi_b - 0) + (0 - \psi_a) = 0 \tag{2.8}$$

2. The algebraic sum of all "through-variable" values into a given node must equal zero. In Figure 2.3, the only node where this rule will apply is node b:

$$\zeta_1 + (-\zeta_2) + (-\zeta_3) = 0 \tag{2.9}$$

The preceding two generalized principles take the form of Kirchhoff's laws when applied to electrical systems. The first (voltage) law appears trivial in the form presented, but in order to apply this law, each of the component terms in the left-hand side of Equation (2.8) has to be expressed as a function of the corresponding through-variables and the system elements. The second (current) law is essentially a statement of the conservation of mass principle. In Equation (2.9), $\zeta_2$ and $\zeta_3$ are given negative signs since, in Figure 2.3, they assume a flow direction that points away from node b. Although each of the boxes labeled E1, E2, and E3 was intended to represent one of the three basic system elements, each in general could also contain a network within itself; within each network, the above two laws would still apply. Thus, starting from the basic system elements, we can construct progressively more complex models by connecting these elements in either series or parallel configurations. And by using the generalized Kirchhoff's laws, it is possible to deduce mathematical expressions that characterize the overall properties of these synthesized networks. Figure 2.4 illustrates the resultant system properties that emerge from simple combinations of resistive and storage elements for electrical, fluidic, thermal, and chemical systems. These results are easily derived using basic circuit analysis.

The expressions that relate combined resistances and compliances to their respective component elements are somewhat different for mechanical systems, as Figure 2.5 shows. Consider case (a), where the two mechanical dashpots (resistances) are placed in parallel. This parallel configuration constrains the motion of the two dashpot plungers by requiring them to

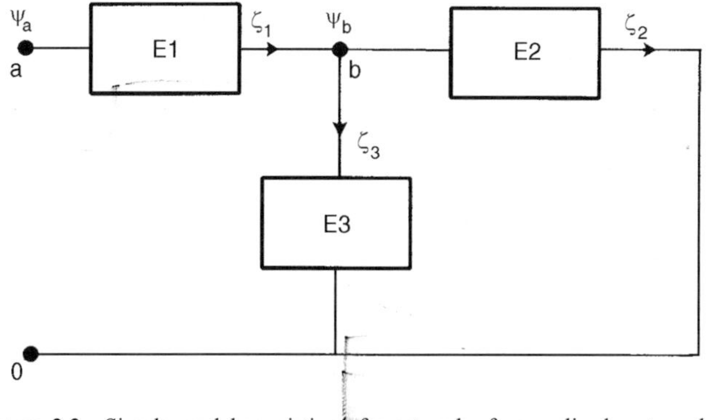


**Figure 2.3** Simple model consisting of a network of generalized system elements.

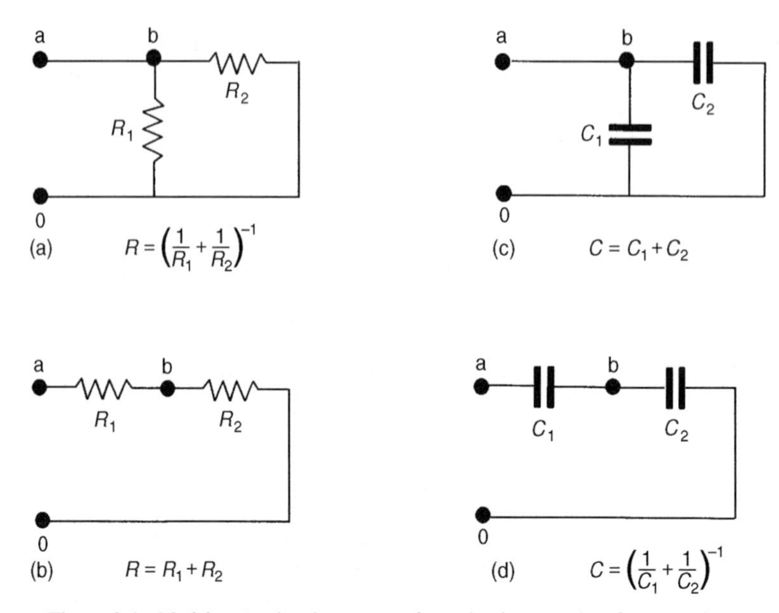


**Figure 2.4** Model properties that emerge from simple networks of system elements in electrical, fluidic, thermal, and chemical systems.

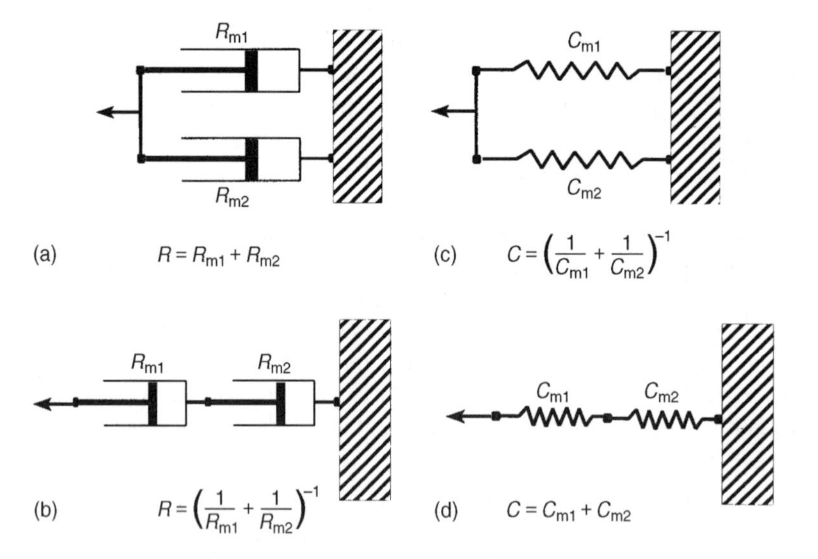


**Figure 2.5** Models of parallel and series combinations of mechanical dashpots (resistances) and springs (compliances).

move at the same velocity. Therefore, the total force $F$ required to extend the two dashpots with resistances $R_{\mathrm{m1}}$ and $R_{\mathrm{m2}}$ at velocity $v$ is

$$F = v(R_{\mathrm{m1}} + R_{\mathrm{m2}}) \tag{2.10}$$

But the combined resistance, $R$, is defined as $F/v$. Thus, Equation (2.10) yields:

$$R = R_{\mathrm{m1}} + R_{\mathrm{m2}} \tag{2.11}$$

This relationship for the two mechanical resistances in parallel is different from the corresponding expression for parallel combinations of the other types of resistances.

Case (c) in Figure 2.5 represents mechanical springs with compliances $C_{m1}$ and $C_{m2}$ connected in parallel. As in case (a), the parallel mechanical arrangement constrains the springs so that they must extend by equal amounts $(=x)$. Thus,

$$x = F_1 C_{m1} = F_2 C_{m2} \tag{2.12}$$

where $F_1$ and $F_2$ represent the corresponding tensions developed in the two springs. But the sum of $F_1$ and $F_2$ yields the total force $F$ required to extend the spring combination by $x$, and since $C = x/F$ by definition, Equation (2.12) leads to

$$\frac{1}{C} = \frac{1}{C_{m1}} + \frac{1}{C_{m2}} \tag{2.13}$$

which again differs from the corresponding situation for capacitances placed in parallel in other systems. Similar considerations apply to series combinations of mechanical dashpots and springs (cases (b) and (d) in Figure 2.5). As such, one has to be cautious in converting models of mechanical systems into their electrical analogs.

## 2.3 LINEAR MODELS OF PHYSIOLOGICAL SYSTEMS: TWO EXAMPLES

In this section, we will derive the mathematical formulations that characterize the input–output properties of two simple physiological models. The first model provides a linearized description of lung mechanics (Figure 2.6). The airways are divided into two categories: the larger or central airways and the smaller or peripheral airways, with fluid mechanical resistances equal to $R_C$ and $R_P$, respectively. Air that enters the alveoli also produces an expansion of the chest-wall cavity by the same volume. This is represented by the connection of the lung ($C_L$) and chest-wall ($C_W$) compliances in series. However, a small fraction of the volume of air that enters the respiratory system is shunted away from the alveoli as a result of the compliance of the central airways and gas compressibility. This shunted volume is very small under normal circumstances at regular breathing frequencies, but becomes progressively more substantial if disease leads to peripheral airway obstruction (i.e., increased $R_P$) or a stiffening of the lungs or chest-wall (i.e., decreased $C_L$ or $C_W$). We account for this effect by

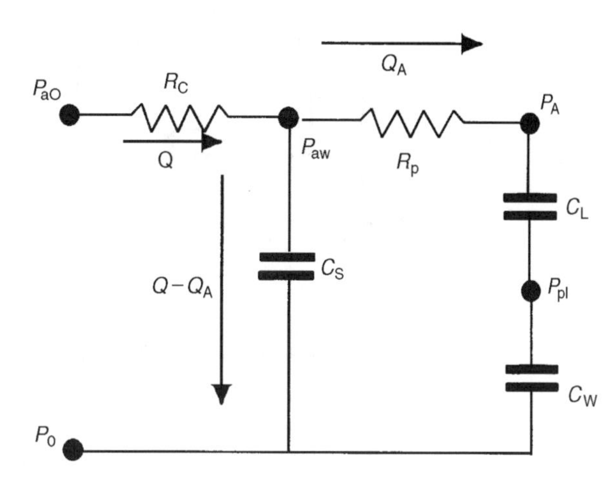


**Figure 2.6** Linear model of respiratory mechanics.

placing a shunt compliance, $C_S$, in parallel with $C_L$ and $C_W$. The pressures developed at the different points of this lung model are: $P_{ao}$ at the airway opening, $P_{aw}$ in the central airways, $P_A$ in the alveoli and $P_{pl}$ in the pleural space (between the lung parenchyma and chest wall). These pressures are referenced to $P_0$, the ambient pressure, which we can set to zero. Suppose the volume flow-rate of air entering the respiratory system is $Q$. Then, the objective here is to derive a mathematical relationship between $P_{ao}$ and $Q$.

From Kirchhoff's Second Law (applied to the node $P_{aw}$), if the flow delivered to the alveoli is $Q_A$, then the flow shunted away from the alveoli must be $Q - Q_A$. Applying, Kirchhoff's First Law to the closed circuit containing $C_S$, $R_P$, $C_L$, and $C_W$, we have

$$R_P Q_A + \left(\frac{1}{C_L} + \frac{1}{C_W}\right) \int Q_A \, dt = \frac{1}{C_S} \int (Q - Q_A) \, dt \tag{2.14}$$

Applying Kirchhoff's First Law to the circuit containing $R_C$ and $C_S$, we have

$$P_{ao} = R_C Q + \frac{1}{C_S} \int (Q - Q_A) \, dt \tag{2.15}$$

Differentiating Equation (2.14) and Equation (2.15) with respect to time, and subsequently reducing the two equations to one by eliminating $Q_A$, we obtain the equation relating $P_{ao}$ to $Q$:

$$\frac{d^2 P_{ao}}{dt^2} + \frac{1}{R_P C_T} \frac{dP_{ao}}{dt} = R_C \frac{d^2 Q}{dt^2} + \left(\frac{1}{C_S} + \frac{R_C}{R_P C_T}\right) \frac{dQ}{dt} + \frac{1}{R_P C_S} \left(\frac{1}{C_L} + \frac{1}{C_W}\right) Q \tag{2.16}$$

where $C_T$ is defined by

$$C_T = \left(\frac{1}{C_L} + \frac{1}{C_W} + \frac{1}{C_S}\right)^{-1} \tag{2.17}$$

The second example that we will consider is the linearized physiological model of skeletal muscle, as illustrated in Figure 2.7. $F_0$ represents the force developed by the active contractile element of the muscle, while $F$ is the actual force that results after taking into account the mechanical properties of muscle. $R$ represents the viscous damping inherent in the tissue, while $C_P$ (parallel elastic element) and $C_S$ (series elastic element) reflect the elastic

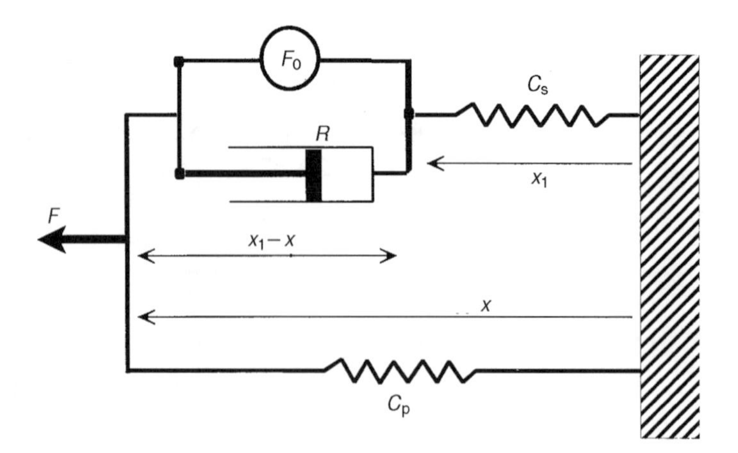


**Figure 2.7** Linear model of muscle mechanics.

storage properties of the sarcolemma and the muscle tendons, respectively. First, consider the mechanical constraints placed on the model components as a result of the parallel configuration. If spring $C_P$ is stretched by an incremental length $x$, the entire series combination of $R$ and $C_S$ will also extend by the same length. Furthermore, the sum of the force transmitted through the two branches of the parallel configuration must equal $F$. However, although the sum of the extensions of $C_S$ and $R$ will have to equal $x$, the individual length contributions from $C_S$ and $R$ need not be equal. Thus, if we assume $C_S$ is stretched a length $x_1$, then the extension in the parallel combination of $R$ and $F_0$ will be $x - x_1$. The velocity with which the dashpot represented by $R$ is extending is obtained by differentiating $x - x_1$ with respect to time, i.e., $d(x - x_1)/dt$.

Using the principle that the force transmitted through $C_S$ must be equal to the force transmitted through the parallel combination of $F_0$ and $R$, we obtain the following equation:

$$\frac{x_1}{C_S} = R\left(\frac{dx}{dt} - \frac{dx_1}{dt}\right) + F_0 \tag{2.18}$$

Then, using the second principle, i.e., that the total force from both limbs of the parallel combination must sum to $F$, we have

$$F = \frac{x_1}{C_S} + \frac{x}{C_P} \tag{2.19}$$

Eliminating $x_1$ from Equation (2.18) and Equation (2.19) yields the following differential equation relating $F$ to $x$ and $F_0$:

$$\frac{dF}{dt} + \frac{1}{RC_S} F = \left(\frac{1}{C_S} + \frac{1}{C_P}\right)\frac{dx}{dt} + \frac{1}{RC_S C_P} x + \frac{F_0}{RC_S} \tag{2.20}$$

Note that in Equation (2.20), under steady-state, isometric conditions (i.e., the muscle length is constrained to be constant), $x = 0$, $dx/dt = 0$ and $dF/dt = 0$, which leads to the result $F = F_0$. Therefore, under steady-state isometric conditions, the force developed by the muscle model will reflect the force developed by the active contractile element of the muscle.

## 2.4 DISTRIBUTED-PARAMETER VERSUS LUMPED-PARAMETER MODELS

The models that we have considered up to this point are known as *lumped-parameter models*. A given property of the model is assumed to be "concentrated" into a single element. For example, in the lung mechanics model (Figure 2.6), the total resistance of the central airways is "lumped" into a single quantity, $R_C$, even though in reality the central airways are comprised of the trachea and a few branching generations of airways, each of which has very different fluid mechanical resistance. Similarly, a single constant, $C_L$, is assumed to represent the compliance of the lungs, even though the elasticity of lung tissue varies from region to region. In order to provide a more realistic characterization of the *spatial* distribution of system properties, it is often useful to develop a *distributed-parameter model*. This kind of model generally takes the form of one or more partial differential equations with time and some measure of space (e.g., length or volume) as independent variables.

A distributed-parameter model can be viewed as a network of many infinitesimally small lumped-parameter submodels. To illustrate this relationship, we will derive the governing differential equation of a distributed-parameter model of the passive cable characteristics of an unmyelinated nerve fiber. As shown in Figure 2.8, the nerve fiber is

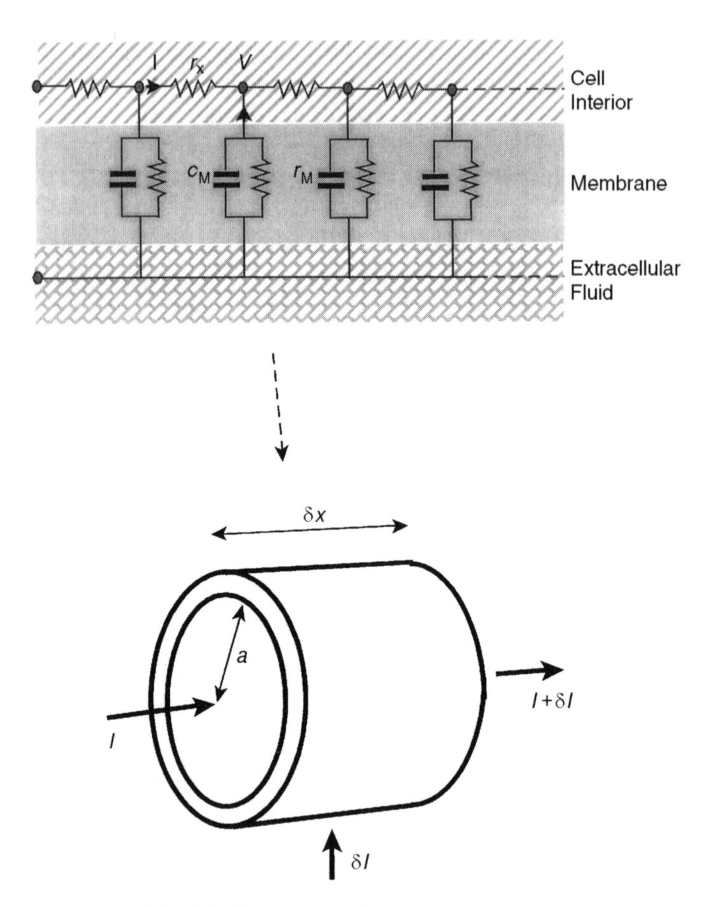


**Figure 2.8** Relationship between the lumped-parameter and distributed-parameter models of the passive cable characteristics of an unmyelinated nerve fiber.

modeled as a network containing serially-connected multiple subunits, each with circuit elements $r_x$, $r_M$ and $c_M$. $r_x$ represents the axial resistance of 1 cm of nerve tissue per cm$^2$ of cross-sectional area, and is given in $\Omega$ cm. $r_M$ and $c_M$ represent the resistance and capacitance of 1 cm$^2$ of nerve membrane surface area, respectively. We assume that the extracellular medium bathing the nerve fiber represents the electrical ground in this model. How do we relate the current passing through to the voltage found at any point in this "cable"?

In the distributed-parameter model, the nerve fiber takes the form of a long cylindrical conductor of radius $a$. We focus our analysis on a small length $\delta x$ of this cable which contains one of these resistance–capacitance subunits. We assume the intracellular voltage to increase by $\delta V$ and the axial current to increase by $\delta I$ over this small segment of cable. The assumption of "increases" instead of "decreases" in $V$ and $I$ merely establishes a sign convention. In reality, the voltage along the nerve fiber drops as current flows out of the nerve fiber through a leaky membrane. If we adopt the stated sign convention in a consistent manner, the final results will show the change in $V$ or $I$ with length to be negative. The voltage increase $\delta V$ occurring over an axial distance of $\delta x$ is related to the axial current by

$$\delta V = -\frac{I r_x \, \delta x}{\pi a^2} \tag{2.21}$$

The right-hand side of Equation (2.21) bears a negative sign since the "increase" in voltage should be associated with a current flowing in a direction *opposite* to that assumed in the figure. Dividing both sides of Equation (2.21) by $\delta x$, and taking the limit as $\delta x$ is made to approach zero, we obtain

$$\frac{\partial V}{\partial x} = -\frac{I r_x}{\pi a^2} \tag{2.22}$$

The membrane current, $\delta I$, is related to the intracellular voltage $V$ by

$$\delta I = -\left(\frac{V}{r_{\mathrm{M}}} \; 2\pi a \; \delta x + c_{\mathrm{M}} \; 2\pi a \; \delta x \; \frac{\partial V}{\partial t}\right) \tag{2.23}$$

Again, dividing both sides of Equation (2.23) by $\delta x$ and taking the limit as $\delta x$ approaches zero, we obtain

$$\frac{\partial I}{\partial x} = -\left(\frac{V}{r_{\mathrm{M}}} \; 2\pi a + c_{\mathrm{M}} \; 2\pi a \; \frac{\partial V}{\partial t}\right) \tag{2.24}$$

Finally, Equation (2.22) and Equation (2.24) can be combined into one equation by differentiating Equation (2.22) with respect to $x$ and substituting for $\partial I/\partial x$ in Equation (2.24):

$$\frac{\partial^2 V}{\partial x^2} = \frac{2 r_x}{a}\left(c_{\mathrm{M}} \; \frac{\partial V}{\partial t} + \frac{V}{r_{\mathrm{M}}}\right) \tag{2.25}$$

Equation (2.25) is known as the one-dimensional cable equation, and describes intracellular voltage along the nerve fiber as a continuous function of length and time.

## 2.5 LINEAR SYSTEMS AND THE SUPERPOSITION PRINCIPLE

All the models we have considered up to this point are *linear systems*. We have shown that these linear systems can be characterized by linear ordinary or partial differential equations. A differential equation is linear when all its terms that contain the output and input variables and its derivatives are of the first degree. For example, the model differential equations that we have derived are of the general form:

$$a_2 \; \frac{d^2 y}{dt} + a_1 \; \frac{dy}{dt} + y = b_2 \; \frac{d^2 x}{dt} + b_1 \; \frac{dx}{dt} + b_0 x \tag{2.26}$$

Since all terms in $y$ and its derivatives, as well as all terms in $x$ and its derivatives, are raised to only the first power (degree), Equation (2.26) is linear. If the coefficients $a_1$, $a_2$, $b_0$, $b_1$ and $b_2$ are constants, then Equation (2.26) is also *time-invariant*. On the other hand, if one or more of these coefficients are functions of time, for example, if $a_1 = t$, Equation (2.26) remains linear but becomes *time-varying*. However, if one or more of the coefficients are functions of $y$ or $x$, Equation (2.26) becomes *nonlinear*.

Let us suppose that the input $x$ of the system described by Equation (2.26) takes on a particular time-course, e.g., $x = x_1(t)$, and the resulting time-course for $y$ is $y_1(t)$. Then it follows that

$$a_2 \; \frac{d^2 y_1}{dt} + a_1 \; \frac{dy_1}{dt} + y_1 = b_2 \; \frac{d^2 x_1}{dt} + b_1 \; \frac{dx_1}{dt} + b_0 x_1 \tag{2.27}$$

When $x$ takes on a different time-course $x_2(t)$ and the resulting time-course for $y$ is $y_2(t)$, the following relationship also holds:

$$a_2 \frac{d^2 y_2}{dt} + a_1 \frac{dy_2}{dt} + y_2 = b_2 \frac{d^2 x_2}{dt} + b_1 \frac{dx_2}{dt} + b_0 x_2 \tag{2.28}$$

By adding terms on both sides of Equation (2.27) and Equation (2.28) with the same coefficients, we have

$$a_2 \frac{d^2(y_1 + y_2)}{dt} + a_1 \frac{d(y_1 + y_2)}{dt} + (y_1 + y_2) = b_2 \frac{d^2(x_1 + x_2)}{dt} + b_1 \frac{d(x_1 + x_2)}{dt} + b_0(x_1 + x_2) \tag{2.29}$$

What Equations (2.27) through (2.29) together imply is that the response of a linear system to the sum of two different inputs is equal to the sum of the responses of the system to the individual inputs. This result can be extended to more than two inputs and is known as the *principle of superposition*. It is a defining property of linear systems and is frequently used as a test to determine whether a given system is linear.

The principle of superposition also implies that the complete solution (i.e., response in $y$) to Equation (2.26) can be broken down into two components, i.e.,

$$y(t) = y_c(t) + y_p(t) \tag{2.30}$$

where $y_c(t)$ is known as the *complementary function* and $y_p(t)$ is called the *particular solution*. $y_c(t)$ is the response of the linear system in Equation (2.26) when the input forcing $x(t)$ is set equal to zero, i.e.,

$$a_2 \frac{d^2 y_c}{dt} + a_1 \frac{dy_c}{dt} + y_c = 0 \tag{2.31}$$

Thus, $y_c(t)$ reflects the component of $y(t)$ that remains the same independently of the form of input $x(t)$. However, the time-course of $y_c(t)$ depends on the initial values taken by $y$ and its derivatives immediately prior to input stimulation. On the other hand, $y_p(t)$ is the response of the same linear system when the input forcing $x(t)$ is a particular function of time. Thus, $y_p(t)$ satisfies the differential equation:

$$a_2 \frac{d^2 y_p}{dt} + a_1 \frac{dy_p}{dt} + y_p = b_2 \frac{d^2 x}{dt} + b_1 \frac{dx}{dt} + b_0 x \tag{2.32}$$

Unlike $y_c(t)$, $y_p(t)$ depends only on the coefficients of Equation (2.32) and the time-course of the input $x(t)$. If the linear system in question is *stable*, then the component of $y(t)$ characterized by $y_c(t)$ will eventually decay to zero while the overall response of the system will be increasingly dominated by $y_p(t)$. For this reason, $y_c(t)$ is also called the *transient response*; $y_p(t)$ is known as the *steady-state response* when $x(t)$ is an input forcing function that persists over time.

## 2.6 LAPLACE TRANSFORMS AND TRANSFER FUNCTIONS

We have shown that linear systems analysis provides the tools that allow us to build increasingly more complex models from smaller and simpler structures that ultimately can be broken down into various combinations of the three basic types of system elements. At each hierarchical level, each structure can be described functionally by its input–output characteristics, which can be expressed mathematically in the form of a differential equation relating the input, $x(t)$, to output, $y(t)$. We now introduce a means of linking the mathematical description of a given system with its block diagram representation.

The Laplace transformation, denoted by $\mathscr{L}[\cdot]$, and its inverse, denoted by $\mathscr{L}^{-1}[\cdot]$, are mathematical operations defined as follows:

$$\mathscr{L}[y(t)] \equiv Y(s) \equiv \int_0^{\infty} y(t)e^{-st}\,dt \tag{2.33a}$$

$$\mathscr{L}^{-1}[Y(s)] \equiv \frac{1}{2\pi j}\int_{\sigma-j\infty}^{\sigma+j\infty} Y(s)e^{st}\,ds \tag{2.33b}$$

where the Laplace variable $s$ is complex, i.e., $s = \sigma + j\omega$, and $j = \sqrt{-1}$. By employing the mathematical operation defined in Equation (2.33a), the function of time, $y(t)$, is converted into an equivalent function, $Y(s)$, in the complex $s$-domain. If we apply this mathematical operation to the time-derivative of $y$, $dy/dt$, we can evaluate the result by performing an integration by parts procedure:

$$\mathscr{L}\left[\frac{dy(t)}{dt}\right] \equiv \int_0^{\infty} \frac{dy(t)}{dt}e^{-st}\,dt = s\int_0^{\infty} y(t)e^{-st}\,dt + \left[y(t)e^{-st}\right]_0^{\infty} \tag{2.34a}$$

Thus,

$$\mathscr{L}\left[\frac{dy}{dt}\right] = sY(s) - y(0) \tag{2.34b}$$

By applying the same transformation to $d^2y/dt^2$, it can be shown that

$$\mathscr{L}\left[\frac{d^2y}{dt^2}\right] = s^2Y(s) - sy(0) - \left(\frac{dy}{dt}\right)_{t=0} \tag{2.35}$$

Conversely, the Laplace transform of the time-integral of $y(t)$, assuming $y(t) = 0$ for $t < 0$, is

$$\mathscr{L}\left[\int_0^t y(t)\,dt\right] = \frac{Y(s)}{s} \tag{2.36}$$

If we apply the Laplace transformation to the linear system represented by Equation (2.26), and use the above expressions in Equation (2.34b) and Equation (2.35) to evaluate the transforms of the derivatives, we will obtain the following result:

$$a_2s^2Y(s) + a_1sY(s) + Y(s) = b_2s^2X(s) + b_1sX(s) + b_0X(s) \tag{2.37a}$$

Equation (2.37a) assumes that the values of $x$ and $y$ and their first time-derivatives are all equal zero at time $t = 0$. Even if any one of these initial values are actually nonzero, they do not affect the functional nature of the dynamics of the linear system being characterized. Equation (2.37a) can be rearranged and presented in the following form:

$$\frac{Y(s)}{X(s)} = \frac{b_2s^2 + b_1s + b_0}{a_2s^2 + a_1s + 1} \tag{2.37b}$$

Equation (2.37b) describes in very compact format how the input to the linear system in question is transformed into its output. For any given input $x(t)$, it is possible to compute its Laplace transform, $X(s)$. Then, multiplying $X(s)$ by the function displayed on the right-hand side of Equation (2.37b), we can obtain the Laplace transform, $Y(s)$, of the system response. Finally, by inverting the transform, we can recover the time-course of the response, $y(t)$. Of course, we could have derived $y(t)$ by solving the differential equation in Equation (2.26).

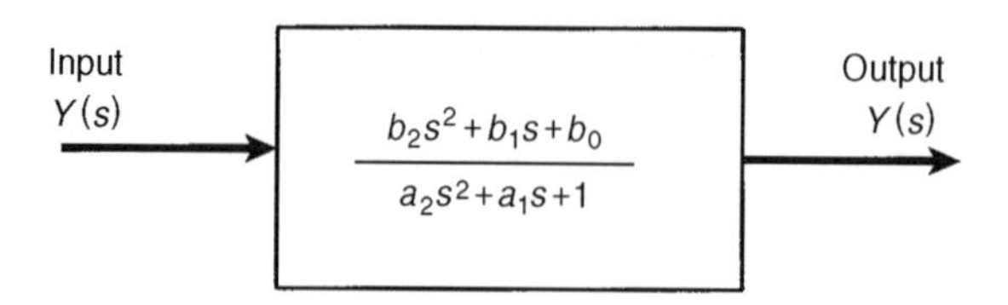


**Figure 2.9** Transfer function representation of the linear system described by Equation (2.26)

However, application of the Laplace transformation converts the differential equation into an algebraic equation, which is generally easier to solve. The ratio of $Y(s)$ to $X(s)$ in Equation (2.37b) is called the *transfer function* of the system in question. Employing this approach allows the convenient representation of the input–output characteristics of any linear system in block diagram form, as illustrated in Figure 2.9 for the example that we have been discussing.

Laplace transforms of various "standard" functions in time have been evaluated and are available in the form of tables, such as that presented in Appendix I. These same tables are also used to convert Laplace expressions back into time-domain functions. One useful technique is to expand the transfer function into simpler partial fraction forms prior to performing the inverse Laplace transformation. For example, in Equation (2.37b), suppose $b_2 = 0$, $b_1 = 0.5$, $b_0 = 1.25$, $a_2 = 0.25$ and $a_1 = 1.25$. If the input were to take the form of a unit step (i.e., $x(t) = 1$ for $t > 0$ and $x(t) = 0$ for $t \leq 0$), then from Appendix I, $X(s) = 1/s$. Substituting these values into Equation (2.37b) would yield

$$Y(s) = \frac{0.5s + 1.25}{s(s+1)(0.25s+1)} = \left(\frac{1}{s} - \frac{1}{s+1}\right) + 0.25\left(\frac{1}{s} - \frac{1}{s+4}\right) \tag{2.38}$$

The time-course of the resulting output, $y(t)$, is obtained by performing the inverse Laplace transformation of the above partial fractions, using Appendix I:

$$y(t) = (1 - e^{-t}) + 0.25(1 - e^{-4t}) \tag{2.39}$$

More details on the methodological aspects of solving differential equations via Laplace transforms can be found in any standard text on applied mathematics or linear systems theory.

## 2.7 THE IMPULSE RESPONSE AND LINEAR CONVOLUTION

Suppose we have a linear system with unknown transfer function $H(s)$. Since, by definition, $Y(s) = H(s)X(s)$, if we can find an input $x(t)$ that has the Laplace transform $X(s) = 1$, then $Y(s) = H(s)$, i.e., the Laplace transform of the resulting response would reveal the transfer function of the unknown system. Appendix I shows that the time-function that corresponds to $X(s) = 1$ is $\delta(t)$, the Dirac delta function or the *unit impulse*. The unit impulse may be considered a rectangular pulse of infinite amplitude but infinitesimal duration. Consider the function $p(t)$ defined such that: $p(t) = a$ for $0 \leq t \leq 1/a$, and $p(t) = 0$ for $t < 0$ and $t > 1/a$, where $a$ is positive constant. Applying the Laplace transformation to $p(t)$ yields

$$\int_0^{\infty} p(t)e^{-st}\,dt = \int_0^{1/a} ae^{-st}\,dt = \frac{1 - e^{-s/a}}{s/a} \tag{2.40}$$

Expanding the exponential term as an infinite series, we obtain

$$\frac{1 - e^{s/a}}{s/a} = \frac{1 - [1 - s/a + (s/a)^2/2 + \cdots]}{s/a} = 1 + (s/a) + \cdots \tag{2.41}$$

Now, if $a \to \infty$, then $p(t) \to \delta(t)$, and Equation (2.40) will yield the Laplace transform of $\delta(t)$:

$$\int_0^{\infty} \delta(t)e^{-st}\, dt = 1 \tag{2.42}$$

Thus, when $X(s) = 1$, the inverse Laplace transformation of $Y(s) = H(s)$ yields the result $y(t) = h(t)$. The system output resulting from the unit impulse input, i.e., the system *impulse response* $h(t)$, is also the inverse Laplace transform of the transfer function $H(s)$.

In the case where $H(s)$ is a known transfer function and $X(s)$ is the Laplace transform of some arbitrary input $x(t)$, we have shown through the example described in Equation (2.38) and Equation (2.39) that the corresponding output $y(t)$ is deduced by performing the inverse Laplace transformation of $Y(s) = H(s)X(s)$. In general, for any $H(s)$ and $X(s)$, we have

$$\begin{aligned} y(t) = \mathscr{L}^{-1}[H(s)X(s)] &= \frac{1}{2\pi j}\int_{\sigma-j\infty}^{\sigma+j\infty} H(s)X(s)e^{st}\, ds \\ &= \frac{1}{2\pi j}\int_{\sigma-j\infty}^{\sigma+j\infty}\int_0^{\infty} h(\tau)e^{-s\tau}\, d\tau \cdot X(s)e^{st}\, ds = \int_0^{\infty}\frac{1}{2\pi j}\int_{\sigma-j\infty}^{\sigma+j\infty} X(s)e^{s(t-\tau)}\, ds \cdot h(\tau)\, d\tau \end{aligned} \tag{2.43}$$

Note that in Equation (2.43), we have reversed the order of integration so that the outer integral is based on the time-variable $\tau$. Consider the inner integral, which is a contour integral made with respect to the complex variable $s$. By definition of the inverse Laplace transform,

$$\frac{1}{2\pi j}\int_{\sigma-j\infty}^{\sigma+j\infty} X(s)e^{s(t-\tau)}\, ds = x(t-\tau) \tag{2.44}$$

Thus, Equation (2.43) simplifies to the following:

$$y(t) = \mathscr{L}^{-1}[H(s)X(s)] = \int_0^{\infty} h(\tau)x(t-\tau)\, d\tau \tag{2.45}$$

The last term in Equation (2.45) represents the *linear convolution* of $h(t)$ and $x(t)$; this could be considered the most important and most fundamental of all mathematical operations in linear systems and signals analysis. Equation (2.45) forms the basis of the following key notions:

- The multiplication operation in the time-domain is equivalent to the convolution operation in the $s$-domain. Similarly, it can be shown that the convolution operation in the time-domain is equivalent to the multiplication operation in the $s$-domain.
- The impulse response, $h(t)$, provides a complete characterization of the dynamic behavior of a given linear system; once $h(t)$ is known, the time-response, $y(t)$, of this system to *any* arbitrary input $x(t)$ can be deduced by *convolving* $h(t)$ with $x(t)$.
- Alternatively, the dynamics of a linear system can also be completely characterized in terms of its transfer function, $H(s)$. Once $H(s)$ is known, the Laplace transform of the system response, $Y(s)$, to any arbitrary input can be deduced by *multiplying* $H(s)$ with the Laplace transform of the input, $X(s)$.

These concepts are illustrated in Figure 2.10.

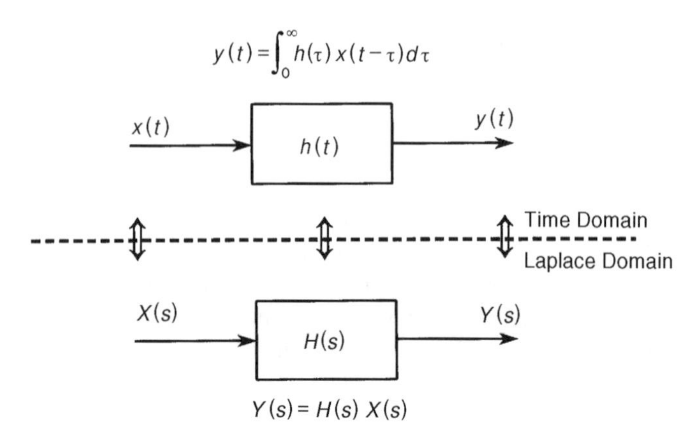


**Figure 2.10** The equivalence between block diagram representations in time and Laplace domains.

In the special case where $x(t)$ takes the form of a unit step input (i.e., $x(t) = u(t)$, where $u(t) = 1$ for $t > 0$ and $u(t) = 0$ for $t \leq 0$), Equation (2.45) yields the following result:

$$y(t) = g(t) = \int_0^t h(\tau)\, d\tau \tag{2.46}$$

which implies that the step response, $g(t)$, of a linear system can be obtained by integrating its impulse response with respect to time.

## 2.8 STATE-SPACE ANALYSIS

To characterize the complexities that are generally found in physiological systems, we often have to resort to the use of differential equations of high order. Analytical solutions of such complicated equations generally are not available, and furthermore, the numerical solution of these high-order differential equations is often fraught with problems of instability. The Laplace transform approach is useful, but this method of solution can become complicated when some of the initial conditions have to assume nonzero values. In such circumstances, *state-space modeling* offers an attractive alternative. A very significant advantage of this approach is that the state-space model can quite easily be extended to characterize *time-varying* and *nonlinear* systems. In addition, problems that are formulated as state-space models are readily amenable to standard parameter estimation techniques, such as Kalman filtering. The use of state-space modeling in system identification (or parameter estimation) is discussed further in Section 7.2.5.

A key premise in state-space analysis is that the dynamics of a given system can be completely characterized by a minimal set of quantities known collectively as the *state*. This means that if the equations describing the system dynamics are known, one can predict the future state of the system given the present state and the complete time-course of the inputs to the system. Suppose we have a linear system that can be described in terms of the following $N$th-order differential equation:

$$\frac{d^N y}{dt^N} + a_{N-1}\frac{d^{N-1}y}{dt^{N-1}} + \cdots + a_1\frac{dy}{dt} + a_0 y = b_0 x(t) \tag{2.47}$$

We define a set of $N$ *state variables* $z_1(t), z_2(t), \ldots, z_N(t)$ such that

$$
\begin{aligned}
z_1(t) &= y(t) \\
z_2(t) &= \frac{dy(t)}{dt} = \frac{dz_1(t)}{dt} \\
&\vdots \\
z_N(t) &= \frac{d^{N-1}y(t)}{dt^{N-1}} = \frac{dz_{N-1}}{dt}
\end{aligned}
\tag{2.48}
$$

Then, using Equation (2.48), we can recast Equation (2.47) into the following form:

$$\frac{dz_N}{dt} = -a_0 z_1 - a_1 z_2 - \cdots - a_{N-1} z_N + b_0 x \tag{2.49}$$

The above equations can be combined to yield the following first-order matrix differential equation:

$$\frac{d\underline{\mathbf{z}}(t)}{dt} = \mathbf{F}\underline{z}(t) + \mathbf{G}x(t) \tag{2.50}$$

where

$$
\underline{\mathbf{z}}(t) = \begin{bmatrix} z_1(t) \\ z_2(t) \\ \vdots \\ z_{N-1}(t) \\ z_N(t) \end{bmatrix}, \quad
F = \begin{bmatrix} 0 & 1 & 0 & \cdot & \cdot & 0 \\ 0 & 0 & 1 & 0 & \cdot & 0 \\ \vdots & \vdots & \vdots & \vdots & \vdots & \vdots \\ 0 & 0 & 0 & \cdot & 0 & 1 \\ -a_0 & -a_1 & \cdot & \cdot & -a_{N-2} & -a_{N-1} \end{bmatrix} \quad \text{and} \quad
G = \begin{bmatrix} 0 \\ 0 \\ \vdots \\ 0 \\ b_0 \end{bmatrix}
\tag{2.51}
$$

It is clear that, basically, what the state-space approach does is to convert the high-order differential equation (Equation (2.48)) into a set of $N$ first-order differential equations (Equation (2.50)).

Finally, we can relate the state vector $\underline{\mathbf{z}}(t)$ to the output $y(t)$ of the system in question through

$$y(t) = \mathbf{C}\underline{\mathbf{z}}(t) + Dx(t) \tag{2.52}$$

where, in this case,

$$\mathbf{C} = [1 \quad 0 \quad \cdots \quad 0 \quad 0] \qquad \text{and} \qquad D = 0 \tag{2.53}$$

Equation (2.52) allows for the possibility that one might not be able to measure the state variables directly, although in the particular example that we have considered, we are able to observe the first state variable, $z_1(t)$. Equation (2.52) is commonly referred to as the *observation equation*, while Equation (2.50) is called the *state equation*.

In linear systems, it is a relatively simple matter to convert a model represented as a transfer function (i.e., in Laplace transform description) into the state-space form, and vice versa. For the sake of illustration, consider the transfer function given by Equation (2.37b).

We define a new intermediate variable, $U(s)$, such that this transfer function, $H(s)$, can be expressed as the product of two components:

$$H(s) = \frac{U(s)}{X(s)} \frac{Y(s)}{U(s)} = \frac{1}{a_2 s^2 + a_1 s + 1} \quad \frac{b_2 s^2 + b_1 s + b_0}{1} \tag{2.54}$$

Now, the first component of this product yields

$$a_2 s^2 U(s) + a_1 s U(s) + U(s) = X(s) \tag{2.55}$$

The inverse Laplace transform of Equation (2.55) is

$$a_2 \frac{d^2 u(t)}{dt^2} + a_1 \frac{du(t)}{dt} + u(t) = x(t) \tag{2.56}$$

As in Equation (2.48), we define the state variables $z_1(t)$ and $z_2(t)$ such that

$$z_1(t) = u(t), \qquad z_2(t) = \frac{dz_1(t)}{dt} = \frac{du(t)}{dt} \tag{2.57}$$

Then, Equation (2.56) can be rewritten

$$\frac{dz_2(t)}{dt} = -\frac{a_1}{a_2} z_2(t) - \frac{1}{a_2} z_1(t) + x(t) \tag{2.58}$$

Equations (2.48) and (2.49) can be combined to form the following matrix state equation:

$$\begin{bmatrix} \dfrac{dz_1(t)}{dt} \\ \dfrac{dz_2(t)}{dt} \end{bmatrix} = \begin{bmatrix} 0 & 1 \\ -\dfrac{1}{a_2} & -\dfrac{a_1}{a_2} \end{bmatrix} \begin{bmatrix} z_1(t) \\ z_2(t) \end{bmatrix} + \begin{bmatrix} 0 \\ 1 \end{bmatrix} x(t) \tag{2.59}$$

We turn next to the other component of $H(s)$. Here, we have

$$Y(s) = b_2 s^2 U(s) + b_1 s U(s) + b_0 U(s) \tag{2.60}$$

Taking the inverse Laplace transform of Equation (2.60), we obtain

$$y(t) = b_2 \frac{d^2 u(t)}{dt^2} + b_1 \frac{du(t)}{dt} + b_0 u(t) \tag{2.61}$$

Then, using Equations (2.57) and (2.58), we obtain the following result:

$$y(t) = \left[ \left( b_0 - \frac{b_2}{a_2} \right) \left( b_1 - b_2 \frac{a_1}{a_2} \right) \right] \begin{bmatrix} z_1 \\ z_2 \end{bmatrix} + b_2 x(t) \tag{2.62}$$

Thus, conversion of the transfer function in Equation (2.37b) into state-space form leads to the state equation given by Equation (2.59) and the observation given by Equation (2.62).

## 2.9 COMPUTER ANALYSIS AND SIMULATION—MATLAB AND SIMULINK

Although the use of Laplace transforms and state-space modeling greatly simplifies the mathematical characterization of linear systems, models that provide adequate representations of realistic dynamical behavior are generally too complicated to deal with analytically. In such complex situations, the logical approach is to translate the system block representation into a computer model and to solve the corresponding problem numerically. Traditionally, one would derive the differential equations that represent the model and develop a program in some basic programming language, e.g., C or Fortran, to solve these equations. However, a variety of software tools are available that further simplify the task of model simulation and analysis. One of these, named SIMULINK®, is currently used by a large segment of the scientific and engineering community. SIMULINK provides a graphical environment that allows the user to easily convert a block diagram into a network of blocks of mathematical functions. It runs within the interactive, command-based environment called MATLAB®. In the discussions that follow, it is assumed that the reader has access to, at least, the Student Versions of MATLAB and SIMULINK, both of which are products of The Mathworks, Inc. (Natick, MA). We also assume that the reader is familiar with the most basic functions in MATLAB. One advantage of employing MATLAB and SIMULINK is that these tools are platform independent; thus, the same commands apply whether one is using SIMULINK on a Windows-based machine or on a Unix-based computer. In the rest of this section, our aim is to give the reader a brief "hands-on" tutorial on the use of SIMULINK by demonstrating in a step-by-step manner how one would go about simulating the linear lung mechanics model discussed in Section 2.3. For more details and advanced topics, the reader is referred to the User's Guides of both SIMULINK (Dabney and Harman, 1998) and MATLAB (Hanselman and Littlefield, 1998).

Let us suppose that we would like to find out how much tidal volume is delivered to a patient in the intensive care unit when the peak pressure of a ventilator is set at a prescribed level. Obviously, the solution of this problem requires a knowledge of the patient's lung mechanics. We assume that this patient has relatively normal mechanics, and the values of the various pulmonary parameters are as follows: $R_C = 1\,\text{cm}\,H_2O\,\text{s}\,L^{-1}$, $R_P = 0.5$ $\text{cm}\,H_2O\,\text{s}\,L^{-1}$, $C_L = 0.2\,L\,\text{cm}\,H_2O^{-1}$, $C_W = 0.2\,L\,\text{cm}\,H_2O^{-1}$, and $C_S = 0.005\,L\,\text{cm}\,H_2O^{-1}$ (see Figure 2.6). We will consider two ways of approaching this problem. The first and most straightforward method is to derive the transfer function for the overall system and use it as a single "block" in the SIMULINK program. The differential equation relating total airflow, $Q$, to the applied pressure at the airway opening, $P_{ao}$, was derived using Kirchhoff's laws and presented in Equation (2.16). Substituting the above parameter values into this differential equation and taking its Laplace transform yields, after some rearrangement of terms, the following expression:

$$\frac{Q(s)}{P_{ao}(s)} = \frac{s^2 + 420s}{s^2 + 620s + 4000} \tag{2.63}$$

To implement the above model, run SIMULINK from within the MATLAB command window (i.e., type `simulink` at the MATLAB prompt). The SIMULINK main block library will be displayed in a new window. Next, click the File menu and select `New`. This will open another window named `Untitled`—this is the working window in which we will build our model. The main SIMULINK menu displays several libraries of standard block functions that we can select from for model development. In our case, we would like to choose a block that represents the transfer function shown in Equation (2.63). Open the `Linear Library` (by double-clicking this block): this will open a new window containing a number of block functions in this library. Select `Transfer Fcn` and drag this block into the working window. Double-click this block to input the parameters of the transfer function. Enter the coefficients of the polynomials in $s$ in the numerator and denominator of the transfer function in the form of a row vector in each case. Thus, for our example, the coefficients for the numerator will be [1 420 0]. These coefficients are ordered in descending powers of $s$. Note that, even though the constant term does not appear in the numerator of Equation (2.63), it is still necessary to include explicitly as zero the "coefficient" corresponding to this term. Once the transfer function block has been set up, the next step is to include a generating source for $P_{ao}$. This can be found in the `Sources` library. In our case, we will select and drag the `sine wave` function generator into the working window. Double-click this icon in order to modify the amplitude and frequency of the `sine wave`: we will set the amplitude to 2.5 cm $H_2O$ (i.e., peak-to-peak swings in $P_{ao}$ will be 5 cm $H_2O$), and the frequency to 1.57 radians $s^{-1}$. which corresponds to $1.57/(2 \cdot \pi) = 0.25$ Hz or 15 breaths $min^{-1}$. Connect the output port of the sine wave block to the input port of the transfer function block with a line. To view the resulting output $Q$, open the `Sinks` library and drag a `Scope` block into the working window. Double-click the `Scope` block and enter the ranges of the horizontal (time) and vertical axes. It is always useful to view the input simultaneously. So, drag another `Scope` block into the working window and connect the input of this block to the line that "transmits" $P_{ao}$. At this point, the model is complete, and one can proceed to run the simulation. However, it is useful to add a couple of features. We would also like to view the results in terms of volume delivered to the patient. This is achieved by integrating $Q$. From the `Linear library`, select the `integrator` block. Send $Q$ into the input of this block and direct the output (Vol) of the block to a third `Scope`. It is also advisable to save the results of each simulation run into a data file that can be examined later. From the `Sinks` library, select and drag the `To File` block into the working window. This output file will contain a matrix of numbers. A name has to be assigned to this output file as well as the matrix variable. In our case, we have chosen to give the file the name `respm1.mat`, and the matrix variable the name `respm1`. Note that `mat` files are the standard (binary) format in which MATLAB variables and results are saved. The first row of matrix `respm1` will contain the times that correspond to each iteration step of the simulation procedure. In our case, we would like to save the time-histories of $P_{ao}$, $Q$ and Vol. Thus, the `To File` block will have to be adjusted to accomodate three inputs. Since this block expects the input to be in the form of a vector, the three scalar variables will have to be transformed into a 3-element vector prior to being sent to the `To File` block. This is achieved with the use of the `Mux` block, found in the `Connections` library. After dragging it into the working window, double-click the `Mux` block and enter 3 for the number of inputs. Upon exiting the `Mux` dialog block, note that there are now three input ports on the `Mux` block. Connect the `Mux` output with the input of the `To File` block. Connect the input ports of the `Mux` block to $P_{ao}$, $Q$ and Vol. The block diagram of the completed SIMULINK model is shown in Figure 2.11a. The final step is to run the simulation. Go to the `Simulation` menu and select `Parameters`. This allows the user to specify the duration over which the simulation will be conducted, as well as the minimum and maximum step sizes for each computational step. The latter will depend on the dynamics of the system in question—in our case, we have chosen both time-steps to be 0.01 s. The user can also select the algorithm for performing integration. Again, the particular choice depends on the problem at hand. For linear systems, `Linsim` is the most appropriate algorithm. However, `Linsim` should not be used for problems involving nonlinearities. For problems that involve vastly different time-constants, the `Gear` or `Adams/Gear` algorithms may be preferable. The user is encouraged to experiment with different algorithms for a given problem. Some algorithms may produce numerically unstable solutions, while others may not. Finally, double-click `Start` in the `Simulation` menu to proceed with the simulation.

In the SIMULINK model of Figure 2.11a, the bulk of our effort in solving the problem was expended on deriving the analytical expression for the transfer function $Q(s)/P_{ao}(s)$ (Equation (2.16)). This clearly is not the approach that we would want to take in general, since

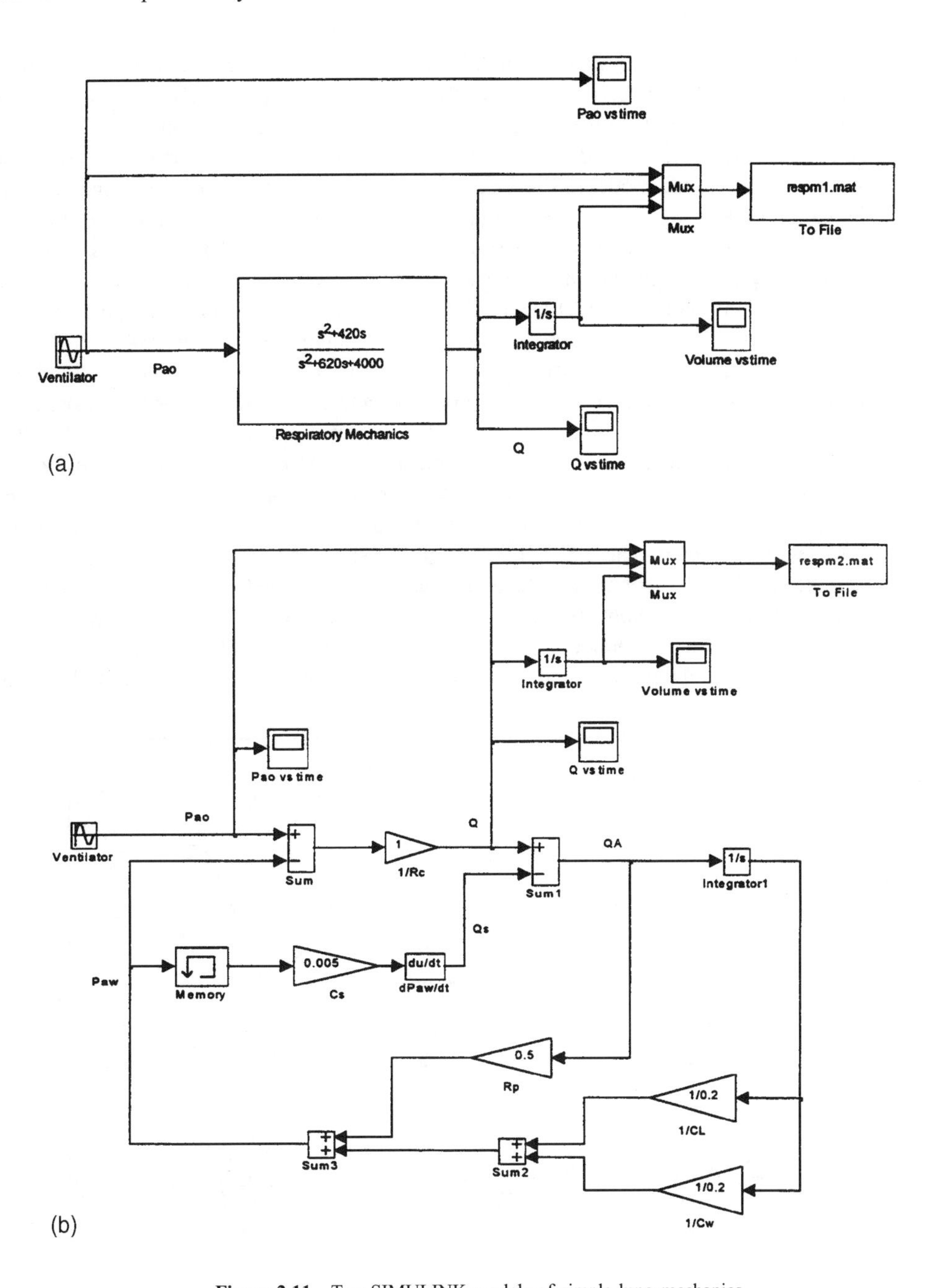


**Figure 2.11** Two SIMULINK models of simple lung mechanics.

one of the reasons for computer simulation is to simplify the task of modeling. Obtaining the solution for $Q(t)$, for given $P_{ao}$ and lung mechanical parameters, leads also to the simultaneous solution of other "internal" variables that might be of interest. These include the alveolar pressure, $P_A$, and the airflow actually delivered to the alveoli, $Q_A$. However, the implementation of the model in Figure 2.11a does not allow access to these internal variables since the system dynamics are "lumped" into a single transfer function block. The SIMULINK implementation of the same model in Figure 2.11b displays a more "open structure." Various segments in the block diagram shown correspond directly with the basic circuit equations (Equations (2.14) and (2.15)) derived from applying Kirchhoff's laws to the model of Figure 2.6. For instance, the double loop containing the gains $1/C_L$, $1/C_W$, $R_P$ and $C_S$ in the SIMULINK diagram of Figure 2.11b represents Equation (2.14). This kind of open architecture also makes it easier to determine how alterations in the parameters, such as what might occur with different lung diseases, are expected to affect overall lung mechanics. However, a common limitation with this approach is the creation of *algebraic loops*. This problem arises when blocks with direct feedthrough are connected together in a loop. For example, in Figure 2.11b, the gains $R_P$ and $C_S$ are connected together with the derivative block in a closed loop. These functions are all feedthrough blocks, so that at any integration step, the simultaneous solution of the two equations represented by this loop is required. This is accomplished iteratively, but numerical ill-conditioning could lead to no convergent solution. For each algebraic loop, SIMULINK reports an error when more than 200 iterations are expended per integration step. We eliminate this problem in our example by adding a

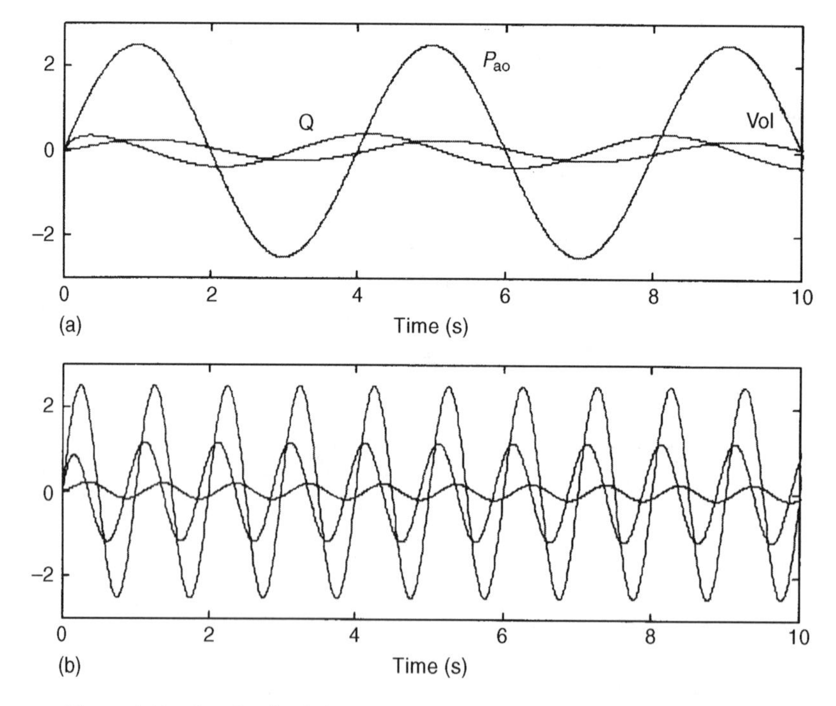


Figure 2.12 Sample simulation results from SIMULINK implementation of lung mechanics model. (a) Predicted dynamics of airflow, $Q$, and volume, Vol, in response to sinusoidal forcing of $P_{ao}$ (amplitude = 2.5 cm $H_2O$) at 15 breaths $min^{-1}$. (b) Predicted dynamics of $Q$ and Vol in response to sinusoidal forcing of $P_{ao}$ (amplitude = 2.5 cm $H_2O$) at 60 breaths $min^{-1}$.

`memory` block to the closed loop in question. The `memory` block simply adds a delay of one integration time-step to the circuit. However, one should be cautioned that this "fix" does not always work and, under certain circumstances, could lead to numerical instability.

Figure 2.12 shows sample simulation results produced by either of the model implementations. As indicated earlier, we assume the ventilator generates a sinusoidal $P_{ao}$ waveform of amplitude 2.5 cm $H_2O$. In Figure 2.12a, the ventilator frequency is set at 15 breaths $min^{-1}$, which is approximately the normal frequency of breathing at rest. At this relatively low frequency, the volume waveform is more in phase with $P_{ao}$; the airflow, $Q$, shows a substantial phase lead relative to $P_{ao}$. This demonstrates that lung mechanics is dominated by compliance effects at such low frequencies. The peak-to-peak change in volume (i.e., tidal volume) is approximately 0.5 L, while peak $Q$ is ~0.4 L $s^{-1}$. When the ventilator frequency is increased fourfold to 60 breaths $min^{-1}$ with amplitude kept unchanged, peak $Q$ clearly increases (to ~1.2 L $s^{-1}$), while tidal volume is decreased (to ~0.4 L). Now, $Q$ has become more in phase with $P_{ao}$ while volume displays a significant lag. Thus, resistive effects have become more dominant at the higher frequency. The changes in peak $Q$ and tidal volume with frequency demonstrate the phenomenon pulmonologists refer to as *frequency dependence* of pulmonary resistance and compliance, i.e., the lungs appear stiffer and less resistive as frequency increases from resting breathing.

The SIMULINK programs displayed in Figure 2.11a and Figure 2.11b have been saved as the model files `respm1.mdl` and `respm2.mdl`, respectively. These have been included in the library of MATLAB script files ("m-files") and SIMULINK model files ("mdl-files") accompanying this book. For a full compilation of all script files, model files, and MATLAB and SIMULINK functions employed in this book, the reader is referred to Appendix II.

## BIBLIOGRAPHY

Blesser, W.B. *A Systems Approach to Biomedicine*. McGraw-Hill, New York, 1969.

Carlson, G.E. *Signal and Linear System Analysis*, 2d ed. Wiley, New York, 1998.

Dabney, J.B., and T.L. Harman. *The Student Edition of SIMULINK: Dynamic Simulation for MATLAB*. Prentice-Hall, Upper Saddle River, NJ, 1998.

Hanselman, D., and B. Littlefield. *The Student Edition of MATLAB: Version 5, User's Guide*. Prentice-Hall, Upper Saddle River, NJ, 1998.

Milsum, J.H. *Biological Control Systems Analysis*. McGraw-Hill, New York, 1966.

## PROBLEMS

**P2.1.** Develop the mechanical equivalent of the electrical analog of respiratory mechanics shown in Figure 2.6.

**P2.2.** Develop the electrical analog of the muscle mechanics model shown in Figure 2.7.

**P2.3.** In emphysematous lungs, the peripheral airways of the diseased regions develop a very high resistance to airflow, and the destruction of the alveolar walls leads to an increased regional compliance. Extend the linear respiratory mechanics model of Figure 2.6 to include a normal and a diseased peripheral lung region, each with its own peripheral resistance and alveolar compliance. Show the electrical analog of the extended model and derive the differential equation relating the airway opening pressure, $P_{ao}$, to the total airflow, $Q$, entering the model. Finally, obtain an expression for the transfer function of the model, with $P_{ao}$ as input and $Q$ as output.

**P2.4.** Figure P2.1 shows a schematic diagram of the 5-element Windkessel model that has been used to approximate the hemodynamic properties of the arterial tree. The model consists of a distensible (as illustrated by the two-ended arrows) aorta and a lumped representation of the rest of the arterial vasculature. The latter is modeled as a simple parallel combination of peripheral resistance, $R_P$, and peripheral compliance, $C_P$. The mechanical parameters pertinent to the aortic portion are: (a) the compliance of the aortic wall, $C_{ao}$, (b) the viscous resistance of the aortic wall, $R_{ao}$; and (c) the inertance to flow through the aorta, $L_{ao}$. Note that resistance to flow in the aorta is considered negligible compared to $R_P$. Construct the electrical analog of this model and derive the transfer function and equivalent state-space model relating aortic pressure, $P_{ao}$, to aortic flow, $Q$.

**P2.5.** A somewhat different version of linear muscle mechanics from that displayed in Figure 2.7 is shown in Figure P2.2. Here, the elastic element, $C_P$, is placed in parallel to the viscous damping element $R$ and the contractile element, and the entire parallel combination is placed in series with the elastic element, $C_S$, and the lumped representation of the muscle mass, $m$. Derive an expression for the transfer function relating the extension of the muscle, $x$, to an applied force, $F$. Convert this transfer function description into the equivalent state-space model.

**P2.6.** Figure P2.3 displays the equivalent circuit of a short length of squid axon according to the Hodgkin–Huxley model of neuronal electrical activity. The elements shown as circles represent voltage sources that correspond to the Nernst potentials for sodium, potassium, and chloride ions. The resistances are inversely proportional to the corresponding membrane conductances for these three types of ions, while $C$ represents membrane capacitance. Derive the Hodgkin–Huxley equation, i.e., the differential equation that relates the net current flowing through the membrane, $I$, to the applied voltage across the membrane, $V$.

**P2.7.** Let us suppose we know that the response of a mechanoreceptor to stretch, applied in the form of a step of magnitude $x_0$ (in arbitrary length units), is

$$V = x_0(1 - e^{-5t})$$

where the receptor potential, $V$, is given in millivolts and time from the start of the step, $t$, is given in seconds. If we assume this system to be linear, deduce an expression for the receptor potential that we would expect to measure in response to the following stretch

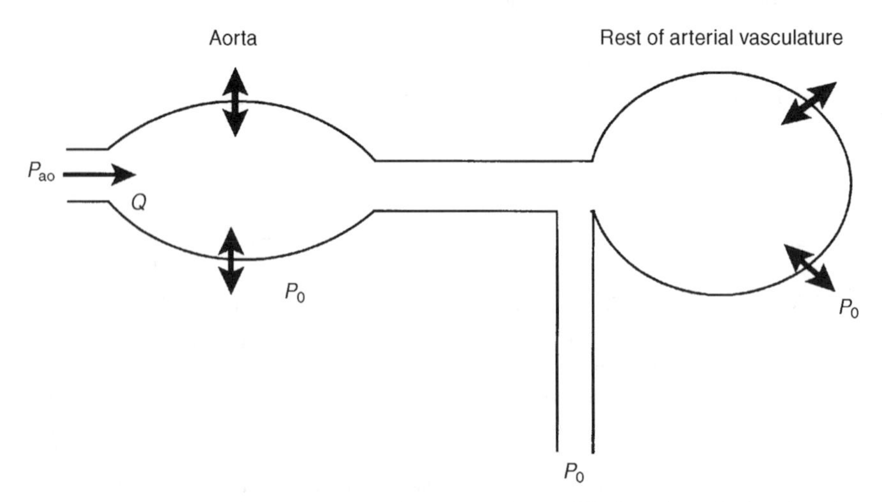


**Figure P2.1** Schematic representation of the 5-element Windkessel model of aortic and arterial hemodynamics.

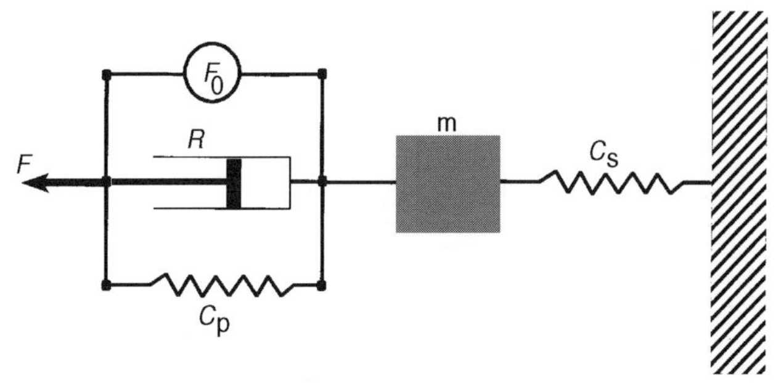


**Figure P2.2** Alternative model of muscle mechanics that includes the effect of muscle mass.

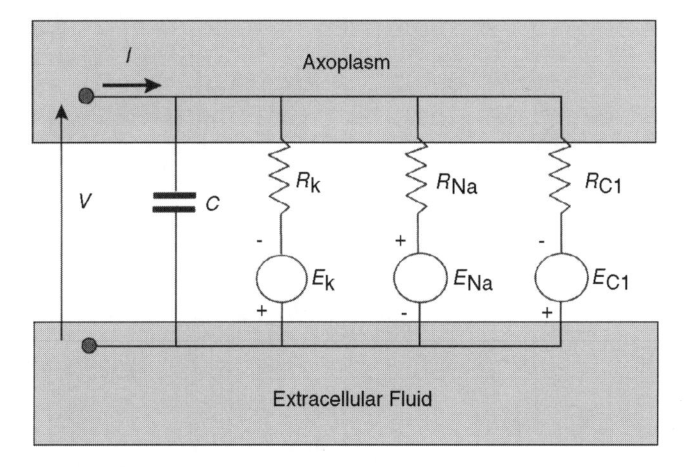


**Figure P2.3** Hodgkin–Huxley model of the electrical properties of nerve membrane.

waveform: $x(t) = 0,\ t < 0;\ x(t) = 4,\ 0 \le t < 0.3;\ x(t) = 9,\ 0.3 \le t < 0.8;\ x(t) = 0,$ $t \ge 0.8$.

**P2.8.** Using Laplace transforms, determine the responses of the mechanoreceptor in Problem P2.7 to a unit impulse and a unit ramp (i.e., $x(t) = t$ for $t > 0$).

**P2.9.** Using the SIMULINK program `respm2.mdl` as the starting point, incorporate into the respiratory mechanics model the effect of inertance to gas flow in the central airways ($L_C$). Assume a value of 0.01 cm $H_2O$ $s^2$ $L^{-1}$ for $L_C$. Keep the values of the other parameters unchanged. Through model simulations, determine how airflow and tidal volume would vary in response to sinusoidal pressure waveforms at the airway opening ($P_{ao}$) applied at 15, 60, 120, 240, 480, and 960 breaths $min^{-1}$. To simulate a subject with emphysema, increase lung compliance ($C_L$) to 0.4 L cm $H_2O^{-1}$ and peripheral airway resistance ($R_P$) to 7.5 cm $H_2O$ s $L^{-1}$. Repeat the simulations at the frequencies listed above. Compare the frequency dependence of the resulting airflow and volume waveforms to those obtained for the normal subject.

# 3

# Static Analysis of Physiological Systems

## 3.1 INTRODUCTION

Although the primary interest in most studies of physiological control systems is in the dynamic aspects, a preliminary investigation of steady-state behavior invariably leads to useful insights. This information can be used subsequently as the basis for further dynamic analysis. Steady-state measurements are generally easier to make in physiological systems, and therefore, this knowledge is usually more complete than knowledge about the dynamics. As such, it is useful to conduct an analysis that can demonstrate how the static characteristics of the various components lead to the steady-state behavior of the overall system. This allows us to verify our working hypothesis of how these components are interconnected and to determine whether there may be other factors that need to be included. Steady-state analysis also allows us to compare the operating characteristics of the system with and without feedback.

Under normal circumstances, physiological control systems generally operate within a relatively narrow range. For instance, body temperature hovers around 37°C, resting arterial blood $P_{CO_2}$ is close to 40 mm Hg, and cardiac output is generally about 5 L min$^{-1}$. These are just a few examples of Walter Cannon's *homeostatic principle* at work. How are these *equilibrium* or *steady-state* values determined? As we discussed in Section 1.3 (Chapter 1), in an engineering control system, one can always introduce a *reference input* which then determines the *set-point* of the system (see Figure 1.1). However, in most physiological systems, it is difficult to identify any explicit "reference input" signal. We also mentioned previously that the equivalent of the "comparator" in engineering control systems is generally not found explicitly in physiology. Instead, negative feedback is *embedded* into the properties of one of the system components. It turns out that it is this "embedded" negative feedback that allows the closed-loop physiological system to determine its steady-state operating level.

## 3.2 OPEN-LOOP VERSUS CLOSED-LOOP SYSTEMS

In Chapter 1, we pointed out in a qualitative way the advantage of employing negative feedback to regulate some selected variable. Here, we will examine this notion quantitatively. We turn back to the example in Chapter 1 of the simple control scheme to regulate the temperature of a room in winter with the use of a fan-heater. The open-loop control scheme is shown in Figure 3.1a. Let us first assume that the environmental temperature is 0°C, and that if the fan-heater is not turned on, the room temperature will also equilibrate to 0°C. We will also assume that the two subsystems involved, the fan-heater and the room, have linear characteristics. For the fan-heater, this takes the form of a constant gain $G_C$, so that if a reference voltage $x_0$ (in volts) is supplied to the fan-heater, the rate at which heat is produced, $u_0$ (in cal s$^{-1}$), will be related to $x_0$ by

$$u_0 = G_C x_0 \tag{3.1}$$

The addition of heat at this rate to the room in the steady state will equal the rate at which heat is lost to the cold exterior at the equilibrium temperature of $y_0$ (in °C). $y_0$ will be related to $u_0$ by

$$y_0 = G_P u_0 \tag{3.2}$$

where $G_P$ is the "gain" of the room. Note that $G_C$ and $G_P$ will have units of cal s$^{-1}$ V$^{-1}$ and °C s cal$^{-1}$, respectively. Thus, for the overall open-loop system, we have

$$y_0 = G_C G_P x_0 \tag{3.3}$$

Now, assume that there is an unexpected change in input voltage to the fan-heater of magnitude $\delta x$. Ignoring transient effects, the resulting room temperature will become

$$y = G_C G_P (x_0 + \delta x) \tag{3.4}$$

Subtracting Equation (3.3) from Equation (3.4), we find that the change in room temperature, $\delta y$, is:

$$\delta y = G_C G_P \, \delta x \tag{3.5}$$

If we extend this result to the case of the general linear open-loop control system, Equation (3.5) states that the change in the output or regulated variable is proportional to the magnitude of the input disturbance. The constant of proportionality is known as the *open-loop gain* (*OLG*), and

$$OLG = G_C G_P \tag{3.6}$$

We turn now to the closed-loop control scheme, illustrated in Figure 3.1b. Here the room temperature is measured and converted into a feedback voltage ($z$), which is subtracted from the reference input. The resulting voltage is used to drive the fan-heater. We assume the reference input required to support a temperature set-point of $y_0$°C is $x_C$ volts. Under set-point conditions, the driving voltage will be $x_C - z$, and consequently, the heat output rate of the fan-heater will be

$$u_0 = G_C (x_C - z) \tag{3.7}$$

As in the open-loop case, the temperature set-point will be related to $u_0$ through Equation (3.2). Therefore, combining Equation (3.7) and Equation (3.2), the relationship between $y_0$ and the driving voltage of the fan-heater will be

$$y_0 = G_C G_P (x_C - z) \tag{3.8a}$$

But $z$ is linearly proportional to $y_0$ through the feedback gain $H$:

$$z = H y_0 \tag{3.9}$$

By eliminating $z$ from Equation (3.8a) and Equation (3.9), and rearranging terms, we obtain the following result:

$$y_0 = \frac{G_C G_P}{1 + G_C G_P H} x_C \tag{3.8b}$$

Comparing Equation (3.8) with Equation (3.3), it is clear that the reference voltage in the open-loop case, $x_0$, will be different from the reference input voltage in the closed-loop case, $x_C$. Now, consider the effect of a disturbance of magnitude $\delta x$ and how it would affect the room temperature.

$$y = \frac{G_C G_P}{1 + G_C G_P H} (x_C + \delta x) \tag{3.10}$$

Subtracting Equation (3.8b) from Equation (3.10), we obtain the change in room temperature, $\delta y$:

$$\delta y = \frac{G_C G_P}{1 + G_C G_P H} \delta x \tag{3.11}$$

Thus, by definition, the *closed-loop gain* (*CLG*) of the feedback system is

$$CLG = \frac{G_C G_P}{1 + G_C G_P H} \tag{3.12a}$$

Comparing Equation (3.6) to Equation (3.12a), it follows that

$$CLG = \frac{OLG}{1 + OLG \cdot H} \tag{3.12b}$$

Since $G_C$, $G_P$, and $H$ are all positive quantities, Equation (3.12b) implies that *CLG* is always be smaller than *OLG*. This means that the incorporation of negative feedback into a control system can lead to a *reduction* of the effect of disturbances on the system. As a result, a closed-loop regulator has a greater ability to maintain the regulated variable within narrower limits than its open-loop counterpart. As well, a closed-loop servomechanism possesses greater inherent capability of tracking its prescribed trajectory in the presence of external noise.

In Equation (3.12a), note that the degree to which feedback reduces the effect of disturbances depends on the factor $G_C G_P H$. The larger this term, the smaller the effect of disturbances on the system output. Since $G_C G_P H$ represents the product of gains of all system components along the closed-loop, it is also known as the *loop gain* (*LG*). Therefore an increased *LG* enhances the effectiveness of the negative feedback. Notice also that, to increase *LG*, we are not necessarily limited to increasing the feedback gain $H$; *LG* can be increased by increasing one or more of the gains $G_C$, $G_P$, and $H$. On the other hand, Equation (3.11) also implies that, although the effect of a constant input disturbance on the operating level of the

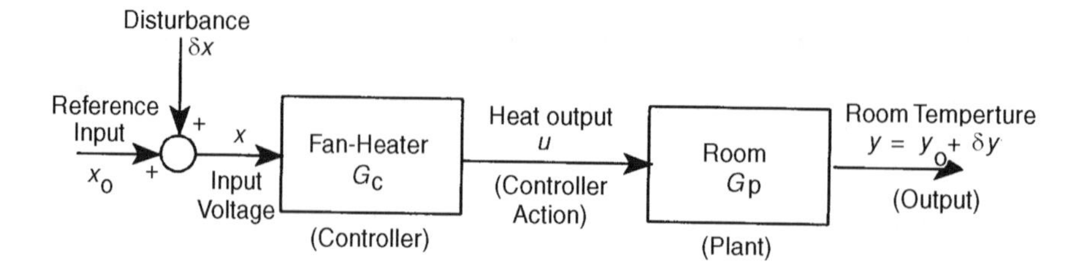


(a)

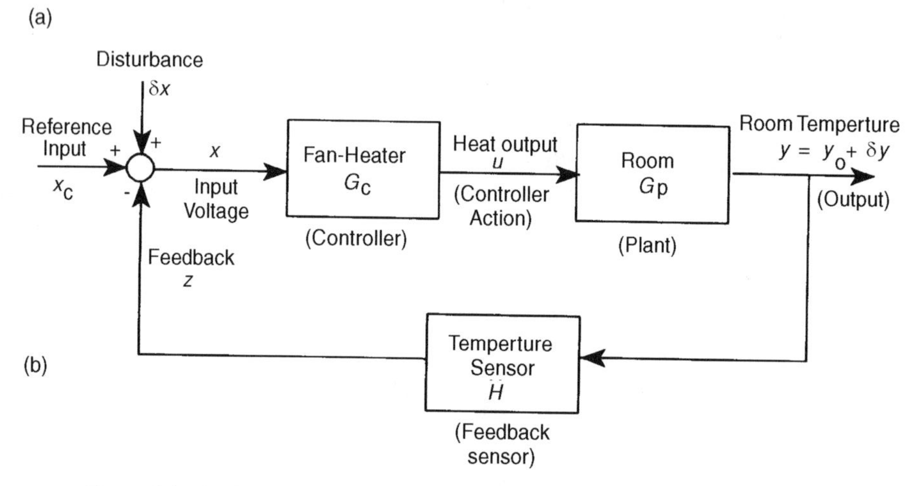


(b)

**Figure 3.1** Analysis of the influence of input fluctuations on the control of room temperature using (a) open-loop and (b) closed-loop schemes.

controlled variable ($y$) is attenuated, it is never completely eliminated unless $LG$ is infinitely high. Therefore, in closed-loop systems with proportional feedback, there will always be a *steady-state error* between the new steady-state operating level (in the presence of the disturbance) and the previous set-point.

## 3.3 DETERMINATION OF THE STEADY-STATE OPERATING POINT

In the previous example, the temperature set-point was a direct function of a reference voltage input. We turn now to physiological control systems, which generally do not have an explicitly controlled set-point. On the other hand, if the primary function of the control system is to regulate some physiological variable, this controlled variable normally will fluctuate within a relatively narrow range. So, although there is no explicit set-point, there is generally a *steady-state operating point*. One could also use the term "equilibrium" loosely to refer to the steady-state operating point. However, in reality, the regulated variable is subject to many cyclical influences, such as circadian rhythms, as well as influences resulting from coupling with other physiological organ systems. Therefore, a true static equilibrium never really exists.

We consider again the simple model of the muscle stretch reflex discussed in Section 1.4 (Chapter 1). However, this time around, we will ignore all dynamic aspects of the model and assume that we know the steady-state characteristics of all the three component blocks.

These are as illustrated in Figure 3.2. At the level of the spinal cord, afferent neural discharge frequency, $f_a$, is converted into efferent discharge frequency, $f_e$, through the linear relationship

$$f_e = G_C f_a \tag{3.13}$$

Since a larger increase in efferent neural frequency leads to a greater contraction of the extensor muscle, the gain of the plant component must be negative. We indicated in Chapter 1 that this is where the negative feedback of the closed-loop control system is "embedded," since there is no physiological component that can be identified as an explicit "comparator." Assuming the amount of contraction is proportional to the increase in efferent frequency, we have the following steady-state muscle characteristic:

$$L = L_0 - G_M f_e \tag{3.14}$$

where $L_0$ is the (hypothetical) muscle length if the efferent nerve is completely silenced. Finally, we assume that the muscle spindle sends afferent neural impulses back to the spinal cord in proportion to the length of the muscle, so that afferent traffic increases when the muscle is stretched:

$$f_a = G_S L \tag{3.15}$$

Given these characteristics and the fact that there is no explicit reference input in this case, what would be the steady-state operating point of this system?

Figure 3.3 shows in graphical form how this equilibrium level is arrived at. Here, we have rearranged the graphs of the static characteristics of the three system components such that they share common axes. The spinal cord ($f_e$ vs. $f_a$) graph shares the same $f_e$ axis as the muscle ($L$ vs. $f_e$) graph and is rotated 90° clockwise, so the $f_a$ axis points downward. The spindle ($f_a$ vs. $L$) graph is rotated 180° clockwise and shares the same axis as the spinal cord

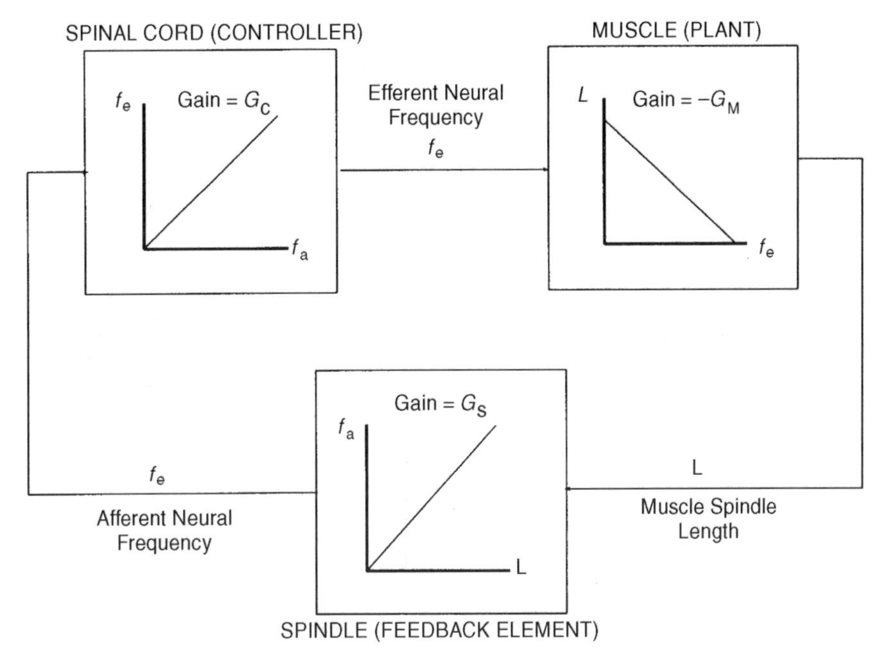


**Figure 3.2** Block diagram displaying the steady-state characteristics of the muscle stretch reflex model components.

graph. If the spindle were to be somehow isolated from the muscle and stretched by some artificial means, then both afferent and efferent neural frequencies would increase with increasing stretch. Thus, if length sensed by the spindle is equivalent to a muscle length of $L_1'$ ($>$ equilibrium muscle length), the afferent neural discharge rate sent to the spinal cord would be $f_{a1}$ ($>$ equilibrium afferent discharge rate). This induces an efferent neural discharge rate of $f_{e1}$ ($>$ equilibrium efferent discharge rate). This increased efferent discharge rate would produce a contraction of the extensor muscle to length $L_1$ ($<$ equilibrium muscle length). On the other hand, if the length sensed by the spindle is equivalent to a muscle length of $L_2'$ ($<$ equilibrium muscle length), the afferent neural discharge rate sent to the spinal cord would be $f_{a2}$ ($<$ equilibrium discharge rate) and the resultant efferent neural discharge rate would be $f_{e2}$ ($<$ equilibrium discharge rate). This decreased efferent discharge rate would produce a relaxation of the extensor muscle to length $L_2$ ($>$ equilibrium muscle length). It follows that at some point between these two extremes, these opposing effects will come to a balance, establishing the steady-state operating level, which is the equilibrium muscle length ($L_3' = L_3 \Rightarrow f_{a3} \Rightarrow f_{e3}$).

A simpler way of determining the steady-state operating level is shown in Figure 3.4. Here, we combine Equations (3.13) and (3.15) so that $f_e$ is expressed as a function of $L$:

$$f_e = G_C G_S L \tag{3.16}$$

This, in essence, collapses two graphs ($f_e$ vs. $f_a$ and $f_a$ vs. $L$) into one. In Figure 3.4, Equation (3.16) is plotted on the same axes as Equation (3.14). However, since $L$ is plotted on the vertical axis, the slope of the line corresponding to Equation (3.16) is $1/G_C G_S$. Figure 3.4 shows that the equilibrium level (labeled "E") is determined by the intersection between the two plots, since it is only at the intersection point that both muscle and spindle+spinal cord relationships are simultaneously satisfied.

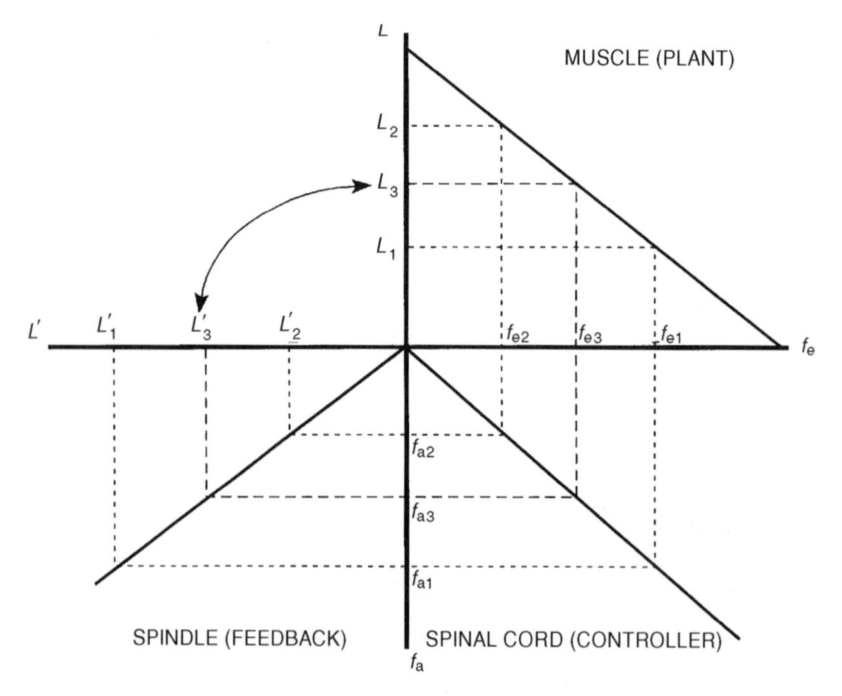


**Figure 3.3** Graphical determination of the steady-state operating point.

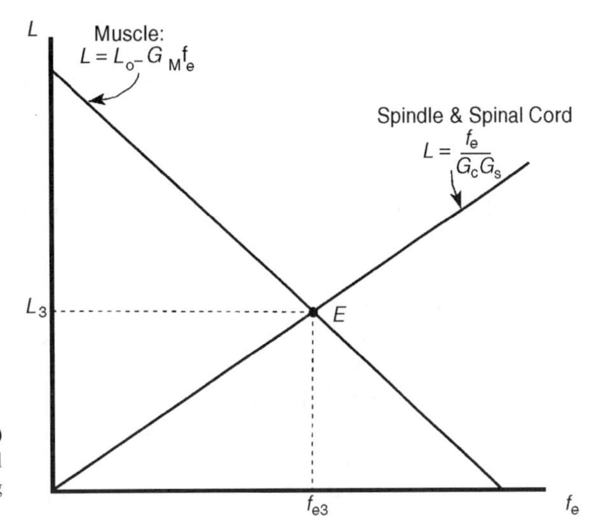


**Figure 3.4** The point of intersection (E) between the "muscle" and "spindle + spinal cord" plots yields the steady-state operating point of the muscle stretch reflex system.

The above considerations point to a third but nongraphical way of arriving at the same answer: simultaneous solution of the algebraic equations represented by Equations (3.13) through (3.15). Eliminating $f_a$ and $f_e$ from these three equations, we obtain the equilibrium solution ($L_3$) for $L$:

$$L_3 = \frac{L_0}{1 + G_M G_C G_S} \tag{3.17}$$

Substituting Equation (3.17) into Equation (3.16) yields the equilibrium solutions ($f_{a3}$ and $f_{e3}$) for the afferent and efferent neural discharge frequencies, respectively:

$$f_{a3} = \frac{G_S L_0}{1 + G_M G_C G_S} \tag{3.18}$$

$$f_{e3} = \frac{G_C G_S L_0}{1 + G_M G_C G_S} \tag{3.19}$$

## 3.4 STEADY-STATE ANALYSIS USING SIMULINK

In systems that contain several components, graphical solution of the steady-state operating point may prove to be somewhat laborious. Furthermore, if these components are nonlinear, simultaneous solution of the corresponding algebraic equations could be difficult. In such situations, it may be useful to solve the problem through numerical means. In this section, we illustrate how a steady-state analysis may be conducted using SIMULINK.

We turn once again to our simple model of the muscle stretch reflex. In the present example, however, we will assume a nonlinear relationship to represent $L$ versus $f_e$ for the muscle component:

$$L = 1 - \frac{f_e^5}{0.5^5 + f_e^5} \tag{3.20}$$

For simplicity, we have scaled all variables involved ($L$, $f_a$, and $f_e$) to their corresponding maximum values, so that scaled variables will range between 0 and 1. To represent the spindle, we assume another nonlinear expression:

$$f_a = 0.6Le^{0.5L} \tag{3.21}$$

Finally, for the spinal cord, we will assume a simple linear gain:

$$f_e = f_a \tag{3.22}$$

For reference purposes, we first deduce the steady-state operating point of this system using the graphical procedure described in the previous section. Figure 3.5 shows, on the same set of axes, a plot of the muscle characteristics (Equation (3.20)) superimposed against a plot of the combined characteristics of the spindle and spinal cord (Equation (3.21) and Equation (3.22) combined). The intersection, labeled E, represents the steady-state operating point of this system; here, $L = 0.58$ and $f_e = 0.47$.

The SIMULINK implementation of the above model, labeled "`msrflx.mdl`", is shown in Figure 3.6. The three system components, representing the spinal cord, muscle, and spindle, are linked together in a closed loop. The spinal cord component takes the form of a simple linear gain function (with gain = 1). To represent the nonlinear muscle characteristics, we employ the `Fcn` block from the `Nonlinear blocks` library. This block allows us to custom-design any mathematical relation using MATLAB-styled expressions. The MATLAB expression corresponding to Equation (3.20) is

```
1 − u^5/(0.5^5 + u^5)
```

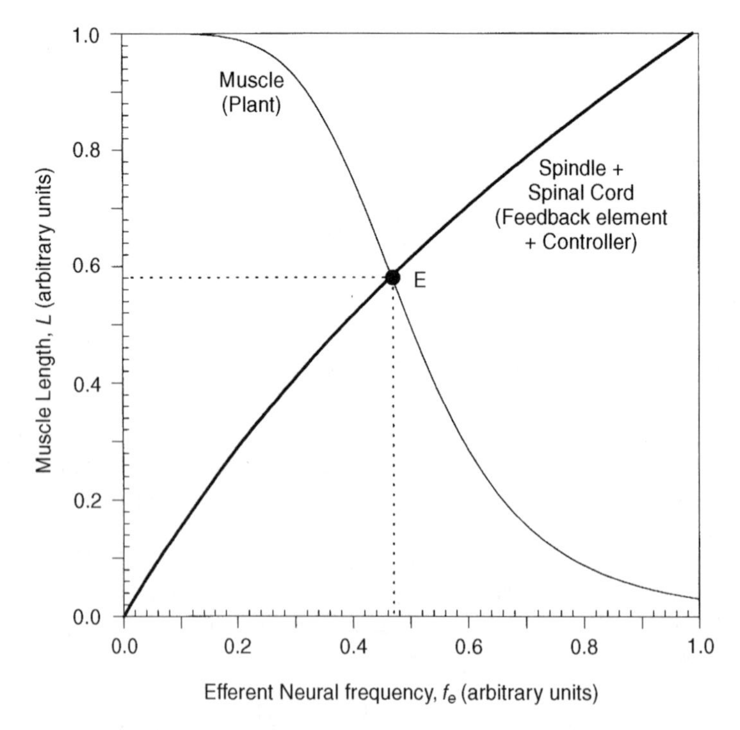


**Figure 3.5** Steady-state solution for muscle stretch reflex model with nonlinear characteristics. The equilibrium point is labeled E.

where "u", the default input variable name, represents $f_e$. The same kind of `Fcn` block is used to represent the spindle characteristics. Here, the corresponding MATLAB expression is

```
0.6*u*exp(0.5*u)
```

where `u` (again, the default input variable name) now represents $L$. Since this SIMULINK implementation solves the system of equations, Equations (3.20) through (3.22), in an iterative fashion, an "initial guess" of the solution has to be made. We achieve this by introducing an initial "impulse" into the closed loop. This is done by using a `Pulse Generator` block from the `Sources` library. The period of the pulse is set to a value (15 s) larger than the duration (i.e., simulation time = 10 s) for which the simulation will be run. The "impulse" is approximated by using a very short pulse duration by setting the duty cycle in the `Pulse Generator` block to 0.75%. The magnitude of the pulse can be set to some arbitrary number; we have chosen a value of unity in this case. A `Scope` block is positioned to display how $L$ would behave as a function of "time" (which translates into iteration number, since this is a steady-state, and not dynamic, analysis). Since we are interested in the final steady-state operating point, we introduce an `XY Graph` block to plot $L$ versus $f_e$ at every iteration. Figure 3.7 shows the results of one simulation, using a total simulation duration of "10 s" and time step of "0.01 s". In Figure 3.7a, $L$ can be seen to start off at its initial condition of 1 and very rapidly converge to its steady-state solution of 0.58. In Figure 3.7b, where $L$ is plotted against $f_e$, the solution begins at point (1,1) and follows a straight line trajectory to end at the final steady-state point where $f_e = 0.47$ and $L = 0.58$. This solution is consistent with the reference solution obtained graphically (see Figure 3.5).

In the sections that follow, we will perform steady-state analyses of three physiological control models. The purpose is not only to show further examples of the analysis procedures that we have been discussing, but also to demonstrate to the reader that steady-state analysis can yield important insights into the integrative physiology of the system in question.

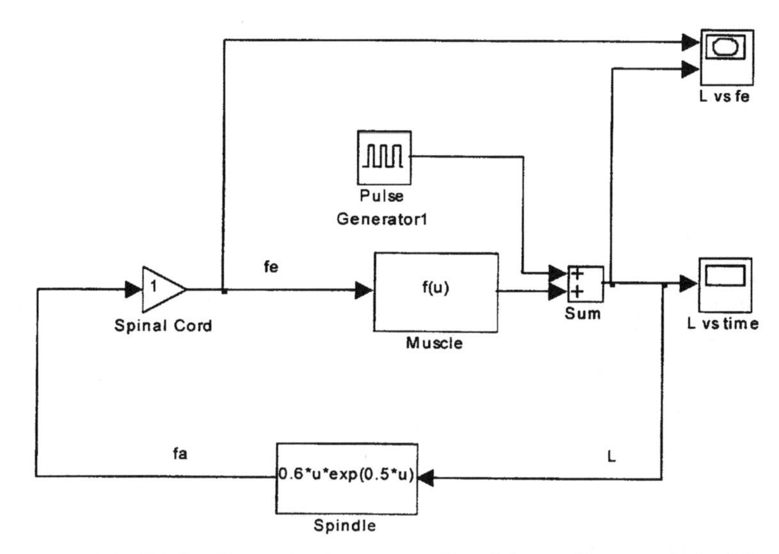


**Figure 3.6** SIMULINK program "msrflx.mdl" used for steady-state analysis of the muscle stretch reflex model.

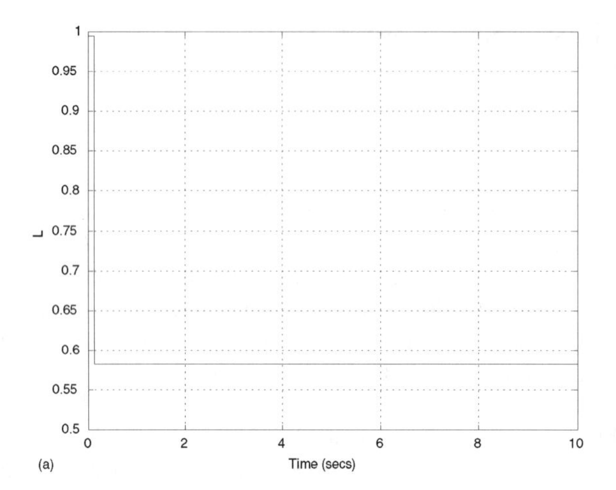


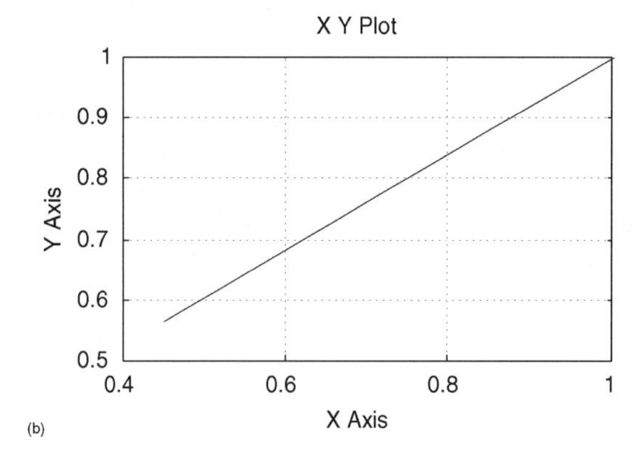


**Figure 3.7** Sample simulation results from the SIMULINK model of Figure 3.6. (a) After short initial transient, $L$ settles quickly to its steady state value of 0.58; (b) $L$ ($Y$-axis) is plotted against $f_e$ ($X$-axis). Starting from initial conditions at (1,1), the system heads for its final steady-state operating point at $f_e = 0.47$ and $L = 0.58$.

## 3.5 REGULATION OF CARDIAC OUTPUT

The fundamental notion underlying cardiac output regulation was best summarized by S.W. Patterson, H. Piper, and E.H. Starling (1914) in what is commonly called *Starling's Law.* They stated in their classic paper that

> *The output of the heart is equal to and determined by the amount of blood flowing into the heart, and may be increased or diminished within very wide limits according to the inflow.*

In other words, in the steady state, the *venous return*, which is determined primarily by the mechanical properties of the systemic circulation, is always equal to the *cardiac output*, which is a function of many factors affecting the pumping ability of the heart.

### 3.5.1 The Cardiac Output Curve

The simplest possible model of the heart and systemic circulation is shown in Figure 3.8. In this simplified model, the component that we will label the "heart" actually incorporates the combined functional characteristics of the right heart, the pulmonary circulation, and the left heart. The "heart" is modeled by assuming its capacitance, $C_H$, to vary between two levels. During diastole (the phase of ventricular relaxation), $C_H = C_D$, while during systole (ventricular contraction), $C_H = C_S$, where $C_D$ is about an order of magnitude larger than $C_S$. During diastole, the heart model is connected to the venous side of the circuit, so that $C_H$ ($= C_D$) is "charged up" by the filling pressure, which is equal to the right atrial pressure (referenced to atmospheric pressure), $P_{ra}$, minus the pleural pressure, $P_{pl}$ (which, in the intact subject, is negative relative to the atmosphere). Thus, at the end of diastole, the volume of blood in the heart would be

$$V_{HD} = C_D(P_{ra} - P_{pl}) \tag{3.23}$$

During systole, the switch S takes on its other position, connecting the variable capacitor to the arterial side of the circuit, allowing the capacitor to "discharge" into the systemic circulation. Therefore, at the end of systole, the volume of blood in the heart becomes:

$$V_{HS} = C_S(P_A - P_{pl}) \approx C_S P_A \tag{3.24}$$

The approximation in Equation (3.24) is valid because $P_A$ is much larger than $P_{pl}$ in magnitude. The difference between the end-diastolic volume and the end-systolic volume is the amount of blood ejected in one beat, i.e., the stroke volume, $SV$:

$$SV = V_{HD} - V_{HS} = C_D(P_{ra} - P_{pl}) - C_S P_A \tag{3.25}$$

But the volume of blood pumped out in each beat multiplied by the number of beats that occurs per unit time ($f$), i.e., the heart rate, must equal the cardiac output, $Q_C$:

$$Q_C = SV \cdot f \tag{3.26}$$

Substituting Equation (3.25) into Equation (3.26), we obtain the following relationship:

$$Q_C = fC_D\left(P_{ra} - \frac{C_S P_A}{C_D} - P_{pl}\right) \tag{3.27}$$

Equation (3.27) states that cardiac output increases proportionally with right atrial pressure (known as "preload") but decreases with increasing arterial pressure (known as "afterload"). Also, note that, since $Q_C$ cannot be negative, cardiac output becomes zero when

$$P_{ra} \leq \frac{C_S P_A}{C_D} + P_{pl} \tag{3.28}$$

Since $C_S \ll C_D$ and $P_{pl}$ is negative, it follows that in the intact subject, cardiac output decreases to zero when $P_{ra}$ takes on slightly negative values. This also means that even when $P_{ra} = 0$, there remains a substantial cardiac output. In Equation (3.27), $Q_C$ increases linearly with $P_{ra}$ without bound. This, of course, cannot be possible physiologically. The degree of diastolic filling is limited beyond a certain point by factors such as increasing stiffening of

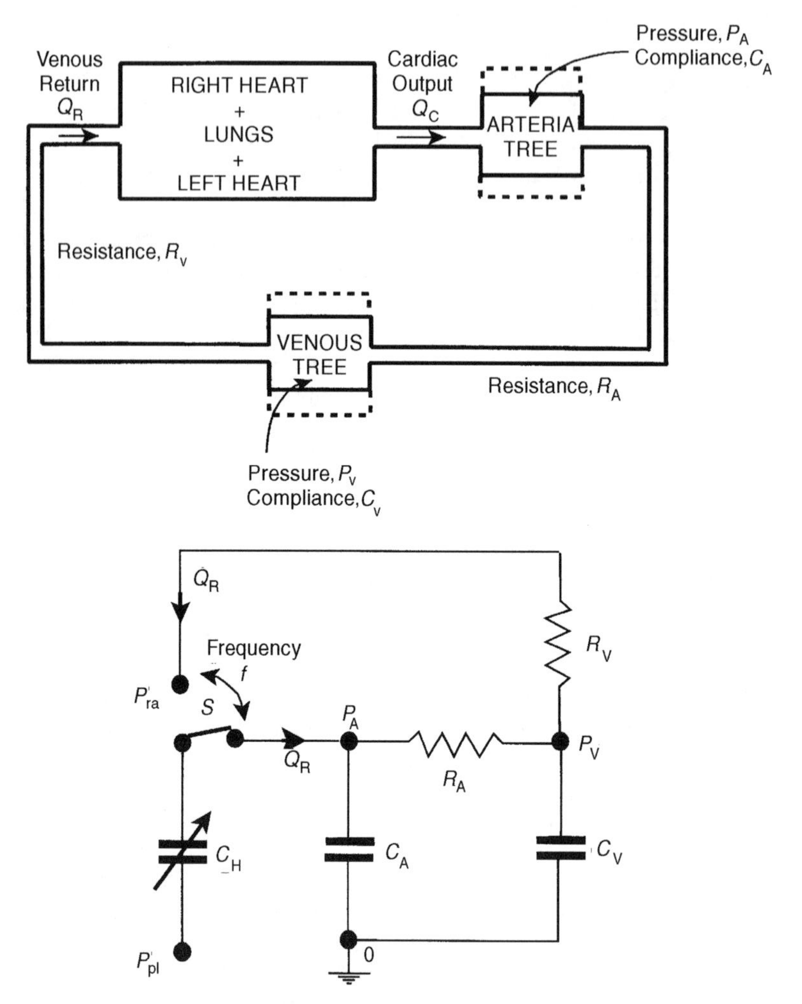


**Figure 3.8** Simplified model of cardiac output regulation.

heart connective tissue and the size restrictions imposed by the pericardial sac. Thus, we impose a threshold limitation on $Q_C$ so that

$$Q_C \leq Q_{C_{max}} = fC_D\left(P^*_{ra} - \frac{C_S P_A}{C_D} - P_{pl}\right) \tag{3.29}$$

where $P^*_{ra}$ is the value of $P_{ra}$ above which $Q_C$ cannot increase any further. The plot of $Q_C$ against $P_{ra}$ is known as the *cardiac output curve* or *cardiac function curve*. Figure 3.9 displays the form of the cardiac output curve, as predicted by our simple model. Note that an increase in heart rate or sympathetic stimulation leads to an elevation in the slope of the curve as well as an increase in $Q_{C_{max}}$. Conversely, parasympathetic stimulation, a decrease in heart rate or the presence of myocardial damage due to heart disease leads to a reduction in slope and a decreased $Q_{C_{max}}$ (Figure 3.9a). Opening the chest wall eliminates the negative intrapleural pressure and therefore shifts the cardiac output curve to the right, without increasing $Q_{C_{max}}$. On the other hand, breathing into a chamber held at negative pressure reduces intrapleural

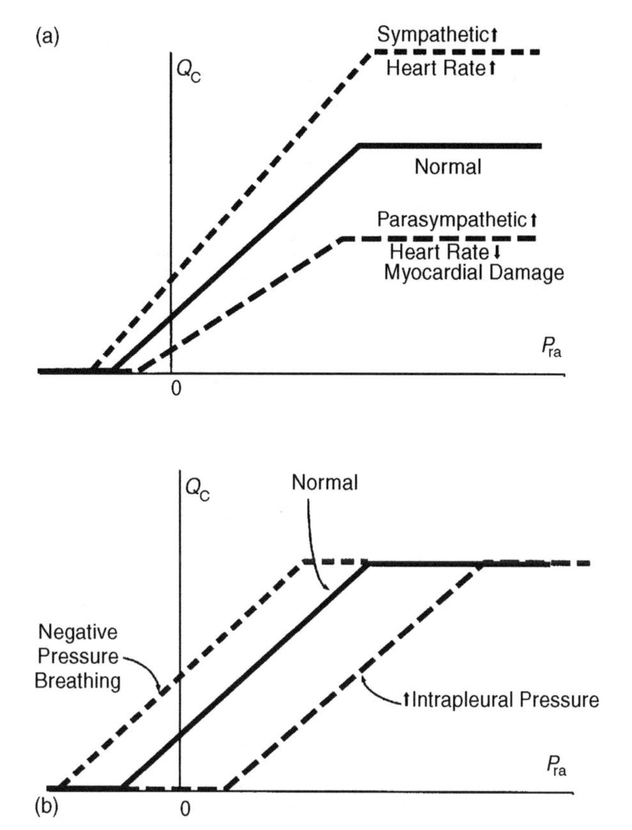


**Figure 3.9** Cardiac output curves: (a) factors that affect slope and position; (b) factors that affect only position.

pressure further, thereby shifting the cardiac output curve to the left (Figure 3.9b). Again, in this case, $Q_{C_{max}}$ is not affected.

### 3.5.2 The Venous Return Curve

We now turn our attention to the systemic circulation part of the model. The resistance and capacitance of the arterial vasculature are modeled as lumped elements $R_A$ and $C_A$, respectively. Similarly, we also lump the resistance and capacitance of the venous vasculature into the elements $R_V$ and $C_V$, respectively. The total resistance of the capillaries is incorporated into $R_A$, while the capacitance of the capillaries is neglected since it is much smaller than $C_A$ or $C_V$. We first consider the hypothetical situation where there is no blood flow. Under this condition, pressures throughout the systemic circulation would be equilibrated. However, due to the finite capacitance of the combined vasculature, this equilibrium pressure would not be zero but would take on a positive value ($\sim$7 mm Hg). This is called the *mean systemic pressure* ($P_{ms}$) or *mean circulatory pressure*. If the volume of blood in the arterial vasculature is $V_A$ and the volume of blood in the venous vasculature is $V_V$, then

$$P_{ms} = \frac{V_A + V_V}{C_A + C_V} \tag{3.30}$$

Now consider the situation where blood is flowing through the systemic circulation at the volumetric rate $Q_R$. Under steady-state conditions, the arterial pressure, $P_A$, and venous pressure, $P_V$, will be related to $Q_R$ through Ohm's law (see Figure 3.8):

$$P_A = Q_R(R_A + R_V) + P_{ra} \tag{3.31}$$

$$P_V = Q_R R_V + P_{ra} \tag{3.32}$$

However, since $V_A$ and $V_V$ remain the same regardless of whether there is blood flow or not, these volumes will be related to $P_A$ and $P_V$, respectively, through the following equations:

$$V_V = C_V P_V \tag{3.33}$$

$$V_A = C_A P_A \tag{3.34}$$

Substituting Equation (3.31) through Equation (3.34) into Equation (3.30) yields the result:

$$P_{ms} = \frac{C_A Q_R (R_A + R_V) + C_A P_{ra} + C_V Q_R R_V + C_V P_{ra}}{C_A + C_V} \tag{3.35a}$$

Rearranging terms in Equation (3.35a), we obtain the following expression that relates $Q_R$ to $P_{ra}$:

$$Q_R = \frac{P_{ms} - P_{ra}}{R_V + \dfrac{R_A C_A}{C_A + C_V}} \tag{3.35b}$$

One should be reminded that Equation (3.35b) describes how $Q_R$ would vary with $P_{ra}$ in the systemic circulation only: this is equivalent to a hypothetical situation in which we have "disconnected" the heart from the systemic circulation and where we are now concerned only with the input–output characteristics of the latter. The significance of Equation (3.35b) is that it tells us what the cardiac output would be, given the mechanical properties of the systemic circulation, the total blood volume (which determines $P_{ms}$), and right atrial pressure. We could also use Equation (3.31) to make this prediction, but we would need to know arterial blood pressure also. For example, if we wanted to know how $Q_R$ would change if $R_A$ were doubled, it would be more difficult to deduce the answer from Equation (3.31), since $P_A$ would also be changed. On the other hand, in Equation (3.35b), $P_{ms}$ is independent of this change, so that if we know $P_{ra}$ and the other mechanical properties of the circulation, $Q_R$ can be determined simply.

The *venous return curve*, described by Equation (3.35b) and illustrated in Figure 3.10, shows that $Q_R$ varies linearly with $P_{ra}$ but with a negative slope, so that as $P_{ra}$ becomes more positive, $Q_R$ decreases. However, the range of $P_{ra}$ over which Equation (3.35b) remains valid is limited. When $P_{ra}$ becomes equal to or higher than $P_{ms}$, no flow occurs. At the other end of the spectrum, when $P_{ra}$ decreases to approximately −4 mm Hg or below, $Q_R$ does not increase any further. This is due to the collapse of the veins in the thoracic cavity when the intramural pressures become lower than intrathoracic pressure. The slope of the linear part of the venous return curve is a function of the mechanical properties of the circulation. Systemic vasoconstriction, which increases peripheral resistance, lowers this slope, while vasodilation increases it (Figure 3.10a). The effect of an arteriovenous (A-V) fistula, which represents a "short-circuiting" of the systemic circulation, is to produce a large increase in slope of the venous return curve. On the other hand, factors that alter $P_{ms}$ act only to shift the venous return curve to the right (increased $P_{ms}$) or left (decreased $P_{ms}$) without altering its slope (Figure 3.10b). An increase in total blood volume (e.g., due to transfusion) would raise $P_{ms}$

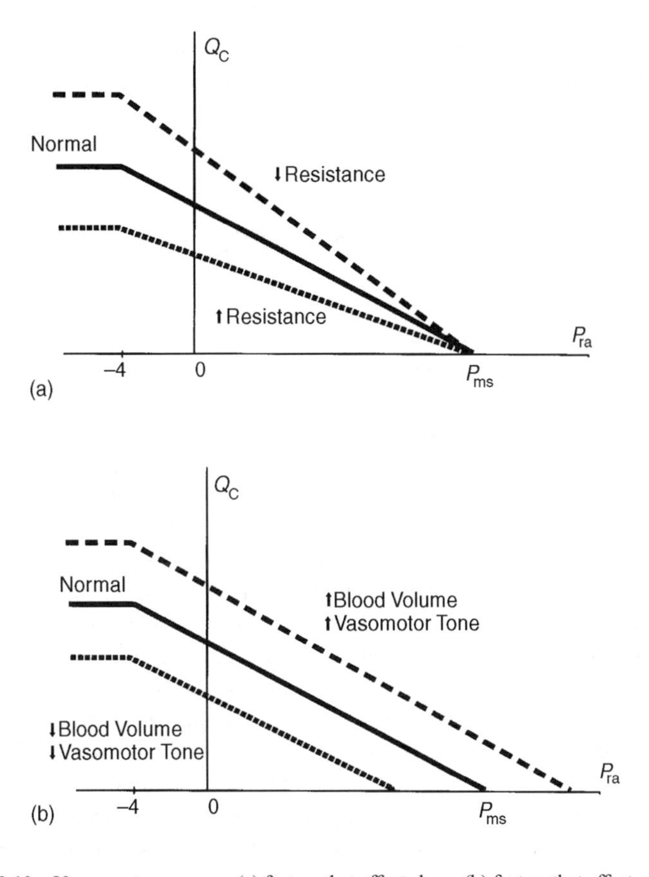


**Figure 3.10** Venous return curves: (a) factors that affect slope; (b) factors that affect position.

and thus shift the venous return curve to the right, while a decrease in blood volume due to hemorrhage would have the opposite effect. Increased vasomotor tone, which decreases the arterial and venous compliances, would also raise $P_{\text{ms}}$.

Equation (3.35b) points to an important model prediction that initially may appear counterintuitive. Since $C_{\text{V}}$ is generally about 18 times as large as $C_{\text{A}}$, Equation (3.35b) may be approximated by

$$Q_{\text{R}} = \frac{P_{\text{ms}} - P_{\text{ra}}}{R_{\text{V}} + \dfrac{R_{\text{A}}}{19}} \tag{3.36}$$

The above equation implies that a change in $R_{\text{V}}$ would have a much larger effect on venous return than a change in $R_{\text{A}}$ of the same magnitude, whereas one would generally think otherwise, since $R_{\text{A}}$ is much larger than $R_{\text{V}}$. The reason for this apparent paradox is that when venous resistance is raised, $P_{\text{V}}$ does not increase much because of the large venous capacitance; thus, the increase in driving pressure is small compared to the increase in resistance, and consequently, blood flow is dramatically reduced. On the other hand, when $R_{\text{A}}$ is increased, $P_{\text{A}}$ also increases substantially because of the relatively small arterial compliance. As a result, venous return is not decreased as much. This kind of initially counterintuitive result is encountered frequently in modeling and underscores the fact that, when several variables are involved in a problem, only the systematic approach inherent in a mathematical model will allow us to make predictions that are consistent with our underlying assumptions.

### 3.5.3 Closed-Loop Analysis: Heart and Systemic Circulation Combined

Starling's law is the consequence of connecting the "heart" and "systemic circulation" components of the model together and allowing the system to operate in closed-loop mode. We assume the following parameter values, which have been chosen so that the model provides a first approximation to the human cardiovascular system under normal resting conditions: $f = 72$ beats min$^{-1}$, $C_D = 0.035$ L mm Hg$^{-1}$, $C_S = 0.0007$ L mm Hg$^{-1}$, $P_{pl} = -4$ mm Hg, $R_A = 19.2$ mm Hg min L$^{-1}$, $R_V = 0.4$ mm Hg min L$^{-1}$, $C_A = 0.028$ L mm Hg$^{-1}$, $C_V = 0.5$ L mm Hg$^{-1}$, $P_{ms} = 7$ mm Hg, $P_A = 100$ mm Hg. Under such conditions, the cardiac output and venous return curves are as shown in Figure 3.11a. The intersection between the two curves yields the steady-state operating point, labeled N. This is established at a cardiac output of 5 L min$^{-1}$ and $P_{ra}$ of 0 mm Hg.

Using this model, can we predict what cardiac output would be during *moderate exercise*? At the onset of exercise, the tensing of the muscles involved plus an increase in venomotor tone produces a decrease in venous compliance, thereby raising $P_{ms}$. Sympathetic stimulation leads subsequently to an increase in heart rate, which elevates the slope of the cardiac output curve, and increased vasomotor tone. Then, local vasodilation of the muscular vascular beds produces a marked decrease in peripheral resistance. In this example, we have assumed $C_A$, $C_V$, $R_A$, and $R_V$ each to decrease by 40% and $f$ to increase 40%. These changes affect the cardiac output and venous return curves in the manner shown in Figure 3.11b, with the new condition being represented in the form of dashed curves. The new steady-state cardiac output is now increased to $\sim 10.5$ L min$^{-1}$, twice the resting value, while $P_{ra}$ remains relatively unchanged (point E). These predictions are consistent with empirical evidence. Note, however, that the bulk of the increase in cardiac output has come about as a result of changes in the systemic circulation: if the latter were to remain at its original state, cardiac output would increase only fractionally by less than 1 L min$^{-1}$ (point E*). This somewhat unexpected result is yet another excellent example of how mathematical modeling can lead us to a conclusion that would have been difficult to predict otherwise.

In Figure 3.11c, the model is used to predict the steady-state values of cardiac output and $P_{ra}$ in heart failure following *myocardial infarction*. To represent the reduction in effectiveness of the heart as a pump, we assume that $C_S$ is increased and $C_D$ is decreased by 30%. Thus, the ratio $C_S/C_D$ is increased from 0.02 to 0.05. According to Equation (3.27), this decreases the slope of the cardiac output curve and shifts the zero-flow intercept to the right by approximately 3 mm Hg. If this were the only effect of heart failure, cardiac output would decrease by 40% to about 3 L min$^{-1}$ (point F*). At the same time, $P_{ra}$ would rise by $\sim 3$ mm Hg. Fortunately, the body generally compensates for this decreased cardiac output by reducing urine output and thus retaining body fluid. This raises blood volume so that $P_{ms}$ increases, thereby shifting the venous return curve to the right. The net effect of these compensatory changes, which usually occur over a week, is to restore cardiac output back toward its normal level. In this example, the steady-state cardiac output following compensation falls just short of 5 L min$^{-1}$. $P_{ra}$ is now $\sim 4$ mm Hg higher than normal.

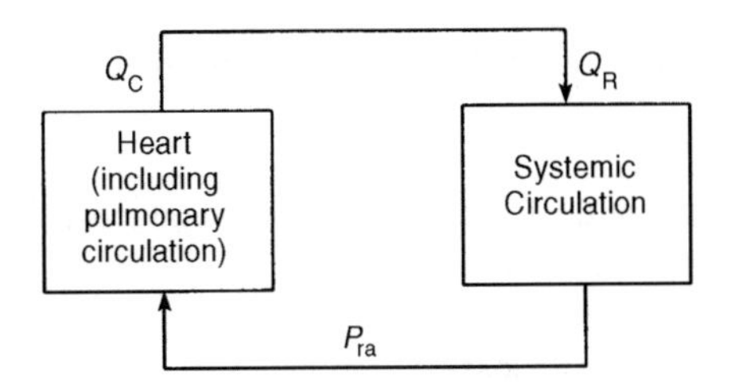


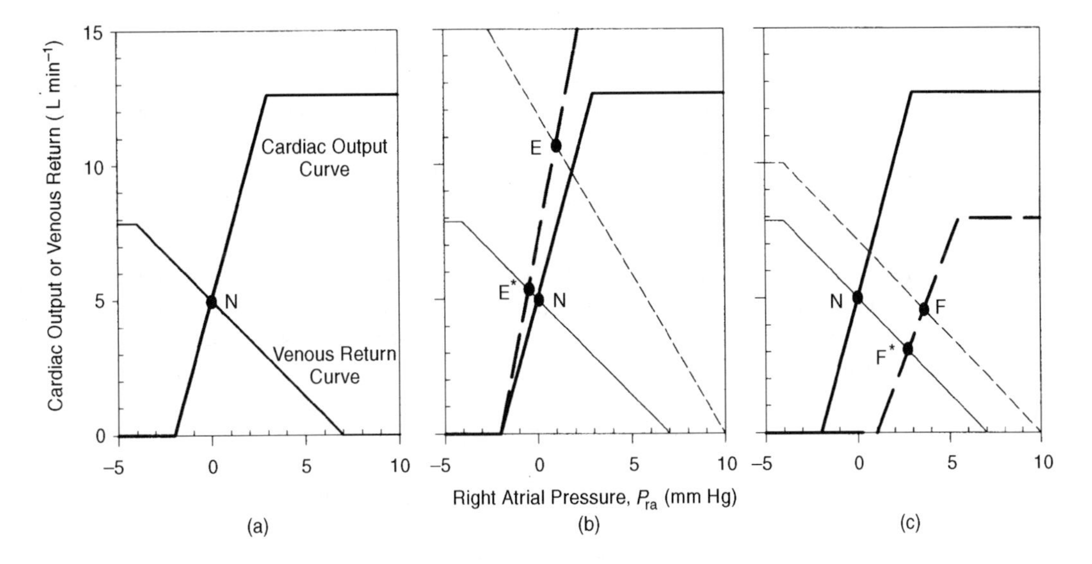


**Figure 3.11** Steady-state closed-loop analysis of cardiac output regulation during (a) normal resting conditions; (b) moderate exercise; and (c) compensated heart failure.

## 3.6 REGULATION OF GLUCOSE

We turn to another example of a physiological control system with negative feedback: the system that regulates blood glucose levels. When plasma glucose levels are elevated, insulin secretion is stimulated. This raises the level of insulin in the blood, which increases the uptake of blood glucose by the tissues. The increased outflow of glucose from the blood and interstitial fluid leads to a decrease in glucose concentration, which subsequently produces a reduction in insulin secretion.

The model we will introduce in this section was first proposed by Stolwijk and Hardy in 1974. We assume the total volume of blood and interstitial fluids to be represented by a single large compartment ($\sim 15\,\mathrm{L}$ in a normal adult), and that the steady-state concentration of glucose in this compartment is $x$ (in units of $\mathrm{mg\,ml^{-1}}$). For this level of $x$ to remain constant, the total inflow of glucose into the compartment must equal the total outflow from the compartment. Figure 3.12 shows a schematic representation of the main processes that affect this balance. Under normal circumstances, glucose enters the blood through absorption from the gastrointestinal tract or through production from the liver. We assume this input flow rate to be $Q_L$ (in mg h$^{-1}$). There are three major ways through which glucose is eliminated from the blood:

- When $x$ is elevated beyond a certain threshold ($\theta$), glucose is excreted by the kidneys at a rate proportional to the gradient between $x$ and $\theta$:

$$\text{Renal Loss Rate} = \mu(x - \theta), \quad x > \theta \tag{3.37a}$$
$$= 0, \quad x \leq \theta \tag{3.37b}$$

- Glucose leaves the blood to enter most cells through facilitated diffusion. In some tissues, the rate of glucose utilization depends only on the extracellular-to-intracellular concentration gradient. In most circumstances, we can ignore the intracellular concentration. Thus, we have

$$\text{Tissue Utilization Rate (Insulin-independent)} = \lambda x \tag{3.38}$$

- In certain types of cells, such as those in muscle and adipose tissue, insulin helps to stimulate this facilitated diffusion process. Therefore, the rate at which glucose is taken up by these cells is proportional to $x$ as well as to the blood insulin concentration, $y$:

$$\text{Tissue Utilization Rate (Insulin-dependent)} = vxy \tag{3.39}$$

In the above equations, $\mu$, $\lambda$, and $v$ are constant proportionality factors.

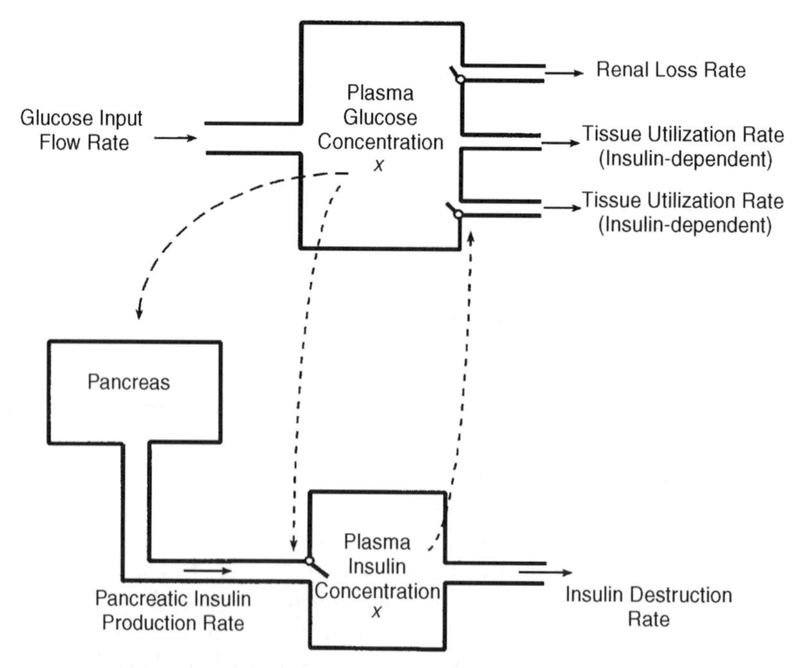


**Figure 3.12** Schematic representation of the processes involved in the regulation of glucose and insulin.

Equating the inflow to the sum of the three outflows, we obtain the following mass balance equations for blood glucose:

$$Q_L = \lambda x + vxy, \qquad x \leq \theta \tag{3.40a}$$

$$= \lambda x + vxy + \mu(x - \theta), \qquad x > \theta \tag{3.40b}$$

Note that in the above equation, a strong *nonlinearity* in the form of the product of $x$ and $y$ is introduced, along with the *thresholding nonlinearity*, which defines one regime above $\theta$ and one below it. Also, the negative feedback in this control system is clearly embedded in the characteristics described by Equations (3.40a) and (3.40b): since $Q_L$ is a constant, an increase in $x$ must lead to a corresponding decrease in $y$, and vice versa.

A similar mass balance can be established for blood insulin. Insulin is produced by the pancreas at a rate dependent on the plasma glucose level. However, if $x$ falls below a certain threshold ($\phi$), insulin production ceases. Thus, we have

$$\text{Insulin Production Rate} = 0, \qquad x \leq \phi \tag{3.41a}$$

$$= \beta(x - \phi), \qquad x > \phi \tag{3.41b}$$

Insulin is destroyed through a reaction involving the insulinase enzyme, at a rate proportional to its concentration in blood:

$$\text{Insulin Destruction Rate} = \alpha y \tag{3.42}$$

Combining Equation (3.41) and Equation (3.42), we obtain the following equation relating the steady-state level of $y$ to that of $x$:

$$y = 0, \qquad x \leq \phi \tag{3.43a}$$

$$= \frac{\beta}{\alpha}(x - \phi), \qquad x > \phi \tag{3.43b}$$

Therefore, aside from the threshold nonlinearity, the insulin response to glucose is basically linear.

The steady-state level of glucose and insulin in the blood under a given set of conditions can be predicted from this model by solving Equations (3.40) and (3.43) simultaneously. As we have shown in the regulation of cardiac output example, graphical analysis is useful in providing not only the steady-state solution but also substantial insight into the overall problem. In Figure 3.13a, steady-state insulin concentration (in milliUnits per ml blood) is plotted against the steady-state blood glucose concentration (in mg per ml). The insulin response to glucose is shown as the bold curve, while the lighter curve reflects the glucose mass balance equation. The parameter values employed in this calculation correspond to the normal adult: $\theta = 2.5\,\text{mg ml}^{-1}$, $\mu = 7200\,\text{ml h}^{-1}$, $\lambda = 2470\,\text{ml h}^{-1}$, $v = 139000\,\text{mU}^{-1}\,\text{h}^{-1}$, $\phi = 0.51\,\text{mg ml}^{-1}$, $\beta = \text{mU ml mg}^{-1}\,\text{h}^{-1}$, $\alpha = 7600\,\text{ml h}^{-1}$, and $Q_L = 8400\,\text{mg h}^{-1}$. The intersection of the glucose and insulin curves yields the steady-state operating point labeled N, where the glucose concentration is $0.81\,\text{mg ml}^{-1}$ and the insulin concentration is $0.055\,\text{mU ml}^{-1}$.

The model is used next to predict the steady-state operating levels of glucose and insulin that would arise from diabetes. In *Type-1* or *insulin-dependent diabetes*, the main defect is in the inability of the islet cells in the pancreas to produce sufficient insulin. The most common form of this disorder begins in childhood and, for this reason, is frequently called *juvenile-onset* diabetes. The other form begins in adulthood and is known as *ketone-prone* diabetes. We can model this condition by lowering the sensitivity of the insulin

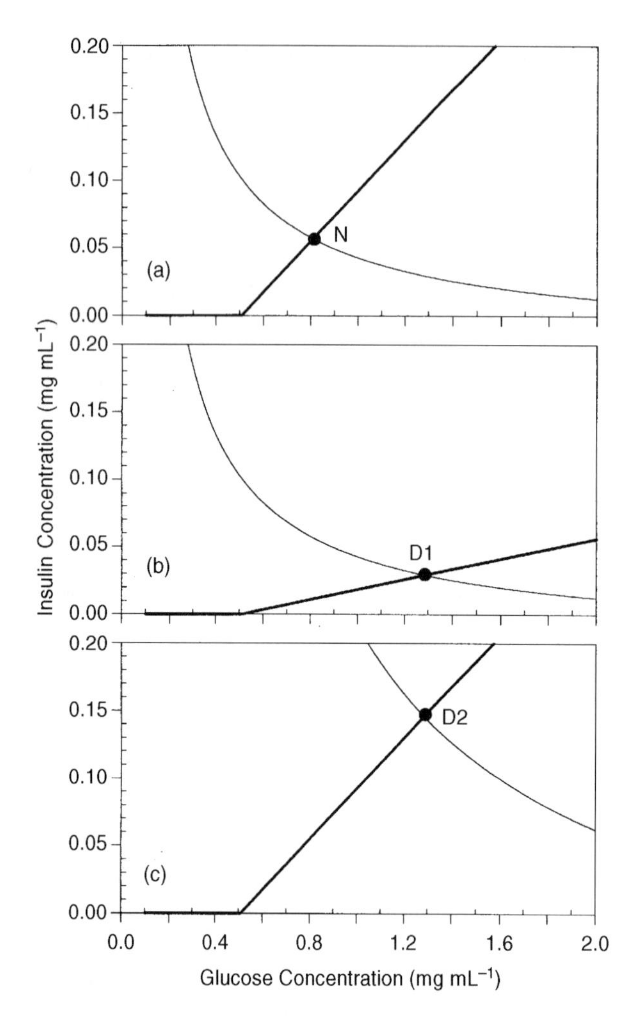


**Figure 3.13** Steady-state analysis of glucose regulation under (a) normal conditions; (b) Type-1 diabetes; and (c) Type-2 diabetes.

response to glucose, $\beta$. Figure 3.13b demonstrates the effect of reducing $\beta$ to 20% of its normal value. The new steady-state operating point is now established at D1, resulting in a highly elevated blood glucose concentration of 1.28 mg ml$^{-1}$ and a depressed plasma insulin concentration of 0.029 mU ml$^{-1}$.

*Type-2 diabetes* is also referred to as *non-insulin-dependent* diabetes since, here, the pancreas may be making normal amounts of insulin. However, for reasons that remain unclear, there is a drastic reduction in the ability of insulin to stimulate glucose uptake by the body tissues. We model this condition by changing the value of the parameter $v$, which is the constant that multiplies the product of $x$ and $y$ in the glucose mass balance equation. The insulin response to glucose may remain normal or may decrease. In Figure 3.13c, however, we have reduced $v$ to 20% of its original value while leaving the insulin curve unchanged. This change produces a shift of the glucose curve away from the origin as well as a steepening in local slopes. The new equilibrium point is established at D2, where the glucose concentration is elevated to 1.29 mg ml$^{-1}$. A somewhat counterintuitive result is that the steady-state insulin concentration now is actually almost three times higher than normal, at a level of 0.146 mU ml$^{-1}$. Thus, in this case, treatment with insulin clearly would not be useful.

## 3.7 CHEMICAL REGULATION OF VENTILATION

The final example that we will consider in this chapter is the chemoreflex regulation of respiration. In normoxic conditions, breathing is controlled almost exclusively by the level of $CO_2$ in the arterial blood. In fact, ventilation is highly sensitive to $P_{aCO_2}$, the partial pressure of $CO_2$ in arterial blood. A rise in $P_{aCO_2}$ by 1 mm Hg from its normal level of approximately 40 mm Hg may increase the ventilatory output by a third of its resting level. However, upon ascent to altitude or during inhalation of a gas mixture containing low $O_2$ content, there is an additional drive to breathe due to hypoxia. This hypoxic drive becomes noticeable when the partial pressure of $O_2$ in arterial blood, $P_{aO_2}$, drops below 70 mm Hg. Since the metabolic consumption rate of $O_2$ and the metabolic elimination rate of $CO_2$ are relatively constant in the steady state, a higher level of ventilation would lead to an increase in $P_{aO_2}$ and a decrease in $P_{aCO_2}$, which in turn would lower ventilation. Therefore, the "negative" part of this negative-feedback system is embedded in the gas exchange characteristics of the lungs. The simple model that we will analyze is depicted in block diagram form in Figure 3.14a. The ventilatory control system is divided into two components: the gas exchanging portion and the respiratory controller. An important distinction between this model and the previous models that we have analyzed is that the components here are either dual-input–single-output (controller) or single-input–dual-output (lungs) systems.

### 3.7.1 The Gas Exchanger

The gas exchanging component involves a combination of many processes that take place in the lungs, vasculature, and body tissues. However, as a first approximation, we will restrict our attention only to gas exchange occurring in the lungs. The operating characteristics of the gas exchanger are obtained by deriving the mass balance equations for $CO_2$ and $O_2$. We begin by considering $CO_2$ exchange, which is depicted schematically in Figure 3.14b. We assume the metabolic $CO_2$ production rate to be $\dot{V}_{CO_2}$; this is the rate at which $CO_2$ is delivered to the lungs from the blood that is perfusing the pulmonary circulation. In the steady state, this must equal the *net* flow of $CO_2$ exiting the lungs in gas phase. The latter is equal to the difference in volumetric fraction (or concentration) of $CO_2$ in the air entering ($F_{ICO_2}$) and leaving ($F_{ACO_2}$) the alveoli multiplied by the *alveolar ventilation*, $\dot{V}_A$. The alveolar ventilation represents that portion of the *total ventilation*, $\dot{V}_E$, that actually participates in the gas exchange process. Part of $\dot{V}_E$ is "wasted" on ventilating the non-gas-exchanging airways in the lungs; this flow is known as "dead space ventilation", $\dot{V}_D$. Thus, we have

$$\dot{V}_A = \dot{V}_E - \dot{V}_D \tag{3.44}$$

and the $CO_2$ mass balance:

$$\dot{V}_{CO_2} = k\dot{V}_A(F_{ACO_2} - F_{ICO_2}) \tag{3.45}$$

In Equations (3.44) and (3.45), the ventilatory flow rates are generally measured in *BTPS* (*body temperature pressure saturated*) units, while the $CO_2$ metabolic production rate is usually expressed in *STPD* (*standard temperature pressure dry*, i.e., at 273 K and 760 mm Hg) units. The constant $k$ allows volumes and flows measured in BTPS units to be converted into STPD units. This conversion is achieved by using the ideal gas equation:

$$\frac{V_{STPD}760}{273} = \frac{V_{BTPS}(P_B - 47)}{310} \tag{3.46a}$$

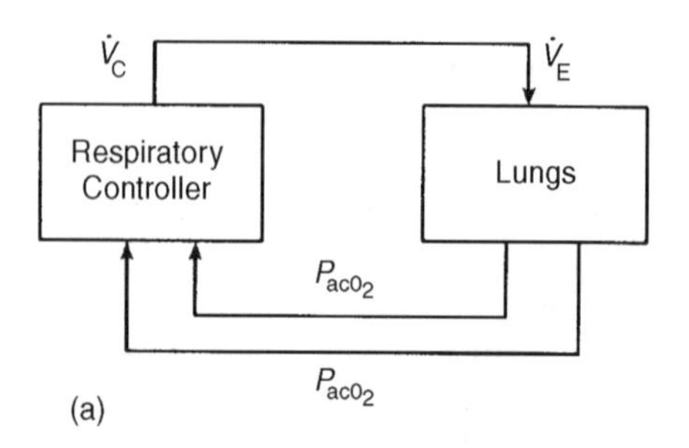


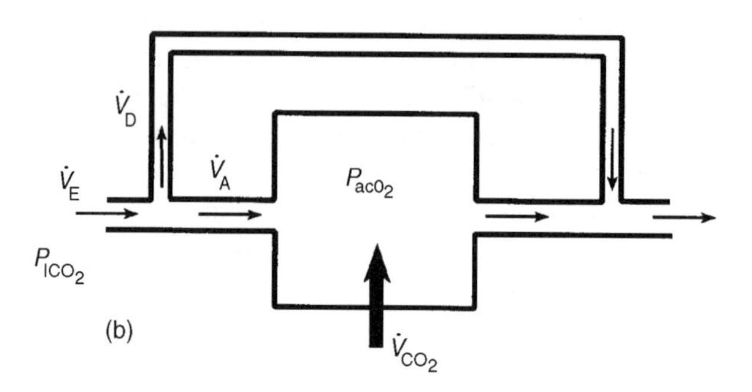


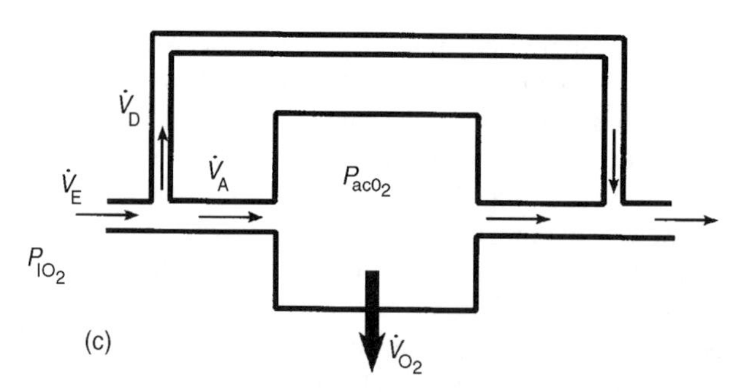


**Figure 3.14** (a) Steady-state model of the chemical regulation of ventilation. (b) Model of steady-state $CO_2$ exchange in the lungs. (c) Model of steady-state $O_2$ exchange in the lungs.

The above equation assumes body temperature to be 37°C or 310 K and a saturated water vapor partial pressure of 47 mm Hg at that temperature. $P_B$ represents the barometric pressure under which the gas exchange process is taking place; at sea level, this is 760 mm Hg, but the value decreases with ascent to high altitude. Upon rearranging Equation (3.46a), we obtain the following expression for $k$:

$$k = \frac{V_{STPD}}{V_{BTPS}} = \frac{P_B - 47}{863} \tag{3.46b}$$

The volumetric fractions, $F_{ICO_2}$ and $F_{ACO_2}$, can be converted into their corresponding partial pressures, $P_{ICO_2}$ and $P_{ACO_2}$, using Dalton's law:

$$P_{ICO_2} = F_{ICO_2}(P_B - 47), \qquad P_{ACO_2} = F_{ACO_2}(P_B - 47) \tag{3.47a,b}$$

Therefore, using Equations (3.46b) and (3.47a,b) in Equation (3.45) yields the following result:

$$P_{ACO_2} = P_{ICO_2} + \frac{863\dot{V}_{CO_2}}{\dot{V}_A} \tag{3.48}$$

Equation (3.48) shows a hyperbolic relation between $P_{ACO_2}$ and $\dot{V}_A$, and for this reason, is commonly referred to as the *metabolic hyperbola*. By employing the same kind of mass balance analysis (see Figure 3.14c), a similar "metabolic hyperbola" can be deduced for $O_2$:

$$P_{AO_2} = P_{IO_2} - \frac{863\dot{V}_{O_2}}{\dot{V}_A} \tag{3.49}$$

The negative sign in Equation (3.49) accounts for the fact that $O_2$ is removed from the lungs by the perfusing blood and, therefore, the alveolar $O_2$ content ($P_{ACO_2}$) will always be lower than the inhaled $O_2$ content ($P_{IO_2}$).

A further assumption that we will make in this model is that the alveolar partial pressures are completely equilibrated with the corresponding arterial blood gas partial pressures, i.e.,

$$P_{aCO_2} = P_{ACO_2}, \qquad P_{aO_2} = P_{AO_2} \tag{3.50a,b}$$

This is approximately true in normals, although for $O_2$, there is an alveolar–arterial gradient of 5 mm Hg or more. However, in patients with lung disease, ventilation–perfusion mismatch can give rise to rather substantial gradients between the alveolar and arterial partial pressures.

Apart from the shared value of $\dot{V}_A$, Equation (3.48) and Equation (3.49) appear to suggest that $CO_2$ and $O_2$ exchange are independent of each other. This, however, is a consequence of limiting our considerations only to the exchange processes that occur in gas phase. For more realistic modeling, it is essential to incorporate the blood-gas dissociation relationships for $CO_2$ and $O_2$, as well as considerations of gas exchange at the level of the body tissues. For instance, $CO_2$ affects the affinity with which $O_2$ is bound to hemoglobin (Bohr effect), and the level of oxygenation affects the blood $CO_2$ concentration at any given partial pressure (Haldane effect). At the level of cellular metabolism, the rate at which $CO_2$ is produced for a given $O_2$ consumption rate depends on the type of nutrient being oxidized. Fortunately, the effects of these complications on the final predictions of the alveolar or arterial partial pressures are not very large.

### 3.7.2 The Respiratory Controller

The controller part of the system includes the chemoreceptors, the neuronal circuits in the lower brain involved in the generation of the respiratory rhythm as well as the neural drive to breathe, and the respiratory muscles. The controller response to $CO_2$ has been shown to be linear over the physiological range. In the absence of vigilance, such as during sleep, the controller output falls rapidly to zero (i.e., central apnea occurs) when $P_{aCO_2}$ decreases slightly below normal awake resting levels. Exposure to hypoxia (i.e., when $P_{aO_2}$ decreases below 100 mm Hg) leads to an increase in the $CO_2$ response slope as well as the ventilatory controller output. Hence, there is a strong interaction between $CO_2$ and $O_2$ at the level of the controller. Cunningham (1974) has modeled the ventilatory controller output ($\dot{V}_C$) as the sum of an $O_2$-independent term and a term in which there is a multiplicative interaction between hypoxia and hypercapnia:

$$\dot{V}_C = \left(1.46 + \frac{32}{P_{aO_2} - 38.6}\right)(P_{aCO_2} - 37), \qquad P_{aCO_2} > 37 \tag{3.51a}$$

$$= 0, \qquad P_{aCO_2} \leq 37 \tag{3.51b}$$

Note that the above expression becomes progressively less valid as $P_{aO_2}$ approaches the asymptotic value of 38.6, in which case $\dot{V}_C$ would become infinitely large. As pointed out below, precautions have to be taken to ensure that $P_{aO_2}$ does not fall below a physiologically realistic range.

### 3.7.3 Closed-Loop Analysis: Lungs and Controller Combined

In the closed-loop situation, the controller output, $\dot{V}_C$, would equal the ventilation, $\dot{V}_E$, driving the gas exchange processes for $CO_2$ and $O_2$, as shown in Figure 3.14a. To obtain the steady-state operating point for the closed-loop system, Equations (3.48) through (3.51) must be solved simultaneously. As we have done previously, it is possible to arrive at the solution through graphical analysis. However, since three variables ($\dot{V}_E$, $P_{ACO_2}$ and $P_{AO_2}$) are involved, both graphical and algebraic methods of solution can be quite laborious. Thus, in this case, we resort to a numerical approach using SIMULINK.

Figure 3.15 displays the layout of the SIMULINK model file "`respss.mdl`" that allows the solution of the steady-state ventilatory control equations. Basically, the program simulates the closed-loop system in "open-loop mode." A `repeating sequence` block (labeled "`VdotEin Input Ramp`") is used to generate a linearly increasing sequence of $\dot{V}_E$ values. Each $\dot{V}_E$ value is fed into Equation (3.48) and Equation (3.49) so that corresponding $P_{ACO_2}$ and $P_{AO_2}$ values are generated. Each pair of $P_{ACO_2}$ and $P_{AO_2}$ values is subsequently used in Equation (3.50) and Equation (3.51) to generate the corresponding ventilatory controller output, $\dot{V}_C$ (labeled "`VdotEout`" in Figure 3.15). The initially low $\dot{V}_E$ values would produce high $P_{ACO_2}$ and low $P_{AO_2}$ levels, which would act on the controller to produce high $\dot{V}_C$ values. However, as $\dot{V}_E$ increases, chemical drive levels would decrease, in turn, decreasing $\dot{V}_C$. The steady-state equilibrium point is established at that combination of $P_{ACO_2}$ and $P_{AO_2}$ values where $\dot{V}_E$ level becomes equal to $\dot{V}_C$. A relational operator block is incorporated to check for this condition and to stop the simulation when the condition is satisfied. The steady-state values of $\dot{V}_E$, $P_{ACO_2}$ and $P_{AO_2}$ are saved to the Matlab workspace in the scalar variables "`vent`", "`paco2`" and "`pao2`", respectively. An important point to note is that we included a `saturation` block to limit the allowable range for $P_{AO_2}$. This ensures that $P_{AO_2}$ would not fall to a point where the $O_2$-dependent term in the controller became infinite or negative.

The results of two SIMULINK simulations are shown in Figure 3.16. In case (a), $P_{IO_2}$ is set equal to 150 mm Hg (i.e., 21% room air) while $P_{ICO_2}$ is set equal to zero. Due to the initially low $\dot{V}_E$ value, $P_{ACO_2}$ and $P_{AO_2}$ are initially ~67 and ~65 mm Hg, respectively, while $\dot{V}_C$ is higher than 20 L min$^{-1}$. As $\dot{V}_E$ increases, $P_{ACO_2}$ decreases while $P_{AO_2}$ rises and $\dot{V}_C$ falls. The simulation is terminated when $\dot{V}_C$ becomes equal to $\dot{V}_E$. This occurs at $\dot{V}_E = \dot{V}_C = 6$ L min$^{-1}$, $P_{ACO_2} = 40$ mm Hg and $P_{AO_2} = 100$ mm Hg. In case (b), we simulate a subject inhaling a gas mixture containing only 15% $O_2$ or, equivalently, a subject ascending to an altitude of 8500 ft. Thus, $P_{IO_2}$ is set equal to 107 mm Hg while $P_{ICO_2}$ is left at zero. As

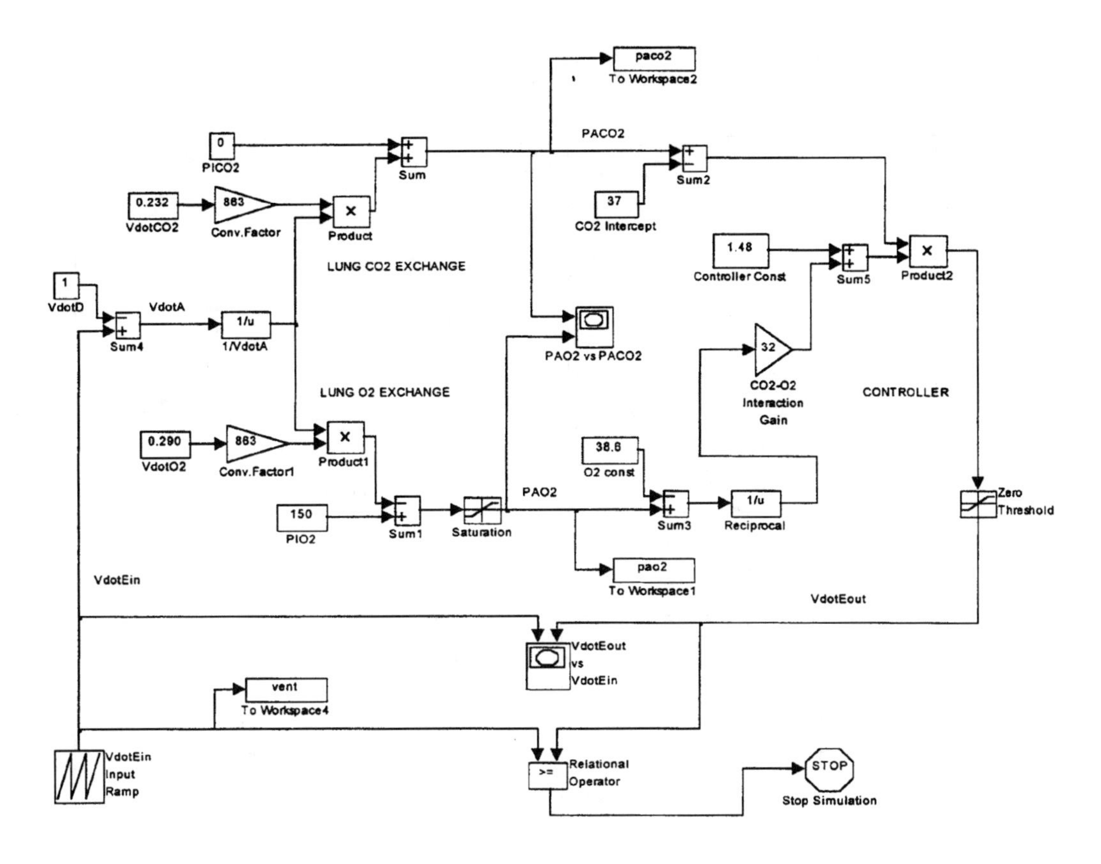


**Figure 3.15** SIMULINK program "respss.mdl" to determine the steady-state operating point of the ventilatory control system.

before, the initial value of $P_{ACO_2}$ is in the high 60s, while $P_{AO_2}$ is consistent with a value lower than 40 mm Hg. However, due to the effect of the saturation block, $P_{AO_2}$ is not allowed to fall below 40. The final equilibrium point is established at $\dot{V}_E = 6.1\,\mathrm{L\,min^{-1}}$, $P_{ACO_2} = 39$ and $P_{AO_2} = 58.3$. These two examples demonstrate quite clearly the negative feedback nature of respiratory control. Although exposure to hypoxia tends to produce an additional drive to breathe, the added ventilation blows off $CO_2$, and consequently, the lower $P_{ACO_2}$ acts to offset the hypoxic-induced drive. As a result, ventilation remains close to its original normoxic level.

The equivalent graphical analyses of cases (a) and (b) are presented in Figure 3.17a and b. The controller responses are depicted as bold curves, while the gas exchange responses are shown as light curves. The steady-state operating points for normoxia and hypoxia are labeled N and H, respectively. The two-dimensional plots do not provide a good sense of the three-dimensional nature of the problem. For instance, the controller plot shown in the ventilation–$P_{ACO_2}$ graph in case (a) represents only the $P_{AO_2}$ value of 100 mm Hg. Similarly, the controller plot shown in the ventilation–$P_{AO_2}$ graph in case (a) assumes $P_{ACO_2}$ to be 40 mm Hg. The same comments apply to the graphs in case (b).

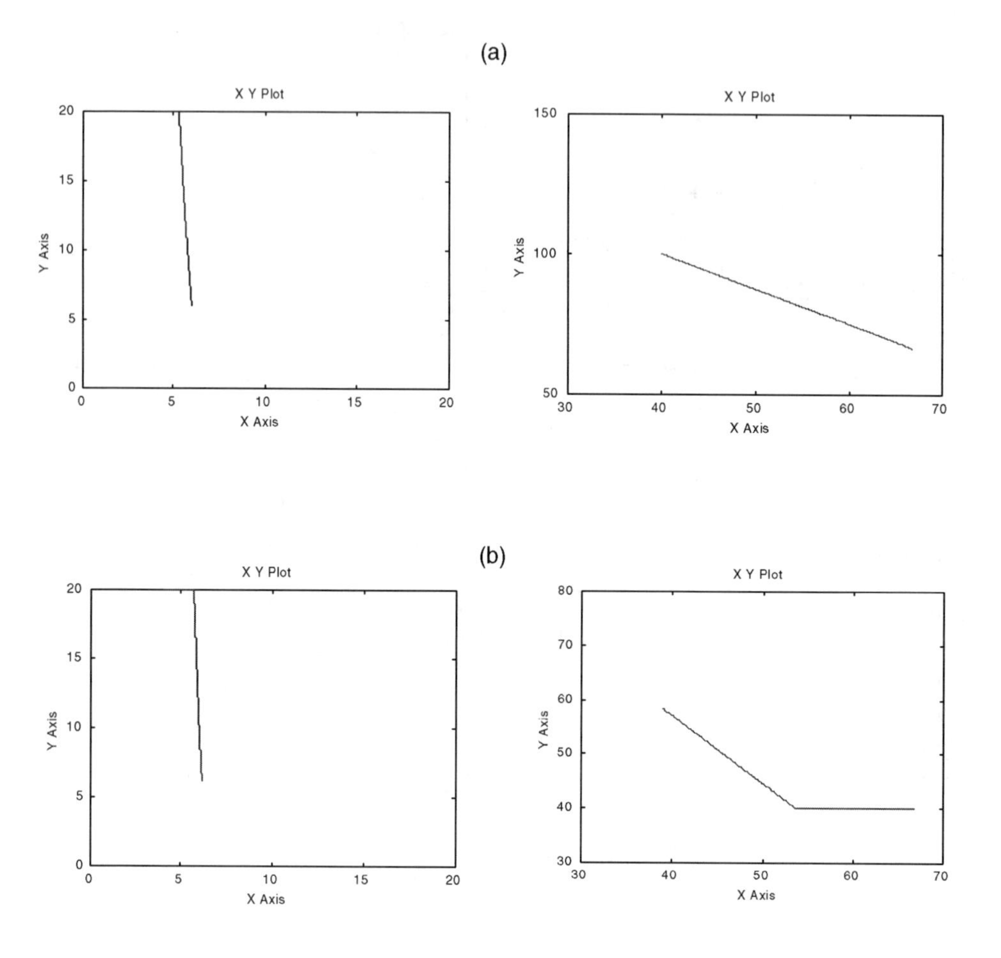


**Figure 3.16** Results of SIMULINK simulations to determine the steady-state operating point during (a) normoxia ($P_{IO_2} = 150\,\text{mm Hg}$) and (b) inhalation of 15% $O_2$ mixture ($P_{OI_2} = 107\,\text{mm Hg}$). Left panels: Ventilatory controller output vs. ventilation in $\text{L min}^{-1}$ (simulation is terminated when they become equal); Right panels: corresponding trajectory of $P_{AO_2}$ vs. $P_{ACO_2}$ in mm Hg.

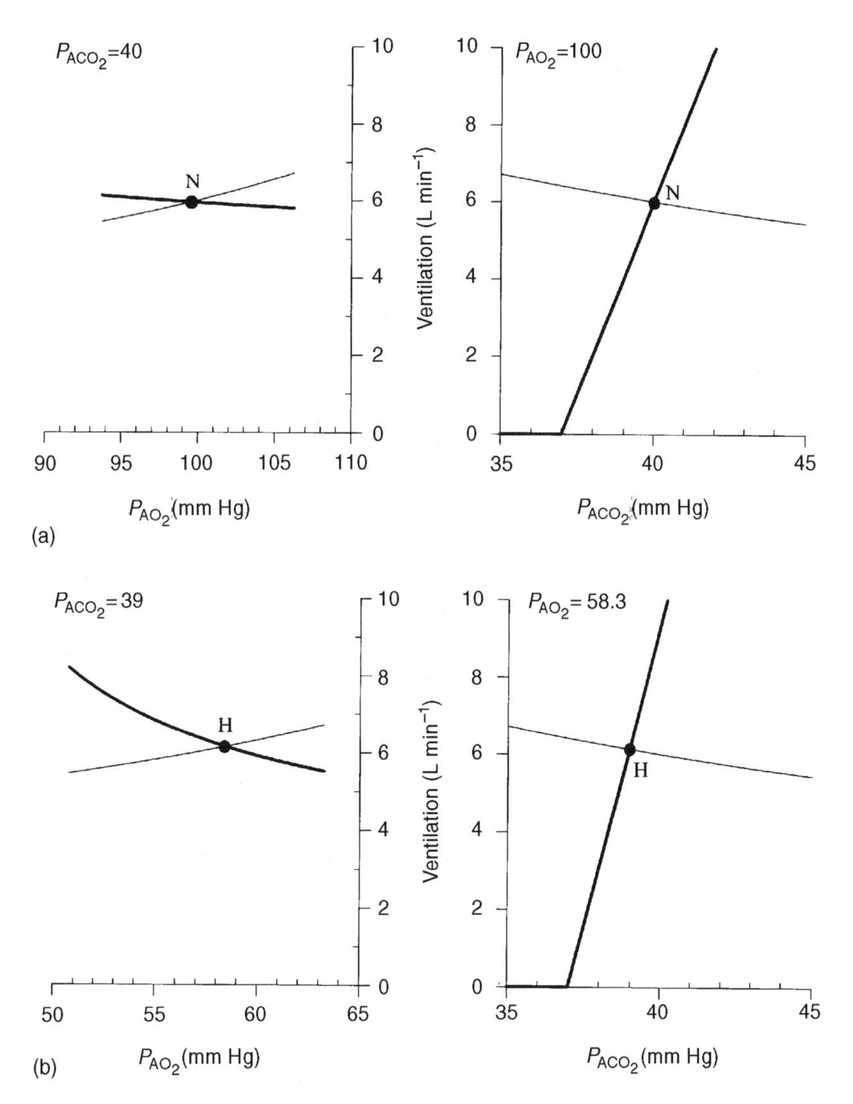


**Figure 3.17** Graphical analysis of the steady-state regulation of ventilation during (a) normoxia ($P_{IO2} = 150$ mm Hg) and (b) exposure to mild hypoxia through inhalation of 15% $O_2$ mixture or ascent to altitude ($\sim$ 8500 ft). The steady-state operating points are labeled N in case (a) and H in case (b).

## BIBLIOGRAPHY


Cunningham, D.J.C. Integrative aspects of the regulation of breathing: A personal view. In: *MTP International Review of Science: Physiology Series One*, Vol. 2, *Respiratory Physiology* (edited by J.G. Widdicombe). University Park Press, Baltimore, 1974; pp. 303–369.

Guyton, A.C., C.E. Jones, and T.G. Coleman. *Circulatory Physiology: Cardiac Output and Its Regulation*, 2d ed. W.B. Saunders, Philadelphia, 1973.

Khoo, M.C.K., R.E. Kronauer, K.P. Strohl, and A.S. Slutsky. Factors inducing periodic breathing in humans: a general model. *J. Appl. Physiol.* **53**: 644–659, 1982.

Milhorn, H.T. *The Application of Control Theory to Physiological Systems*. W.B. Saunders, Philadelphia, 1966.

Patterson, S.W., H. Piper, and E.H. Starling. The regulation of the heart beat. *J. Physiol.* (*London*) **48**: 465, 1914.

Stolwijk, J.E., and J.D. Hardy. Regulation and control in physiology. In: *Medical Physiology*, 13th ed. (edited by V.B. Mountcastle). C.V. Mosby, St. Louis, 1974; pp. 1343–1358.

## PROBLEMS

**P3.1.** Assume that the block diagram of a temperature-regulating space-suit to be worn by an astronaut for a mission to Mars is as shown in Figure P3.1. The variable $x$ represents the external temperature while $y$ represents the temperature inside the space-suit. $G_C$ is the steady-state gain of the heating/cooling device (controller) built into the space-suit, while $G_P$ represents the steady state gain associated with the thermal characteristics of the astronaut. $H$ is the gain with which the internal temperature is fed back to the controller. The operating internal temperature ($y$) is allowed to range from 60°F to 100°F.

Assume that $G_C = 2$, $G_P = 1$, and $H = 7$.

(a) What range of external temperatures can this space-suit be used for, if it is deployed in open-loop mode?

(b) What is the permissible range of external temperatures when the space-suit is deployed in closed-loop mode?

(c) Based on the results obtained in (a) and (b), what can you conclude about the effect of negative feedback in this device?

**P3.2.** Figure P3.2 shows the block diagram of a sophisticated biomedical device for regulating the dosage of anesthetic gases being delivered to a patient during surgery. Note that the plant and controller are themselves feedback control systems.

(a) Derive an expression for the *open-loop gain* of the overall control system.

(b) Derive an expression for the *closed-loop gain* of the overall control system.

(c) If $G_1 = 1$, $G_2 = 2$, $H_1 = 1$, and $H_2 = 2$, what is the *loop-gain* of the overall system?

**P3.3.** The cardiac output curve of a heart that has been transplanted into a patient is given in tabular form as follows:

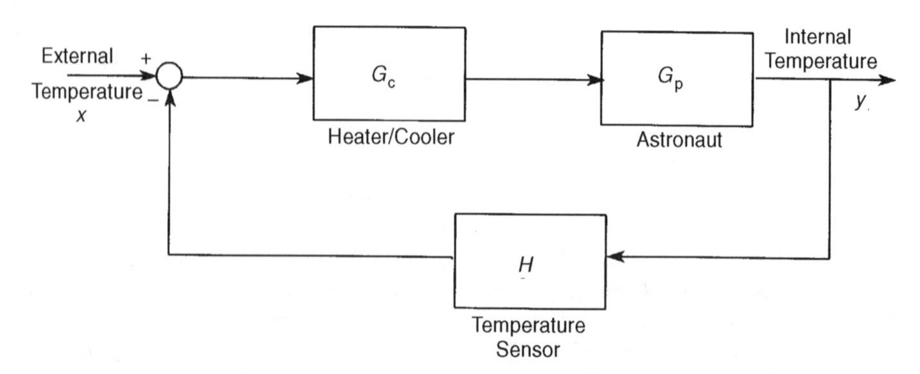


**Figure P3.1** Block diagram of the temperature control system of a space-suit.

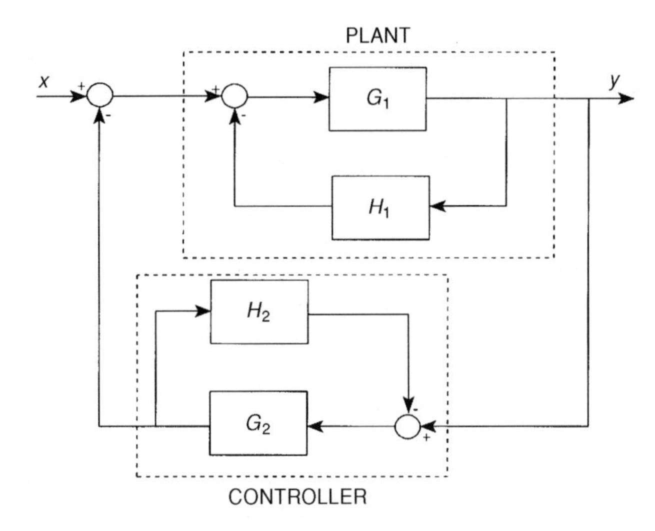


**Figure P3.2** Block diagram of the control system of a hypothetical biomedical device.

| Right Atrial Pressure (mm Hg) | 0 | 1 | 2 | 3 | 4 | 5 | 6 | 7 | 8 | 9 | 10 | 11 | 12 | 13 |
|---|---|---|---|---|---|---|---|---|---|---|---|---|---|---|
| Cardiac Output ($L\,min^{-1}$) | 0 | 0.3 | 1.0 | 2.5 | 4.8 | 7.0 | 9.0 | 10.5 | 11.0 | 11.3 | 11.5 | 11.6 | 11.6 | 11.6 |

Suppose the venous return characteristics of the patient's systemic circulation can be expressed in the form of the following equations:

$$
\begin{aligned}
Q_R &= 14, & P_{ra} &\leq 0 \\
&= 14 - 2P_{ra}, & 0 &< P_{ra} < 7 \\
&= 0, & P_{ra} &\geq 7
\end{aligned}
$$

(a) Deduce the patient's steady-state right atrial pressure ($P_{ra}$) and cardiac output, assuming the transplant operation has been successful.

(b) What would be the steady-state values for cardiac output and $P_{ra}$ if the total circulatory resistance were to be doubled?

(c) To counteract the increased circulatory resistance in (b), suppose a sufficient quantity of blood is transfused into the patient so that mean systemic pressure is raised by 5 mm Hg. What would be the new steady-state values for $P_{ra}$ and cardiac output?

**P3.4.** Assume the metabolic hyperbola for $CO_2$ given by Equation (3.48), where the steady-state $CO_2$ production rate is 200 ml min$^{-1}$ and the inspired $CO_2$ concentration is zero. Also, assume a dead-space ventilation rate of 1 L min$^{-1}$. Now, suppose the steady-state ventilatory response to $CO_2$ is given by Equation (3.51), where $P_{aO_2}$ is set equal to 100 mm Hg.

(a) What are the steady-state values of ventilation and $P_{aCO_2}$?

(b) The onset of sleep shifts the $CO_2$ response curve to the right, so that the apneic threshold is increased from 37 to 42 mm Hg. How would this affect the steady-state values of ventilation and $P_{aCO_2}$?

(c) How would inhalation of a gas mixture containing 7% $CO_2$ in air affect the steady-state ventilation and $P_{aCO_2}$ during sleep?

**P3.5.** Rising suddenly from a reclining to standing position sometimes causes a feeling of faintness due to a decrease in blood flow to the brain. However, in the normal person, this is quickly compensated for by adjustments in the circulation. Although cardiac output and venous return curves reflect *steady-state responses*, they remain useful for providing a qualitative picture of the sequence of events accompanying the change in posture. Explain how the cardiac output and venous return curves are affected at each stage of the response. Also, describe how cardiac output and right atrial pressure are changed.

(a) Rising suddenly causes extra blood to be stored in the veins of the legs.

(b) The drop in blood pressure is sensed by the baroreceptors, which lead to an increase in generalized sympathetic outflow. This increases heart rate and cardiac contractility as well as peripheral resistance.

(c) Finally, venoconstriction restores mean systemic pressure back toward its normal level.

**P3.6.** Using the model of glucose regulation in Section 3.6, to estimate the steady-state glucose and insulin levels in a patient with an abnormally high gain in the insulin response to glucose. Assume $\beta$ to be twice as large as its normal value.

**P3.7.** Develop a SIMULINK program that will solve for the steady-state levels of glucose and insulin, assuming the glucose regulation model discussed in Section 3.6. Employ the approach that was adopted in the SIMULINK model "`respss.mdl`." Use the program to simulate the conditions displayed in Figure 3.13.

# 4

# Time-Domain Analysis of Linear Control Systems

## 4.1 LINEARIZED RESPIRATORY MECHANICS: OPEN-LOOP VERSUS CLOSED-LOOP

In the previous chapter, we considered how feedback can change the steady-state behavior of physiological systems. In this chapter, we will explore the basic concepts and analytical techniques used to quantify the dynamics of *linearized* physiological models. We will perform extensive mathematical analyses of models with first-order and second-order dynamics. These are models that one can employ as "first approximations" to a number of physiological systems. They are useful in demonstrating the methods of analysis and concepts that can be applied, while allowing the mathematics to remain at a manageable, nondistracting level.

We consider a simplified version of the linearized lung mechanics model discussed in Section 2.3. Instead of the several regional resistances and compliances, this model contains only one resistance ($R$) and one compliance ($C$) element which represent, respectively, the overall mechanical resistive and storage properties of the respiratory system. Thus, $R$ represents a combination of resistance to airflow in the airways, lung tissue resistance, and chest-wall resistance. $C$ represents the combined compliance of lung tissue, chest wall, and airways. In addition, however, we will also add an inductance element, $L$, that represents fluid inertance in the airways. The electrical analog of this model is displayed in Figure 4.1. Our task is to predict how the alveolar pressure, $P_A$, will respond dynamically to different pressure waveforms ($P_{ao}$) applied at the airway opening.

Applying Kirchhoff's First Law (see Section 2.2) to the model, we find that the pressure drop across the entire model must be equal to the sum of all the pressure drops across each of the circuit elements. Thus,

$$P_{ao} - P_0 = L\frac{dQ}{dt} + RQ + \frac{1}{C}\int Q\,dt \tag{4.1}$$

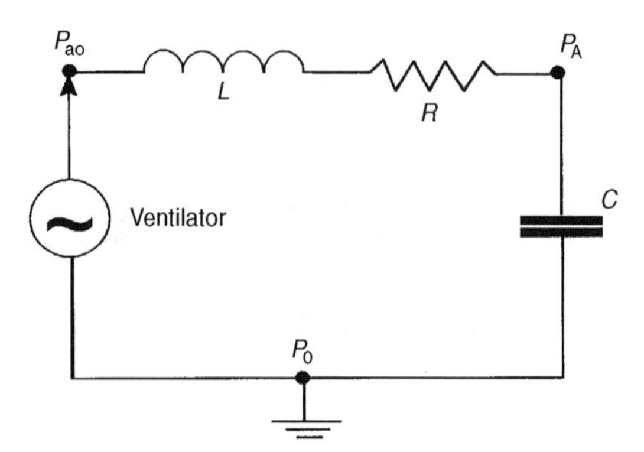


**Figure 4.1** Electrical analog of lung mechanics model.

In Equation (2.9), $Q$ represents the airflow rate. A similar expression can be derived to relate $P_A$ to $Q$:

$$P_A - P_0 = \frac{1}{C}\int Q\,dt \tag{4.2}$$

We will reference all pressures to the ambient pressure (i.e., set $P_0 = 0$). Combining Equations (4.1) and (4.2), and eliminating $Q$ from both equations, we obtain

$$P_{ao} = LC\,\frac{d^2P_A}{dt^2} + RC\,\frac{dP_A}{dt} + P_A \tag{4.3}$$

Equation (4.3) describes the dynamic relationship between $P_{ao}$ and $P_A$. Applying the Laplace transform to this second-order differential equation yields the transfer function of the model:

$$\frac{P_A(s)}{P_{ao}(s)} = \frac{1}{LCs^2 + RCs + 1} \tag{4.4}$$

This transfer function is displayed schematically in Figure 4.2a. Note that, since $P_A$ is entirely dependent on $P_{ao}$, this depicts an *open-loop* configuration.

Let us now consider an alternative situation where we would like to be able to attenuate the changes in $P_A$ as much as possible, for a given set of lung mechanical parameters and a given imposed change in $P_{ao}$. In the clinical setting, this is desirable, since large fluctuations

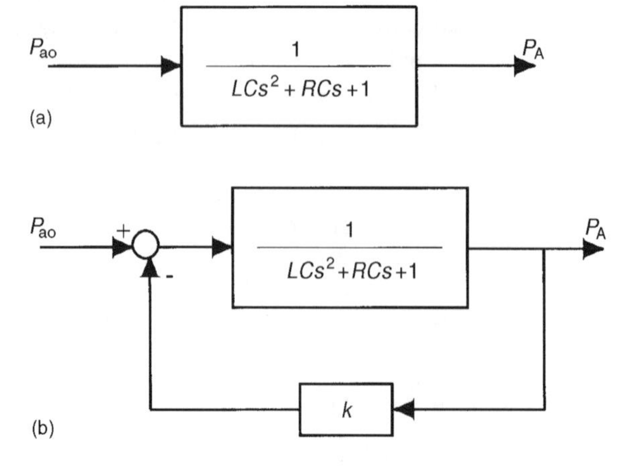


**Figure 4.2** (a) Lung mechanics model—open-loop configuration. (b) Lung mechanics model—closed-loop configuration.

in $P_A$ can cause pulmonary barotrauma or damage to lung tissue. In order to "control" $P_A$, it is necessary to measure this variable and feed the measurement back to the controller. In practice, this can be achieved by measuring the pressure in the mid-esophagus with the use of an esophageal balloon, since fluctuations in esophageal pressure have been demonstrated to closely reflect fluctuations in alveolar pressure. Thus, we assume the arrangement shown in Figure 4.2b, where $P_A$ is measured and a scaled representation of this measurement is fed back and subtracted from the input, $P_{ao}$. This clearly is a *closed-loop* configuration and the type of control scheme is known as *proportional feedback*, since the feedback variable is proportional to the system output. Reanalysis of the new block diagram yields the following result:

$$\frac{P_A(s)}{P_{ao}(s) - kP_A(s)} = \frac{1}{LCs^2 + RCs + 1} \tag{4.5a}$$

By rearranging terms in Equation (4.5a), we can derive the following expression for the overall transfer function of the closed-loop system:

$$\frac{P_A(s)}{P_{ao}(s)} = \frac{1}{LCs^2 + RCs + (1 + k)} \tag{4.5b}$$

Equations (4.4) and (4.5b) can be generalized to represent both the open-loop and closed-loop conditions:

$$\frac{P_A(s)}{P_{ao}(s)} = \frac{1}{LCs^2 + RCs + \lambda} \tag{4.6}$$

where $\lambda = 1$ for the open-loop case, and $\lambda = 1 + k$ for the closed-loop case.

## 4.2 OPEN-LOOP AND CLOSED-LOOP TRANSIENT RESPONSES: FIRST-ORDER MODEL

In the range of spontaneous breathing frequencies, studies with reasonably realistic models of respiratory mechanics, such as that by Jackson and Milhorn (1973), have demonstrated that airway fluid inertance plays a virtually insignificant role in determining lung pressures and airflow. Thus, under these conditions, we can ignore inertance effects by setting $L$ to zero. The transfer function in Equation (4.6) then becomes

$$\frac{P_A(s)}{P_{ao}(s)} = \frac{1}{\tau s + \lambda} \tag{4.7}$$

where $\tau = RC$.

### 4.2.1 Impulse Response

We can obtain the impulse response, $h_1(t)$, of the first-order system in Equation (4.7) by setting $P_{ao}(s)$ to 1, since we are assuming the input to take the form of a unit impulse. We also multiply both numerator and denominator of the right-hand side of Equation (4.7) by $1/\tau$ to reduce it to the standard form:

$$P_A(s) = \frac{1/\tau}{s + \lambda/\tau} \tag{4.8}$$

Using the table of Laplace transforms in Appendix I, it can be seen that the impulse response is

$$h_1(t) = \frac{1}{\tau} e^{-(\lambda/\tau)t} \tag{4.9}$$

Thus, the impulse response under both open-loop and closed-loop conditions is a simple exponential. Note that the peak of the impulse response is a function only of $\tau$, which depends on the system parameters $R$ and $C$, but not of $\lambda$, i.e., it is the same value under open-loop and closed-loop conditions. However, the time constant of the exponential is $\tau/\lambda$. Without proportional feedback, this time constant is $\tau$, since $\lambda$ is unity. However, with proportional feedback, $\lambda > 1$; therefore, the closed-loop impulse response decays faster. Theoretically, the "response time" of the system can be made infinitely fast if the feedback gain, $k$, is raised to an infinitely high level. A comparison of open-loop and closed-loop responses is shown in Figure 4.3a for the case where $R = 1\,\text{cm}\,H_2O\,\text{s}\,L^{-1}$, $C = 0.1\,\text{L}\,\text{cm}\,H_2O^{-1}$, and $\lambda = 2$ (i.e., $k = 1$).

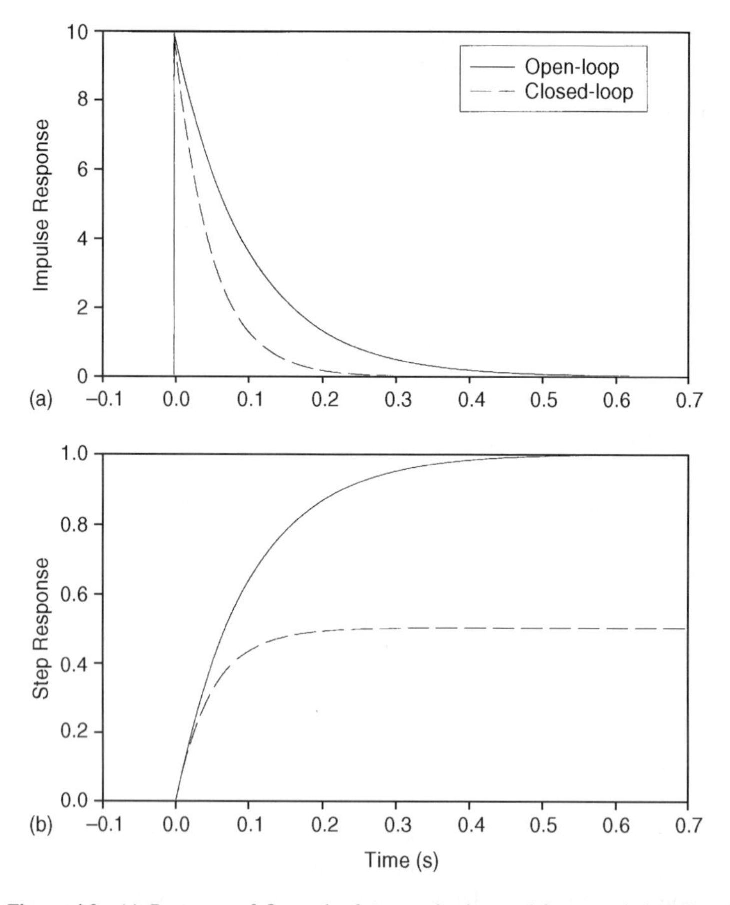


**Figure 4.3** (a) Response of first-order lung mechanics model to a unit impulse. (b) Response of first-order lung mechanics model to a unit step. Solid and dashed lines represent the model responses in open-loop and closed-loop modes. Parameter values used: $R = 1$ cm $H_2O\,s\,L^{-1}$, $C = 0.1\,L$ cm $H_2O^{-1}$, $\lambda = 2$.

### 4.2.2 Step Response

To deduce the response, $g_1(t)$, of the first-order model to a unit step, we set $P_{ao}(s)$ to $1/s$ and rearrange Equation (4.7) to obtain

$$P_A(s) = \frac{1/\tau}{s(s + \lambda/\tau)} \tag{4.10}$$

Again, the corresponding response in the time-domain can be found by using Appendix I:

$$g_1(t) = \frac{1}{\lambda}(1 - e^{-(\lambda/\tau)t}) \tag{4.11}$$

As in the case for the impulse response, the time constant for the step response is decreased when proportional feedback is introduced, i.e., the closed-loop system responds faster. Whereas the peak amplitude of the impulse response is not affected by feedback, the steady-state magnitude of the closed-loop step response is inversely proportional to $\lambda$. Thus, the greater the feedback gain, $k$, the smaller the steady-state value of the closed-loop step response. This can also be expressed as the *steady-state error*, $\varepsilon_1$, defined as the final $(t \to \infty)$ difference between the input (which is the unit step function) and the closed-loop step response. Thus, in this case, $\varepsilon_1$ increases as $k$ and $\lambda$ increase:

$$\varepsilon_1|_{t\to\infty} = 1 - \frac{1}{\lambda} \tag{4.12}$$

Figure 4.3b compares the step responses of the first-order open-loop and closed-loop respiratory mechanics model for the same parameter values as in Figure 4.3a. In the open-loop case, there is no steady-state error.

## 4.3 OPEN-LOOP VERSUS CLOSED-LOOP TRANSIENT RESPONSES: SECOND-ORDER MODEL

We now turn to the more general situation, which covers a larger range of respiratory frequencies. As the rates of change of airflow become larger, so will the effects derived from fluid inertance, $L$. This brings us back to the second-order model represented by Equation (4.6).

### 4.3.1 Impulse Responses

To deduce the impulse response, $h_2(t)$, of the second-order system, we set $P_{ao}(s) = 1$ in Equation (4.6), which becomes

$$P_A(s) = \frac{1/LC}{s^2 + (R/L)s + \lambda/LC} \tag{4.13}$$

To determine the inverse Laplace transform of Equation (4.13), it is necessary to evaluate the roots of the quadratic function in $s$. If we denote the roots by $\alpha_1$ and $\alpha_2$, then

$$\alpha_{1,2} = -\frac{R}{2L} \pm \sqrt{\frac{R^2}{4L^2} - \frac{\lambda}{LC}} \tag{4.14}$$

Depending on the values of the model parameters, the roots $\alpha_1$ and $\alpha_2$ can be imaginary, complex, real and equal, or real and different. As we will demonstrate below, these roots determine whether the model behavior is oscillatory, underdamped, critically damped, or overdamped. We will consider each of these cases individually.

***4.3.1.1. Undamped Behavior.*** The roots $\alpha_1$ and $\alpha_2$ are imaginary when $R = 0$, so that Equation (4.13) becomes

$$P_A(s) = \frac{1/LC}{s^2 + \lambda/LC} \tag{4.15}$$

Thus, the roots are

$$\alpha_1 = j\sqrt{\frac{\lambda}{LC}} \quad \text{and} \quad \alpha_2 = -j\sqrt{\frac{\lambda}{LC}} \tag{4.16a,b}$$

The inverse Laplace transform of Equation (4.15) is

$$h_2(t) = \frac{1}{\sqrt{\lambda LC}} \sin\left(\sqrt{\frac{\lambda}{LC}}\, t\right) \tag{4.17}$$

Equation (4.17) implies that the response of the model to an impulsive change in $P_{ao}$ is a sustained oscillation. In the *open-loop* configuration ($\lambda = 1$), the amplitude and the angular frequency of this oscillation are equal in magnitude, with both assuming values of $(LC)^{-1/2}$. However, in the *closed-loop* situation, $\lambda > 1$, which lowers the amplitude of the oscillation but increases its frequency. These responses are shown graphically in Figure 4.4a. In the example displayed, we have assumed the following parameter values: $L = 0.01\ \mathrm{cm\,H_2O\,s^2\,L^{-1}}$ and $C = 0.1\ \mathrm{L\,cm\,H_2O^{-1}}$. Under open-loop conditions, these parameter values produce an oscillation of frequency $(1000)^{1/2}/(2\pi)$ Hz, or approximately, 5 Hz. As in Section 4.2.1, we again assume the feedback gain, $k$, is set equal to unity, so that $\lambda = 2$. Then, the closed-loop oscillation amplitude will be $1/\sqrt{2}$ times, or approximately 71%, the oscillation amplitude in the open-loop case. At the same time, the oscillation frequency will be $\sqrt{2}$ times the corresponding value under open-loop conditions, or approximately 7 Hz.

***4.3.1.2. Underdamped Behavior.*** The sustained oscillatory responses in the previous section are, of course, highly unrealistic, since they require that $R$ be reduced to zero. Consider now the situation when $R$ is nonzero but small, so that

$$\frac{R^2}{4L^2} < \frac{\lambda}{LC} \tag{4.18}$$

The term within the square-root operation in Equation (4.14) will become negative, and consequently, the characteristic roots $\alpha_1$ and $\alpha_2$ will be complex. Equation (4.13) then becomes

$$P_A(s) = \frac{1/LC}{\left(s + \dfrac{R}{2L}\right)^2 + \left(\dfrac{\lambda}{LC} - \dfrac{R^2}{4L^2}\right)} \tag{4.19}$$

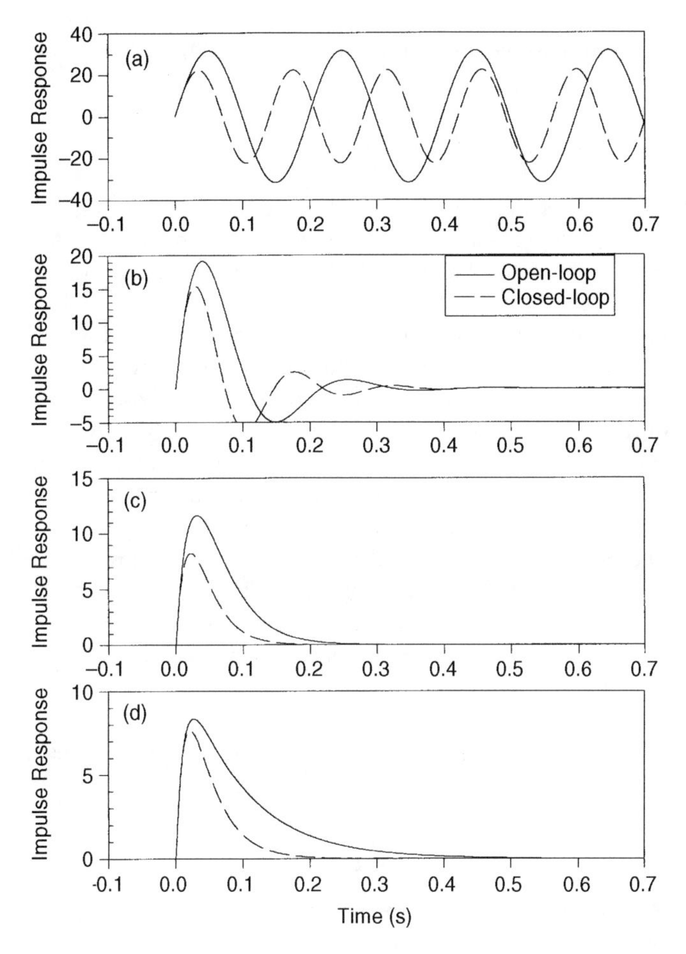


**Figure 4.4** Responses of the second-order lung mechanics model to a unit impulse under open-loop (solid lines) and closed-loop (dashed lines) modes: (a) undamped responses; (b) underdamped responses; (c) critically damped responses; and (d) overdamped responses.

which is easily converted to the standard form

$$P_A(s) = \frac{1/LC}{\left(s + \dfrac{R}{2L}\right)^2 + \gamma^2} \tag{4.20}$$

where

$$\gamma = \frac{R}{2L}\sqrt{\left(\frac{4L\lambda}{R^2C} - 1\right)} > \frac{R}{2L} \tag{4.21}$$

From Equation (4.14), the characteristic roots of the denominator are clearly

$$\alpha_{1,2} = -\frac{R}{2L} \pm j\gamma \tag{4.22}$$

Applying the inverse Laplace transform to Equation (4.20), we obtain the following impulse response:

$$h_2(t) = \frac{1}{LC\gamma} e^{-(R/2L)t} \sin(\gamma t) \tag{4.23}$$

The above result shows that, in the underdamped situation, the model responds to a unit impulse with dynamics that can be described as a damped sinusoid. Note that, in the limit when $R$ decreases to zero, Equation (4.23) degenerates into the sustained oscillation represented by Equation (4.17).

How does the incorporation of negative feedback affect this underdamped response? In the closed-loop situation, $\lambda$ becomes larger than unity, which increases $\gamma$ relative to the open-loop case. This, in turn, reduces the amplitude of the damped oscillations but increases their frequency. However, the exponential decay term is unaffected by $\lambda$. A graphical comparison of underdamped impulse responses under open-loop versus closed-loop conditions is displayed in Figure 4.4b. In this example, the values of $L$ and $C$ are the same as those employed in Section 4.3.1.1. The value of $R$ used here is 0.5 (cm $H_2O$) s $L^{-1}$. With these parameter values, $\gamma = 19.4$ in the open-loop case; thus, the frequency of the damped oscillation is approximately 3 Hz. With the incorporation of negative feedback ($k = 1$, so that $\lambda = 2$), $\gamma \approx 37.1$ so that the damped oscillation frequency becomes approximately 6 Hz. At the same time, the amplitude of the damped oscillation in the closed-loop case is significantly lower than that in the open-loop case.

***4.3.1.3. Critically Damped Behavior.*** If $R$ is increased further until the following condition becomes valid:

$$\frac{R^2}{4L^2} = \frac{\lambda}{LC} \tag{4.24}$$

$\gamma$ will become zero and Equation (4.13) will reduce to

$$P_A(s) = \frac{1/LC}{\left(s + \dfrac{R}{2L}\right)^2} \tag{4.25}$$

Thus, in this case, the characteristic roots will be real and equal, as shown below:

$$\alpha_{1,2} = -\frac{R}{2L} \tag{4.26}$$

The inverse Laplace transform of Equation (4.25) yields the impulse response of the model:

$$P_A(t) = \frac{1}{LC} te^{-t/\tau_c} \tag{4.27}$$

where

$$\tau_c = \frac{2L}{R} = \sqrt{\frac{LC}{\lambda}} \tag{4.28}$$

Note that the second part of Equation (4.28) follows directly from the equality condition expressed in Equation (4.24).

The above results demonstrate that, in the critically-damped mode, all oscillatory behavior disappears. How is the response affected by the introduction of negative feedback?

Equation (4.28) shows quite clearly that, in the closed-loop configuration where $\lambda > 1$, the single time constant for the exponential decay is shorter compared to the open-loop case when $\lambda = 1$. Thus, as was the case for the first-order model, proportional feedback increased the speed of response of the system. This comparison is displayed graphically in Figure 4.4c. However, compared to the open-loop case, we see from Equation (4.24) that $R$ has to be increased to a higher value before the damped oscillatory behavior disappears and critical damping is achieved in the closed-loop system.

***4.3.1.4. Overdamped Behavior.*** When $R$ increases above the point at which critical damping occurs, the following inequality will take effect:

$$\frac{R^2}{4L^2} > \frac{\lambda}{LC} \tag{4.29}$$

Under these circumstances, the characteristic roots of Equation (4.14) become real and different:

$$\alpha_1, \alpha_2 = -\frac{R}{2L}(1 \pm \mu) \tag{4.30}$$

where

$$\mu = \sqrt{1 - \frac{4L\lambda}{R^2 C}} \tag{4.31}$$

It follows from the inequality expressed in Equation (4.29) that $\mu$ must lie between zero and unity in Equation (4.31). The resulting expression for $P_A(s)$ becomes

$$P_A(s) = \frac{\dfrac{1}{LC}}{\left(s + \dfrac{R}{2L}(1-\mu)\right)\left(s + \dfrac{R}{2L}(1+\mu)\right)} \tag{4.32}$$

Consequently, the inverse transform of Equation (4.32) yields

$$h_2(t) = \frac{1}{\mu RC}\left(e^{-t/\tau_1} - e^{-t/\tau_2}\right) \tag{4.33}$$

where

$$\tau_1 = \frac{2L}{R(1-\mu)} \quad \text{and} \quad \tau_2 = \frac{2L}{R(1+\mu)} \tag{4.34a,b}$$

Thus, in the overdamped system, the impulse response is composed of two exponential decay contributions with larger time constant $\tau_1$ and smaller time constant $\tau_2$.

To compare the overdamped impulse responses in the closed-loop versus open-loop cases, we assume the values of $L$ and $C$ employed previously: $L = 0.01\ \mathrm{cm\,H_2O\,s^2\,L^{-1}}$ and $C = 0.1\ \mathrm{L\,cm\,H_2O^{-1}}$. To ensure that the condition described by Equation (4.29) is met in both open-loop and closed-loop conditions, we set $R = 1\ \mathrm{cm\,H_2O\,s\,L^{-1}}$. Since the "tails" of the impulse responses will be dominated by the contribution with the longer time constant, we will compare only the values of $\tau_1$ for open-loop versus closed-loop conditions. Applying Equation (4.34a), we find that in the open-loop situation, $\tau_1$ is approximately 0.09 s, while in the closed-loop condition, it is approximately 0.04 s. This comparison is shown in Figure

4.4d. Therefore, as the previous cases considered, closing the loop here also increases the speed of response of the system.

### 4.3.2 Step Responses

To determine the response of our lung mechanics model to a unit step change in $P_{ao}$, we could apply the same approach that was employed for calculating the step response of the first-order model. However, for illustrative purposes, we will proceed along a somewhat different path by making use of the results that were derived for the impulse response of the second-order model. The basic principle employed here is the equivalence between multiplication in the Laplace domain and convolution in the time domain (Section 2.7). Thus, the step response, represented in the Laplace domain as

$$P_A(s) = \frac{1}{LCs^2 + RCs + \lambda}\frac{1}{s} \tag{4.35}$$

can be evaluated in the time domain from

$$g_2(t) = \int_0^t h_2(\sigma)u(t-\sigma)\,d\sigma \tag{4.36}$$

where

$$\begin{aligned} u(t) &= 1, \qquad t > 0 \\ &= 0, \qquad t \leq 0 \end{aligned} \tag{4.37}$$

and $h(t)$ represents the impulse response of the model. Inserting Equation (4.37) into Equation (4.36), the step response can be evaluated as follows:

$$g_2(t) = \int_0^t h(\sigma)\,d\sigma \tag{4.38}$$

The expression shown in Equation (4.38) implies that the step response can be evaluated by integrating the impulse response with respect to time.

***4.3.2.1. Undamped Behavior.*** Integrating Equation (4.17) with respect to time, we obtain:

$$g_2(t) = -\frac{1}{\lambda}\cos\left(\sqrt{\frac{\lambda}{LC}}\,t\right) + A \tag{4.39a}$$

where $A$ is an arbitrary constant. Imposing the initial condition $P_A(0) = 0$ on Equation (4.39a), we obtain the step response for undamped conditions:

$$g_2(t) = \frac{1}{\lambda}\left[1 - \cos\left(\sqrt{\frac{\lambda}{LC}}\,t\right)\right] \tag{4.39b}$$

As in the case for the impulse response, the step input elicits a sustained oscillation when there is no resistance in the system. Closing the loop increases the frequency of the oscillation but decreases its amplitude. These responses are shown in Figure 4.5a.

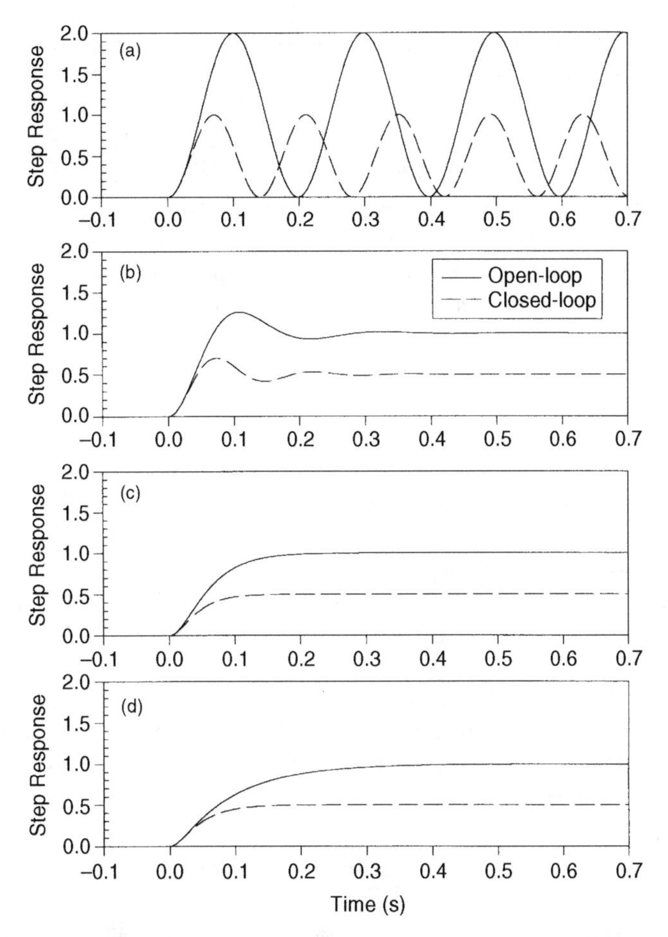


**Figure 4.5** Responses of the second-order lung mechanics model to a unit step under open-loop (solid lines) and closed-loop (dashed lines) modes: (a) undamped responses; (b) underdamped responses; (c) critically damped responses; and (d) overdamped responses.

***4.3.2.2. Underdamped Behavior.*** As in the undamped case, we obtain the step response here by convolving the impulse response described in Equation (4.23) with a unit step. This turns out to be the same as integrating the impulse response with respect to time:

$$g_2(t) = \frac{1}{LC\gamma}\int_0^t e^{-(R/2L)\sigma}\ \sin(\gamma\sigma)\ d\sigma \tag{4.40}$$

where $\gamma$ is given by Equation (4.21). Performing integration by parts and imposing the initial condition that $P_A(0) = 0$, we obtain the following expression for the underdamped step response:

$$g_2(t) = \frac{1}{\lambda}\left(1 - e^{-(R/2L)t}\ \cos\gamma t - \frac{R}{2L\gamma}\ e^{-(R/2L)t}\ \sin\gamma t\right) \tag{4.41}$$

As can be seen from Figure 4.5b, the damped oscillatory characteristics of this response are the same as those of the impulse response in both open-loop and closed-loop modes.

However, in the steady state, the oscillations become fully damped out and the response settles to the constant level given by

$$g_2(t \to \infty) = \frac{1}{\lambda} \tag{4.42}$$

In Equation (4.42), note that in the open-loop case where $\lambda = 1$, the response in $P_A$ settles down to a value of 1, i.e, the same as the unit step in $P_{ao}$. With the loop closed, however, where $\lambda > 1$, the steady state value of $P_A$ is less than unity. Thus, as it was for the first-order model, the underdamped step response for the second-order model shows a *steady-state error*, $\varepsilon_2$, given by

$$\varepsilon_2|_{t \to \infty} = 1 - \frac{1}{\lambda} \tag{4.43}$$

***4.3.2.3. Critically Damped Behavior.*** We obtain the critically damped response to the unit step by integrating Equation (4.27) with respect to time. After imposing the initial condition $P_A(0) = 0$, we have

$$g_2(t) = \frac{1}{LC}\left(\tau_c^2 - \tau_c(\tau_c + t)e^{-t/\tau_c}\right) \tag{4.44}$$

where $\tau_c$ is defined by Equation (4.28). The step responses for open-loop and closed-loop conditions are displayed in Figure 4.5c. As in the case for the corresponding impulse responses, closing the loop leads to a smaller $\tau_c$ and thus faster speed of response. In the steady state, as $t \to \infty$, Equation (4.44) becomes

$$g_2(t \to \infty) = \frac{\tau_c^2}{LC} = \frac{1}{\lambda} \tag{4.45}$$

where the second part of the above equation is derived by using Equation (4.28) to substitute for $\tau_c$. Thus, the steady-state response to a unit step and the corresponding steady-state error in the critically damped mode are the same as those in the underdamped mode.

***4.3.2.4. Overdamped Behavior.*** For the overdamped response to the unit step, we integrate Equation (4.27) with respect to time and impose the initial condition $P_A(0) = 0$ to obtain

$$g_2(t) = \frac{1}{\mu RC}\left[\tau_1(1 - e^{-t/\tau_1}) - \tau_2(1 - e^{-t/\tau_2})\right] \tag{4.46}$$

As in the critically damped case, introducing negative feedback increases the speed of response. In the steady state, as $t \to \infty$, we obtain the following result:

$$g_2(t \to \infty) = \frac{\tau_1 - \tau_2}{\mu RC} \tag{4.47a}$$

By substituting for $\tau_1$ and $\tau_2$ in Equation (4.47a) and employing the definition of $\mu$ given in Equation (4.31), it can be shown that this equation reduces to

$$g_2(t \to \infty) = \frac{1}{\lambda} \tag{4.47b}$$

Closing the loop gives rise to a steady-state error of the same magnitude as in the previous step responses. Open- and closed-loop overdamped responses to the unit step are compared in Figure 4.5d.

## 4.4 DESCRIPTORS OF IMPULSE AND STEP RESPONSES

### 4.4.1 Generalized Second-Order Dynamics

The impulse and step responses of both first-order and second-order lung mechanics models have demonstrated that when proportional feedback is introduced, alveolar pressure changes resulting from perturbations in $P_{ao}$ (the input) are attenuated. The resulting fluctuations in $P_A$ also respond more quickly to changes in $P_{ao}$ under closed-loop conditions.

In the various impulse and step responses derived for the lung mechanics model in Section 4.3, it should be pointed out that although the model contained three physiological parameters ($L$, $C$, and $R$), these parameters always appeared in combination with one another, e.g., $LC$ and $RC$. Indeed, the transfer function $P_A(s)/P_{ao}(s)$ contains only two free parameters for a given value of $k$, the feedback gain; thus, more than one combination of $R$, $C$, and $L$ may produce the same dynamics. In this section, we will present the system equations for a generalized second-order model that is characterized by the same dynamics as the lung mechanics model. In the generalized model, the second-order dynamics are governed by two independent parameters. A third parameter, the steady-state input–output gain, $G_{SS}$, is also introduced. In the particular example of the lung mechanics model, $G_{SS}$ turned out to be unity. In addition, we generalize the input and output to be $x(t)$ and $y(t)$. Then, denoting the Laplace transforms of $x(t)$ and $y(t)$ by $X(s)$ and $Y(s)$, respectively, we can convert Equation (4.5a) to

$$\frac{Y(s)}{X(s) - kY(s)} = \frac{G_{SS}}{LCs^2 + RCs + 1} \tag{4.48}$$

We can generalize Equation (4.48) further by introducing two new parameters to substitute for the three redundant parameters, $R$, $L$, and $C$. It will soon become obvious that these two new parameters provide a highly intuitive description of the dynamic properties of the model. We begin by considering the undamped open-loop system ($k = 0$). As we had shown earlier, the responses to unit impulse or step took the form of a sustained oscillation. In fact, the angular frequency of the oscillation represents the highest frequency at which the system will "resonate." This frequency is commonly referred to as the *natural frequency*, $\omega_n$. From Equation (4.17), we find that $\omega_n$ is defined by

$$\omega_n = \frac{1}{\sqrt{LC}} \tag{4.49}$$

The second new parameter that we will introduce is $\zeta$, defined as

$$\zeta = \frac{R}{2}\sqrt{\frac{C}{L}} \tag{4.50}$$

Substituting Equations (4.49) and (4.50) into Equation (4.48) and rearranging terms, it can be easily shown that the overall transfer function for the model now becomes

$$\frac{Y(s)}{X(s)} = \frac{G_{SS}\omega_n^2}{s^2 + 2\zeta\omega_n s + (1 + kG_{SS})\omega_n^2} \tag{4.51}$$

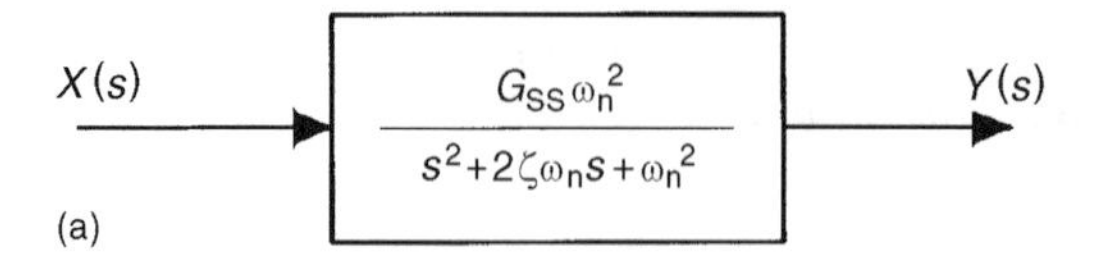


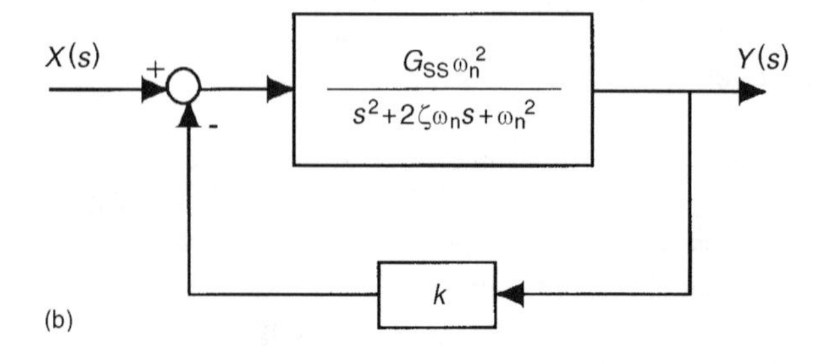


**Figure 4.6** (a) Generalized second-order open loop model. (b) Generalized second-order closed-loop model.

The open-loop and closed-loop versions of this generalized system are depicted schematically in Figures 4.6a and 4.6b, respectively.

Whether the resulting impulse or step responses are undamped, underdamped, critically damped or overdamped depends on the roots of the denominator in Equation (4.51). It can be seen that this, in turn, depends on the value of the parameter, $\zeta$. Note that when $\zeta = 0$, the impulse or step response will be a sustained oscillation. In the open-loop case, when $0 < \zeta < 1$, the impulse and step responses will show damped oscillatory behavior. However, when $\zeta \geq 1$, these responses will assume an exponential form. It is clear that $\zeta$ represents the amount of "damping" inherent in the system and, for this reason, it is commonly referred to as the *damping factor* or *damping ratio*.

***4.4.1.1. Undamped Dynamics.*** In the case when $\zeta = 0$, Equation (4.51) becomes

$$\frac{Y(s)}{X(s)} = \frac{G_{SS}\omega_n^2}{s^2 + (1 + kG_{SS})\omega_n^2} \tag{4.52}$$

The inverse Laplace transform of Equation (4.52) yields an oscillatory solution for the impulse response, $h_2(t)$:

$$h_2(t) = \frac{G_{SS}\omega_n}{\sqrt{1 + kG_{SS}}} \sin\left(\sqrt{1 + kG_{SS}}\,\omega_n t\right) \tag{4.53}$$

The step response, $g_2(t)$, which is also oscillatory, is obtained by integrating $h_2(t)$ with respect to time:

$$g_2(t) = \frac{G_{SS}}{1 + kG_{SS}}\left(1 - \cos\left(\sqrt{1 + kG_{SS}}\,\omega_n t\right)\right) \tag{4.54}$$

Note from Equation (4.54) that the step response oscillates around the constant level $G_{SS}/(1 + kG_{SS})$.

***4.4.1.2. Underdamped Dynamics.*** In the underdamped mode, when $\zeta^2 < 1 + kG_{SS}$, the denominator in Equation (4.51) can be rearranged so that the following form is obtained:

$$\frac{Y(s)}{X(s)} = \frac{G_{SS}\omega_n^2}{(s + \zeta\omega_n)^2 + \omega_n^2(1 + kG_{SS} - \zeta^2)} \tag{4.55}$$

The impulse response corresponding to Equation (4.55) is

$$h_2(t) = \frac{G_{SS}\omega_n}{\sqrt{1 + kG_{SS} - \zeta^2}}\, e^{-\omega_n \zeta t} \sin\left(\omega_n \sqrt{1 + kG_{SS} - \zeta^2}\, t\right) \tag{4.56}$$

while the step response is

$$g_2(t) = \frac{G_{SS}}{1 + kG_{SS}} \left(1 - \frac{e^{-\zeta\omega_n t}}{\sqrt{1 + kG_{SS} - \zeta^2}} \sin\left(\omega_n \sqrt{1 + kG_{SS} - \zeta^2}\, t + \theta\right)\right) \tag{4.57}$$

where

$$\theta = \tan^{-1}\left(\sqrt{\frac{1 + kG_{SS} - \zeta^2}{\zeta}}\right)$$

***4.4.1.3. Critically Damped Dynamics.*** The roots of the denominator become real and equal when $\zeta^2 = 1 + kG_{SS}$. At this point, all oscillatory dynamics disappear and the system becomes "critically damped." Equation (4.51) becomes

$$\frac{Y(s)}{X(s)} = \frac{G_{SS}\omega_n^2}{(s + \zeta\omega_n)^2} \tag{4.58}$$

The impulse response that corresponds to Equation (4.58) is

$$h_2(t) = G_{SS}\omega_n^2 t e^{-\omega_n t} \tag{4.59}$$

while the step response is

$$\begin{aligned} g_2(t) &= \frac{G_{SS}}{\zeta}\left[\frac{1}{\zeta} - \left(\frac{1}{\zeta} + \omega_n t\right) e^{-\zeta\omega_n t}\right] \\ &= \frac{G_{SS}}{\sqrt{1 + kG_{SS}}}\left[\frac{1}{\sqrt{1 + kG_{SS}}} - \left(\frac{1}{\sqrt{1 + kG_{SS}}} + \omega_n t\right) e^{-\zeta\omega_n t}\right] \end{aligned} \tag{4.60}$$

***4.4.1.4. Overdamped Dynamics.*** When $\zeta^2$ exceeds $1 + kG_{SS}$, the roots of the denominator of Equation (4.51) become real and different, so that the corresponding impulse and step responses now become

$$h_2(t) = \frac{G_{SS}\omega_n}{2\sqrt{\zeta^2 - 1 - kG_{SS}}}\left(e^{-\omega_n(\zeta - \sqrt{\zeta^2 - 1 - kG_{SS}})t} + e^{-\omega_n(\zeta + \sqrt{\zeta^2 - 1 - kG_{SS}})t}\right) \tag{4.61}$$

$$g_2(t) = \frac{G_{SS}}{2\sqrt{\zeta^2 - 1 - kG_{SS}}}\left(\frac{1 - e^{-\omega_n(\zeta - \sqrt{\zeta^2 - 1 - kG_{SS}})t}}{\zeta - \sqrt{\zeta^2 - 1 - kG_{SS}}} + \frac{1 - e^{-\omega_n(\zeta + \sqrt{\zeta^2 - 1 - kG_{SS}})t}}{\zeta + \sqrt{\zeta^2 - 1 - kG_{SS}}}\right) \tag{4.62}$$

***4.4.1.5. Steady-State Error.*** In the underdamped, critically damped and overdamped modes of the generalized second-order system, the step response attains the same steady-state value. In Equations (4.57) and (4.60), we can deduce this final value easily by letting $t$ tend to infinity. The same can be done for the overdamped mode in Equation (4.62), except that a little algebra will be needed to obtain the following expression for the steady-state response:

$$g_2(t \to \infty) = \frac{G_{SS}}{1 + kG_{SS}} \tag{4.63}$$

The steady-state error is deduced by subtracting the steady-state response from the input value, which is unity since the unit step was employed. Under open-loop circumstances ($k = 0$), the steady-state error would be

$$\varepsilon_2|_{\text{open-loop}} = 1 - G_{SS} \tag{4.64}$$

Note that, in the special case when $G_{SS} = 1$, as in the example considered in Section 4.3.2, the open-loop steady-state error is zero. However, when $G_{SS}$ assumes other values, the open-loop steady-state error can be quite large. In the closed-loop case, the steady state error is given by

$$\varepsilon_2|_{\text{closed-loop}} = 1 - \frac{G_{SS}}{1 + kG_{SS}} \tag{4.65}$$

In the special case where $G_{SS} = 1$, the steady-state error becomes $k/(1 + k)$, as was previously shown in the example in Section 4.3.2.

### 4.4.2 Transient Response Descriptors

The first-order and second-order impulse and step responses we have discussed constitute the simplest approximations to the corresponding time-domain dynamics of real physiological systems. To characterize more realistic impulse and step responses, one could in principle extend the modeling analysis to higher-order models. But as this process continues, the mathematics rapidly become less and less tractable. Furthermore, the number of parameters needed to describe these responses will also increase. In some situations, it may be necessary to compare the dynamic behavior of one system with that of another. Alternatively, one may need to compare the dynamic characteristics of the same system under different conditions. In order to do this, it is possible to first estimate the impulse and/or step responses, and then extract certain descriptors from these responses empirically. Subsequently, standard statistical analyses, such as the Student $t$-test, can be employed to determine whether the two sets of dynamic responses are significantly different from one another. The descriptors discussed below are among the most commonly used in systems analysis.

***4.4.2.1. Impulse Response Descriptors.*** These descriptive features of the impulse response are illustrated in Figure 4.7a. The most direct feature is the *peak amplitude*, which simply measures the maximum (or minimum, if the response is predominantly negative) value of the impulse response. Thus,

$$\text{Peak Amplitude} = \max[h(t)] \qquad \text{or} \qquad |\min[h(t)]| \tag{4.66}$$

The *area* under the impulse response function represents the integral of $h(t)$ over time, which in turn yields the *steady-state gain*, $G_{SS}$, of the system:

$$G_{SS} = \int_{-\infty}^{\infty} h(t)\, dt \tag{4.67}$$

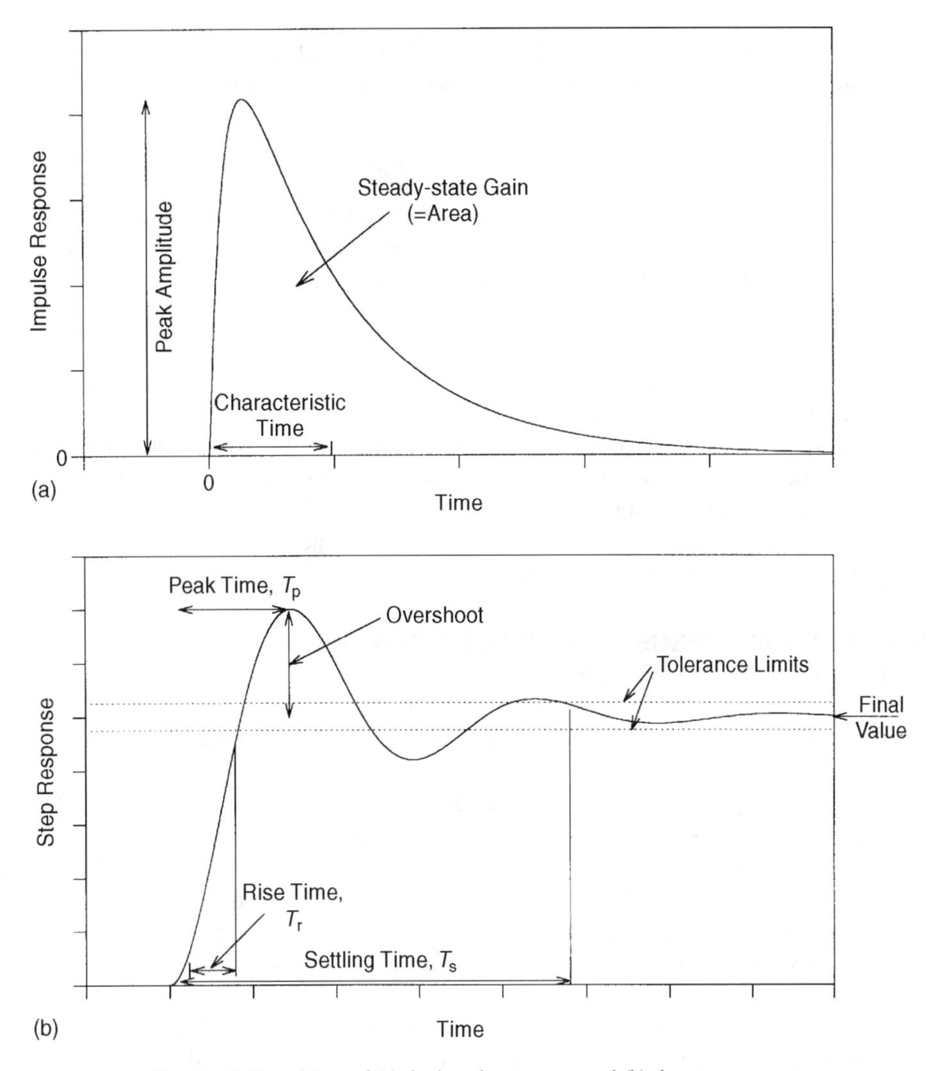


**Figure 4.7** Descriptors of (a) the impulse response and (b) the step response.

Finally, the *characteristic time*, $T_c$, provides a measure of the approximate latency following which the bulk of the impulse response occurs. Alternatively, $T_c$ may also be thought of in the following way. If the impulse response function is represented as a two-dimensional mass, then $T_c$ will be the location (on the time axis) at which the center of mass acts:

$$T_c = \frac{\int_{-\infty}^{\infty} t\, |h(t)|\, dt}{\int_{-\infty}^{\infty} |h(t)|\, dt} \tag{4.68}$$

***4.4.2.2. Step Response Descriptors.*** The most commonly used descriptors of the step response are shown in Figure 4.7b. As mentioned previously, the *final value* of the response is the steady-state level achieved by the system in question. If the input is a *unit step*, this final value will yield the steady-state gain, $G_{SS}$. If the peak value of the step response is larger than the final value, the *overshoot* will be the difference between this peak value and the final value. Frequently, this overshoot is expressed in percentage terms:

$$\text{Percent Overshoot} = \frac{\text{Peak Response} - \text{Final Value}}{\text{Final Value}} \times 100\% \tag{4.69}$$

The time taken for the step response to achieve its peak value is known as the *peak time*, or $T_p$ as illustrated in Figure 4.7b. Aside from peak time, there are two other measures of speed of response. One is the *rise time*, $T_r$, defined as

$$T_r = t_{90\%} - t_{10\%} \tag{4.70}$$

where $t_{90\%}$ = time at which response first achieves 90% of its final value, $t_{10\%}$ = time at which response first achieves 10% of its final value.

The other measure of speed of response is the settling time, $T_s$, defined as the time taken for the step response to settle within $\pm\delta\%$ of the final value. The upper and lower levels of this band of values, i.e. $100 + \delta\%$ and $100 - \delta\%$ of the final value, define the tolerance limits within which the step response will remain at all times greater than $T_s$. The values of $\delta$ generally employed range from 1% to 5%.

## 4.5 OPEN-LOOP VERSUS CLOSED-LOOP DYNAMICS: OTHER CONSIDERATIONS

### 4.5.1 Reduction of the Effects of External Disturbances

In our previous discussions of the first-order and second-order models of lung mechanics, we showed that one clear consequence of introducing negative feedback into the control scheme is *an increase in speed of system response*. A second major effect of closing the loop is the reduction in overall system gain. For both first-order and second-order models, closing the loop led to a significant reduction of the final values in the unit step responses (see Figures 4.3b and 4.5). This result is consistent with the conclusion that we arrived at in Section 3.2, although those considerations were based entirely on steady-state conditions. As we had pointed out in that section, what is most advantageous about this reduction in overall system gain is the enhanced ability of the closed-loop system to attenuate the impact of external disturbances. To emphasize the importance of this point, we will consider a simple example here.

Figures 4.8a and 4.8b illustrate the open-loop and closed-loop versions of a generalized linear control system. $D(s)$ represents the Laplace transform of an external disturbance that contributes "noise" directly and additively to the output. Thus, in the open-loop case,

$$Y(s) = G(s)X(s) + D(s) \tag{4.71}$$

which clearly shows that 100% of the external disturbance is reflected in the output. However, in the closed-loop case, we have

$$Y(s) = G(s)[X(s) - H(s)Y(s)] + D(s) \tag{4.72a}$$

which, upon rearranging terms, becomes:

$$Y(s) = \frac{G(s)}{1 + G(s)H(s)} X(s) + \frac{1}{1 + G(s)H(s)} D(s) \tag{4.72b}$$

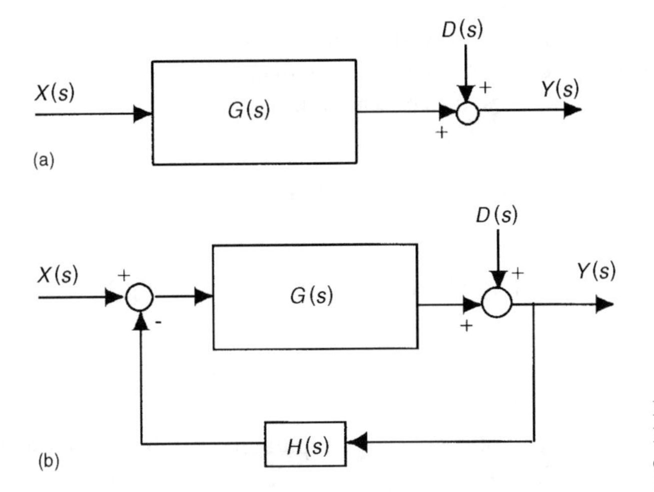


**Figure 4.8** (a) Generalized linear open-loop system. (b) Generalized linear closed-loop system.

Since we have explicitly incorporated negative feedback into the equations, the common denominator in Equation (4.72b) satisfies the following condition:

$$|1 + G(s)H(s)| > 1 \tag{4.73}$$

As such, the effect of $D(s)$ on $Y(s)$ will be attenuated and can be further attenuated as we increase the magnitude of the product $G(s)H(s)$, which is the *loop gain* (*LG*) of the closed-loop system.

### 4.5.2 Reduction of the Effects of Parameter Variations

There are situations, particularly when dealing with the artificial control of some physiological variable, where there may be a need to decide upon a range of the input $x(t)$ signal in order to closely regulate variations in the output $y(t)$. This can only be done if we have a very good idea of the characteristics of the feedforward subsystem $G(s)$. However, this may not always be possible, as we may have erroneous estimates of $G(s)$ or $G(s)$ may actually be time-varying. These variations in the system parameters will have an impact on the controlled output.

First, consider the open-loop case. Assume that there is a small change in the transfer characteristics of $G(s)$, which we will denote by $\Delta G(s)$. Then, the effect on the output will be

$$Y(s) + \Delta Y(s) = [G(s) + \Delta G(s)]X(s) + D(s) \tag{4.74}$$

Eliminating the equivalent expression for $Y(s)$, we can derive the following:

$$\Delta Y(s) = \Delta G(s)X(s) \tag{4.75}$$

The above result shows that the variation in $G(s)$ is directly reflected in the output. Now, consider the corresponding result as we apply the same type of analysis to the closed-loop system:

$$Y(s) + \Delta Y(s) = D(s) + [G(s) + \Delta G(s)]\{X(s) - H(s)[Y(s) + \Delta Y(s)]\} \tag{4.76}$$

Again, we expand Equation (4.76) and eliminate $Y(s)$ from both sides of the equation. We also eliminate the term containing the product of differences $\Delta G(s)$ and $\Delta Y(s)$, since we have assumed these differences to be small. These steps lead to the following result:

$$\Delta Y(s) = \frac{\Delta G(s)}{[1 + G(s)H(s)]^2} X(s) \tag{4.77}$$

Thus, in the closed-loop case, the effect of $\Delta G(s)$ is reduced by a factor of $[1 + G(s)H(s)]^2$.

### 4.5.3 Integral Control

In spite of the many advantages of employing proportional feedback control, one problem that can be highly aggravating in some applications is the existence of the steady-state error. We will demonstrate in this section that the steady-state error can be eliminated completely by employing integral control. To understand how this can be achieved, consider the proportional control and integral control systems shown in Figures 4.9a and 4.9b, respectively.

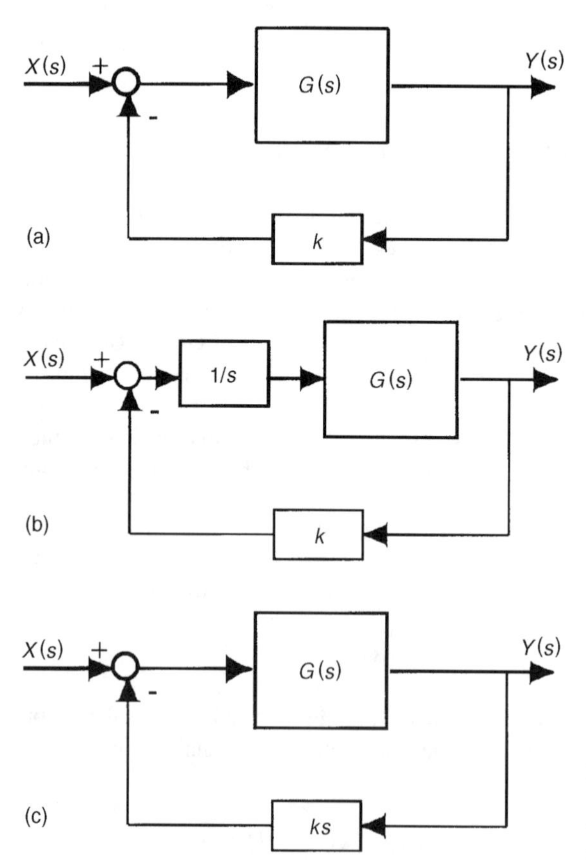


**Figure 4.9** Different closed-loop control schemes: (a) proportional feedback control; (b) integral control; (c) derivative feedback control.

First consider the proportional control system shown. This represents a generalization of the particular first-order and second-order models discussed in Sections 4.2 and 4.3. The Laplace transform of the difference (error) between the input and output is given by

$$E(s) = X(s) - Y(s) = \left(1 - \frac{G(s)}{1 + kG(s)}\right) X(s) \tag{4.78}$$

The steady-state error, $e(t \to \infty)$, can be deduced from the above equation by using the unit step input (i.e., setting $X(s) = 1/s$) and evaluating the result via the Final Value Theorem for Laplace transforms:

$$e(t \to \infty) = \lim_{s \to 0} sE(s) \tag{4.79}$$

Applying Equation (4.79) to Equation (4.78), we obtain

$$e(t \to \infty) = \frac{1 + (k-1)G_{SS}}{1 + kG_{SS}} \tag{4.80}$$

where $G_{SS}$ here represents the steady-state value of $G(s)$. Note that we can minimize the steady-state error by setting $k$ equal to unity, in which case:

$$e(t \to \infty)_{\min} = \frac{1}{1 + G_{SS}} \tag{4.81}$$

Now consider the case for integral control, in which the error signal is integrated prior to being used to drive the actuator (plant) portion of the closed-loop system. In this case, the Laplace transform of the difference between input and output is

$$E(s) = X(s) - Y(s) = \left(1 - \frac{G(s)}{s + kG(s)}\right) X(s) \tag{4.82}$$

Using Equation (4.79), and assuming the input to be a unit step ($X(s) = 1/s$), we obtain

$$e(t \to \infty) = \frac{(k-1)G_{SS}}{kG_{SS}} \tag{4.83}$$

In this case, we can eliminate the steady-state error completely by setting $k$ to unity. However, this advantage of not having a steady-state error is derived at the expense of speed of system response.

For a more intuitive explanation of why there is always a steady-state error in proportional feedback control but not in integral control, consider both cases when $k = 1$. In proportional control, if the steady-state error ($e(t \to \infty) = x(t \to \infty) - y(t \to \infty)$) is zero, this would also make the error signal that drives the actuator/plant zero, which in turn implies that the steady-state output $y(t \to \infty)$ would become zero. This result would be incompatible with the prior assertion that $e(t \to \infty)$ is zero, since $x(t \to \infty)$ equals unity. Thus, for the proportional feedback system, *a steady-state error must exist* in order for the system to produce a nonzero output. Now consider the integral control scheme. Assume that, before time zero, both input and output are zero. When the unit step takes effect at the input, $y(t)$ will initially remain at zero and consequently, there will be a large error signal that feeds into the integrator. However, with time, as $y(t)$ increases toward its final value, this error signal will diminish. On the other hand, the output of the integrator will remain high since it represents the accumulation of all previous values of the error signal. Finally, when $y(t \to \infty)$ attains the same value as $x(t \to \infty)$, the steady-state error will become zero, and the integrator output, which drives the actuator/plant, will cease increasing but remain at its final positive value so that $y(t \to \infty)$ will be unchanged (and equal to $x(t \to \infty)$).

### 4.5.4 Derivative Feedback

Instead of feeding back a signal directly proportional to the system output, how would closed-loop dynamics be different if the feedback signal were proportional to the *time-derivative* of the output? Consider the control scheme illustrated in Figure 4.9c and, for the sake of simplicity, let us assume in this example that

$$G(s) = \frac{1}{\tau s + 1} \tag{4.84}$$

Then,

$$\frac{Y(s)}{X(s) - ksY(s)} = \frac{1}{\tau s + 1} \tag{4.85}$$

From Equation (4.85), we derive the following expression for the overall system transfer function:

$$\frac{Y(s)}{X(s)} = \frac{1}{\tau' s + 1} \tag{4.86}$$

where

$$\tau' = \tau + k \tag{4.87}$$

The unit step response corresponding to Equations (4.86) and (4.87) is

$$g_1(t) = 1 - e^{-(t/\tau + k)} \tag{4.88}$$

It is clear from this result that derivative feedback increases the effective time constant and therefore produces a more sluggish response. In other words, derivative feedback increases system damping.

To determine how derivative feedback affects steady-state error, we derive from Equation (4.86) the following expression:

$$E(s) = X(s) - Y(s) = \left(1 - \frac{1}{\tau' s + 1}\right) X(s) \tag{4.89}$$

Then, using Equation (4.79), the steady-state error is found to be

$$e(t \to \infty) = \lim_{s \to 0} \left[ s\left(1 - \frac{1}{\tau' s + 1}\right)\frac{1}{s}\right] = 0 \tag{4.90}$$

Thus, derivative feedback of the kind shown in Figure 4.9c leads to the elimination of the steady-state error.

There is a popular variant of this type of control known as "velocity feedback," in which the feedback signal consists of the sum of a term proportional to the output and a term proportional to the derivative of the output. In this case, there will in general be a steady-state error. However, the steady-state error can be attenuated by increasing the gain of the forward block, $G_{SS}$. In the limit, when $G_{SS} \to \infty$, the steady-state error will become zero.

## 4.6 TRANSIENT RESPONSE ANALYSIS USING MATLAB

If the form of the transfer function of a given model is known, the response of the system to standard inputs, such as the unit impulse or unit step, as well as any arbitrary input waveform, can be deduced easily in MATLAB. The following MATLAB command lines (also found in the script file `tra_llm.m`) demonstrate how transient response analysis can be applied to the linearized lung mechanics model that we have been discussing.

Assuming that the parameter values of $L$, $R$, $C$, and $k$ in Equation (4.5b) have been preassigned, we begin by setting up the transfer function, `Hs`, of the model:

```
>> num = [1]
>> den = [L*C R*C 1 + k];
>> Hs = tf(num,den;
>> t=[0:0.005:0.8];
```

The first two lines assign values to the various terms in the numerator (`num`) and denominator (`den`) of `Hs`. In the case of the denominator, these values are assigned in the order of descending powers of $s$. The fourth line simply generates a time vector covering the duration of the response that we will examine.

The impulse response is computed and plotted using the following command lines:

```
>> x = impulse (Hs, t);
>> plot(t,x)
```

The command lines that follow produce a plot of the unit step response:

```
>>  y = step (Hs, t);
>> plot(t,y)
```

Finally, the response of this system to an input, $u$, of arbitrary time-course can be computed using the `lsim` function:

```
>> [u, t] = gensig('square',0.5,5,0.005);
>> y = lsim (Hs, u, t);
>> plot(t,y)
```

In this example, the "arbitrary input" is a square wave of period 0.5 s, lasting up to time $t = 5$ s (in time steps of 0.005 s), generated with the function "`gensig`."

## 4.7 SIMULINK APPLICATION: DYNAMICS OF NEUROMUSCULAR REFLEX MOTION

Up to this point, we have limited our analyses to simple models with only first-order or second-order dynamics. This was done intentionally to demonstrate the methodology employed in classical time-domain analysis without letting the mathematical details become too intractable and distracting. To extend this kind of analysis to more complex (and more realistic) physiological models, it becomes progressively more convenient to employ the methods of computer simulation. In this section, we will demonstrate an example of time-domain analysis using SIMULINK.

### 4.7.1 A Model of Neuromuscular Reflex Motion

Examination of the dynamics of neuromuscular reflex motion can yield valuable insight into the status of patients who have neurological disorders. The model that we will consider assumes the following test. The patient is seated comfortably and his shoulder and elbow are held by adjustable supports so that the upper arm remains in a fixed horizontal position throughout the test. The subject's forearm is allowed to move only in the vertical plane. At the start of the experiment, he is made to flex his arm by pulling on a cord that has been attached to a cuff on his wrist. The cord runs around a pulley system and supports a sizeable weight. The initial angle between the forearm and upper arm is 135°. The subject is not given any specific instructions about maintaining this angle, except to relax his arm as much as possible while supporting the weight. Then, at time $t = 0$, an electromagnetic catch is switched off so that an additional weight is abruptly added to the original load. Changes in angular motion, $\theta(t)$, of the forearm about the elbow are recorded during and after the quick release of the weight. The mathematical model used to interpret the results of this test is based on the work of Soechting et al. (1971).

***4.7.1.1. Limb Dynamics.*** Figure 4.10a shows a schematic diagram of the forearm, with the black filled circle representing the elbow joint. $M_x$ represents the change in external moment acting on the limb about the elbow joint; in this experiment, $M_x$ would be a step. $M$ represents the net muscular torque exerted in response to the external disturbance. Neglecting the weight of the forearm itself, application of Newton's Second Law yields the following equation of motion:

$$M_x(t) - M(t) = J\ddot{\theta} \tag{4.91}$$

where $J$ is the moment of inertia of the forearm about the elbow joint.

***4.7.1.2. Muscle Model.*** Although this reflex involves both the biceps and triceps muscles, we will assume for simplicity that the net muscular torque in response to $M_x$ is generated by a single equivalent muscle model, illustrated in Figure 4.10b. Note that in this mechanical analog, $M$ is treated as if it were a "force," although it is actually a torque. Accordingly, the "displacements" that result are in fact angular changes, $\theta$ and $\theta_1$. As such, the muscle stiffness parameter, $k$, and the viscous damping parameter, $B$, have units consistent with this representation. The equations of motion for the muscle model are:

$$M(t) = k(\theta - \theta_1) \tag{4.92}$$

and

$$M(t) = M_0(t) + B\dot{\theta}_1 \tag{4.93}$$

where $M_0(t)$ is the torque exerted by the muscle under isometric conditions. $M_0(t)$ is represented as a function of time, since it is dependent on the pattern of firing of the alpha motorneurons.

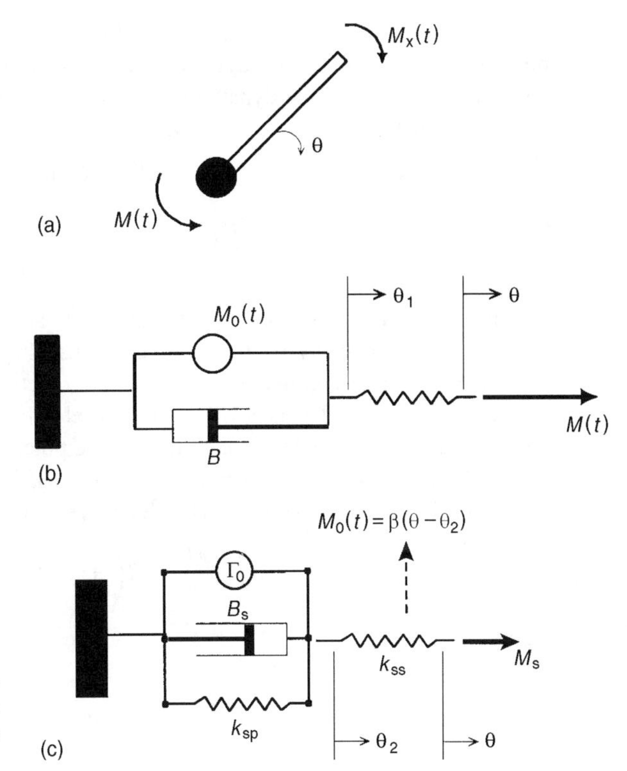


**Figure 4.10** Components of the neuromuscular reflex model: (a) limb dynamics; (b) muscle model; (c) muscle spindle model.

***4.7.1.3. Plant Equations.*** By combining Equations (4.91) through (4.93), we obtain an equation of motion that characterizes the dynamics of the plant, i.e., describing how $\theta$ would change due to the torque exerted by the external disturbance $M_x$ and the resulting muscular response:

$$\frac{BJ}{k}\dddot{\theta} + J\ddot{\theta} + B\dot{\theta} = M_x(t) - M_0(t) \tag{4.94}$$

***4.7.1.4. Muscle Spindle Model.*** This model describes the dynamics by which changes in $\theta$ are transduced at the level of the muscle spindles into afferent neural signals. The latter travel to the spinal cord, which sends out efferent signals to the contractile machinery of the muscle to generate $M_0(t)$. We assume that the neural output of the spindle is proportional to the amount by which its nuclear bag region is stretched, so that ultimately

$$M_0(t) = \beta(\theta - \theta_2) \tag{4.95}$$

Figure 4.10c shows the mechanical analog of the muscle spindle model. $k_{sp}$ and $B_s$ are parameters that represent the elastic stiffness and viscous damping properties, respectively, of the pole region of the spindle, while $k_{ss}$ represents the elastic stiffness of the nuclear bag region. $\Gamma_0$ represents the contractile part of the pole region, which allows the operating length of the spindle to be reset at different levels, using the gamma motorneuronal pathways. We will assume $\Gamma_0$ to be constant at the equilibrium length of the spindle, so that this parameter does not play a role in the dynamics of changes about this equilibrium length. With this consideration in mind, the dynamics of the muscle spindle model may be characterized by the following equations:

$$M_s = K_{ss}(\theta - \theta_2) \tag{4.96}$$

and

$$M_s = B_s\dot{\theta}_2 + k_{sp}\theta_2 \tag{4.97}$$

Another important factor that must be taken into account is the fact that, although $\theta$ is sensed virtually instantaneously by the spindle organs, there is a finite delay before this feedback information is finally converted into corrective action at the level of the muscle. This total delay, $T_d$, includes all lags involved in neural transmission along the afferent and efferent pathways as well as the delay taken for muscle potentials to be converted into muscular force. Eliminating the intermediate variables, $M_s$ and $\theta_2$, from Equations (4.95) through (4.97), we obtain the following equation for the feedback portion of the stretch reflex model:

$$M_0 + \frac{M_0}{\tau} = \beta\left(\dot{\theta}(t - T_d) + \frac{\theta(t - T_d)}{\eta\tau}\right) \tag{4.98}$$

where

$$\tau = \frac{B_s}{k_{ss} + k_{sp}} \tag{4.99}$$

and

$$\eta = \frac{k_{ss} + k_{sp}}{k_{sp}} \tag{4.100}$$

***4.7.1.5. Block Diagram of Neuromuscular Reflex Model.*** Taking the Laplace transforms of Equations (4.94) and (4.98), we obtain the following equations that are represented schematically by the block diagram shown in Figure 4.11:

$$\theta(s) = \frac{M_x(s) - M_0(s)}{s\left(\frac{BJ}{k}s^2 + Js + B\right)} \tag{4.101}$$

and

$$M_0(s) = \beta\frac{\tau s + 1/\eta}{\tau s + 1}e^{-sT_d}\theta(s) \tag{4.102}$$

### 4.7.2 SIMULINK Implementation

The SIMULINK implementation of the neuromuscular reflex model is depicted in Figure 4.12. This program has been saved as the file "`nmreflex.mdl`." Note that the model parameters appear in the program as variables and not as fixed constants. This gives us the flexibility of changing the parameter values by entering them in the MATLAB command window or running a MATLAB m-file immediately prior to running the SIMULINK program. In this case, we have chosen the latter path and created an m-file called

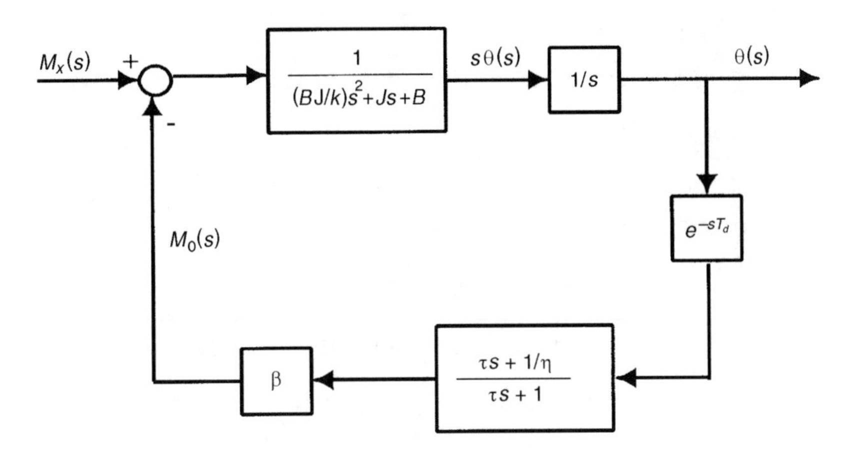


**Figure 4.11** Block diagram of neuromuscular reflex model.

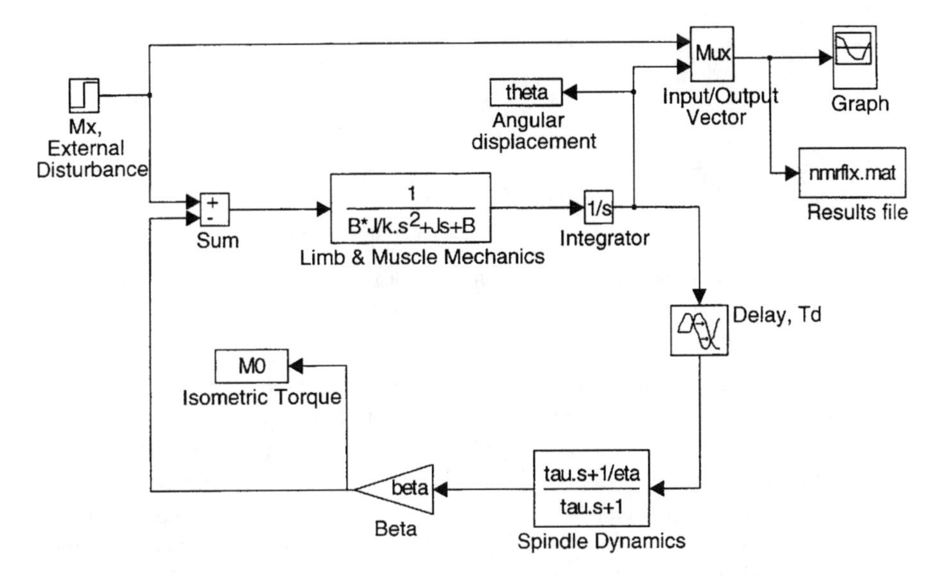


**Figure 4.12** SIMULINK implementation of neuromuscular reflex model.

"`nmr_var.m`" that specifies the parameter values. The nominal parameter values used in the simulation are as follows: $J = 0.1\,\text{kg m}^2$, $k = 50\,\text{N m}$, $B = 2\,\text{N m s}$, $T_d = 0.02\,\text{s}$, $\tau$ ("`tau`" in Figure 4.12) = 1/300 s, $\eta$ ("`eta`" in Figure 4.12) = 5, and $\beta$ ("`beta`" in Figure 4.12) = 100. These values are consistent with the average physiological equivalents found in normal adult humans.

Figure 4.13 displays the results of three simulation runs with "`nmreflex.mdl`" using the nominal parameter values mentioned above. The upper panel shows the time-course of the external disturbance, $M_x$, which is a step increase of 5 N m in the moment applied to the forearm. The solid tracing in the lower panel represents the corresponding response in $\theta$, the angular displacement of the forearm, when $\beta$ was set equal to 100. Note that positive values of $\theta$ correspond to increases in the angle of flexion between the forearm and the upper arm. There is a slight overshoot in $\theta$, followed by an almost undetectable oscillation before the

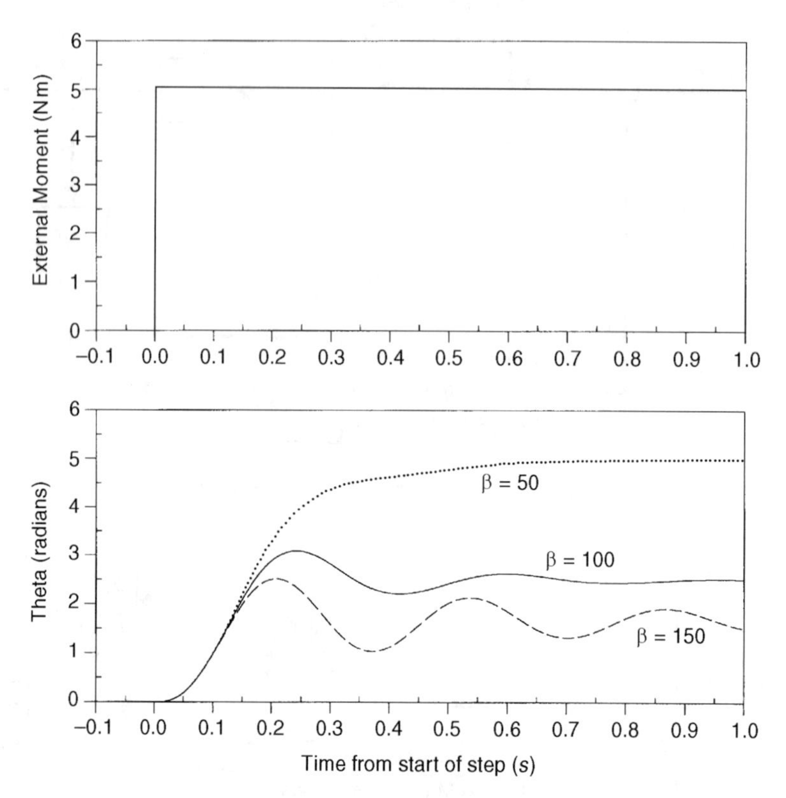


**Figure 4.13** Sample results of simulations using the SIMULINK implementation of the neuromuscular reflex model.

steady-state value of approximately 0.25 radian is attained. Note that $\beta$ represents the overall gain of the reflex arc. When $\beta$ was increased to 150, the response was a damped oscillation, but the steady-state value achieved by $\theta$ became smaller than that obtained with the nominal value of $\beta$. In the third simulation, $\beta$ was decreased to half the nominal value (i.e., 50). This produced an overdamped response and also resulted in a larger end-value for $\theta$. These results reiterate the point that increased feedback gain leads to better attenuation of the effects of imposed disturbances—higher values of $\beta$ produced smaller ending values for $\theta$. On the other hand, the responses also become more oscillatory. This issue of instability will be discussed further in Chapter 6.

## BIBLIOGRAPHY

Dorf, R.C., and R.H. Bishop. *Modern Control Systems*, 7th ed. Addison-Wesley, Reading, MA, 1995.

Dorny, C.N. *Understanding Dynamic Systems*. Prentice-Hall, Englewood Cliffs, NJ, 1993.

Jackson, A.C., and H.T. Milhorn. Digital computer simulation of respiratory mechanics. *Comput. Biomed. Res.* **6**: 27–56, 1973.

Kuo, B.C. *Automatic Control Systems*, 4th ed. Prentice-Hall, Englewood Cliffs, NJ, 1994.

Milhorn, H.T. *The Application of Control Theory to Physiological Systems*. W.B. Saunders, Philadelphia, 1966.

Milsum, J.H. *Biological Control Systems Analysis*. McGraw-Hill, New York, 1966.

Shahian, B., and M. Hassul. *Control System Design using MATLAB*. Prentice-Hall, Englewood Cliffs, NJ, 1993.

Soechting, J.F., P.A. Stewart, R.H. Hawley, P.R. Paslay, and J. Duffy. Evaluation of neuromuscular parameters describing human reflex motion. *Trans. ASME, Series G* **93**: 221–226, 1971.

Strum, R.D., and D.E. Kirk. *Contemporary Linear Systems using MATLAB*. PWS Publishing Co., Boston, MA, 1994.

## PROBLEMS

**P4.1.** Figure P4.1 shows the block diagram of a simplified model of eye-movement control. $J$ represents the moment of inertia of the eyeball about the axis of rotation, while $B$ represents the viscous damping associated with the rotational movement of the eye. The target angular position of the eye, $\theta_{ref}$, is set by the higher centers. $G$ is a gain that converts the controlling signal into the torque exerted by the extraocular muscles. Information about the angular position of the eye, $\theta$, is fed back to the controller with unity gain. Velocity information is also fed back with variable gain, $k_v$ ($> 0$). Deduce expressions for the responses of this system to a unit step change in $\theta_{ref}$ when:

(a) there is no feedback at all;
(b) there is only position feedback ($k_v = 0$);
(c) both position and velocity feedback exist.

**P4.2.** Determine the response in angular displacement of the eye in Figure P4.1 if the target input $\theta_{ref}$ were to follow the trajectory of a unit ramp, i.e., $\theta_{ref} = t$ ($t > 0$). How would this ramp response be affected if the velocity feedback gain, $k_v$, were made negative?

**P4.3.** The following transfer function is one of the simplest linear approximations to the pure time delay, $T$:

$$H(s) = \frac{1 - \frac{Ts}{2}}{1 + \frac{Ts}{2}}$$

Determine the open-loop and closed-loop responses for the system shown in Figure P4.2 when the input is a unit step.

**P4.4.** Many types of physiological receptors exhibit the property of rate sensitivity. Carbon dioxide ($CO_2$) receptors have been found in the lungs of birds and reptiles, although it remains unclear whether such receptors are also found in human lungs. Figure P4.3 shows a highly simplified model of the way in which ventilation may be controlled by these intrapulmonary receptors following denervation of the carotid bodies. The feedforward

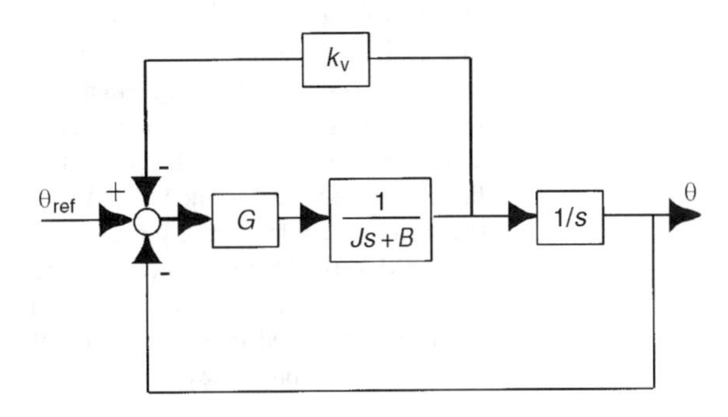


**Figure P4.1** Simple model of eye-movement control.

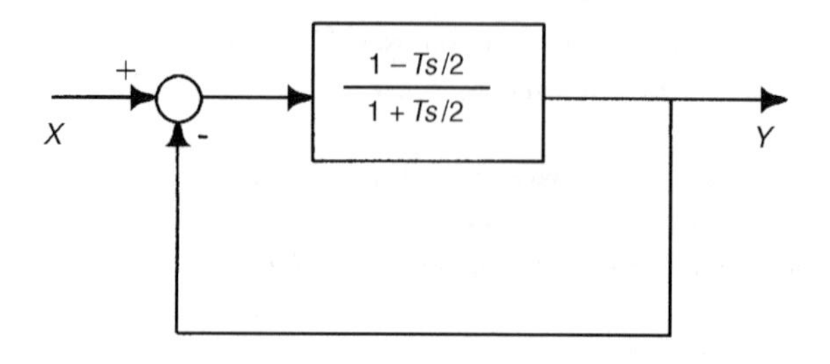


Figure P4.2 Closed-loop system containing time-delay approximation.

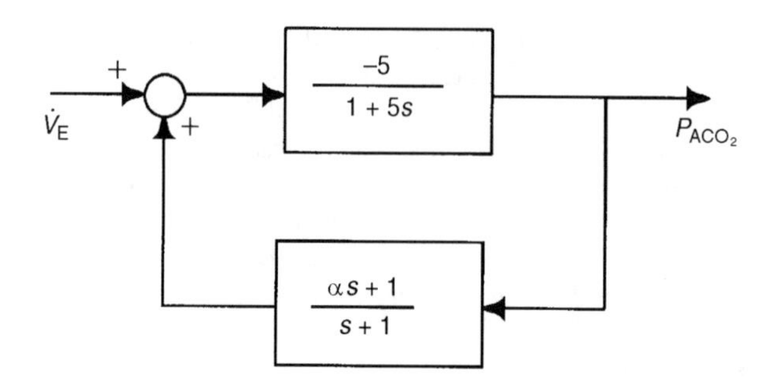


Figure P4.3 Simplified model of ventilatory control with intrapulmonary $CO_2$ receptor feedback.

element in the closed-loop system represents the gas exchange processes of the lungs, while the feedback element represents the dynamic characteristics of the intrapulmonary $CO_2$ receptors. The parameter $\alpha$ determines how rate-sensitive these receptors are. Determine the responses of this system to a large hyperventilatory sigh (which may be approximated by an impulse function) when: (a) $\alpha = 0$ (there is no rate sensitivity), (b) $\alpha = \frac{1}{2}$, and (c) $\alpha = 2$. (Note that the feedback element belongs to a class of systems known as lag-lead (when $\alpha < 1$) or lead-lag (when $\alpha > 1$) systems).

**P4.5.** Develop a SIMULINK program that simulates the linearized lung mechanics model shown in Figure 4.2b. Using the same parameter values as those given in Section 4.3, verify that you can obtain the impulse and step responses shown in Figures 4.4 and 4.5 for both open-loop and closed-loop circumstances.

**P4.6.** The SIMULINK program "`glucose.mdl`" is a dynamic version of the glucose regulation model discussed in Section 3.6. Determine the time-courses of the concentrations of glucose and insulin in response to the steady infusion of glucose at the rate of 80 000 mg h$^{-1}$ for a period of 1 hour. The values of the other parameters are as given in Section 3.6. Compare these time-courses to the corresponding cases where the insulin production parameter, $\beta$, has been reduced to 20% of its nominal value.

**P4.7.** The degree of spasticity in patients with neuromuscular disorders can be quantified with the use of the "pendulum test." In this clinical procedure, the subject sits relaxed on a table with his lower leg initially supported by the medical examiner so that the knee joint is fully extended. The examiner abruptly releases the lower leg so that it swings freely until it finally comes to rest in the vertical position. The trajectory of the swing, as measured by the change in angle of knee flexion can reveal information about the neuromuscular stretch reflex. Modify the SIMULINK program "`nmreflex.mdl`" so that it can be used to simulate this test. Note that the major difference between the pendulum test and the procedure described in Section 4.6 is that, here, the externally applied moment does not remain constant but varies according to the angular displacement of the lower leg, since it is a function of the weight of the lower leg and the moment arm between the center-of-gravity of the lower leg and the knee joint. Assume the same parameter values used in "`nmreflex.m`", except for the following: moment of inertia of the lower leg about the knee joint = 0.25 kg m$^2$; length of lower leg = 40 cm, weight of lower leg = 5 kg. Determine how the trajectory of the lower leg would change with different values of stretch reflex gain $\beta$.

# 5

# Frequency-Domain Analysis of Linear Control Systems

## 5.1 STEADY-STATE RESPONSES TO SINUSOIDAL INPUTS

While the impulse and step functions are useful test signals to use in the characterization of linear systems, it is difficult to find naturally occurring signals that approximate these highly idealized waveforms. Moreover, abrupt steps and impulsive changes are difficult to generate as test signals. On the other hand, periodic phenomena are a common occurrence in physiology. Since it is possible, using the *Fourier series*, to decompose any periodic signal into its sinusoidal components, sine waves represent a highly useful class of basic test inputs. Furthermore, sinusoidal changes are generally much easier to approximate in practice relative to other periodic and most nonperiodic signals.

### 5.1.1 Open-Loop Frequency Response

Consider the linearized respiratory mechanics model (open-loop case) discussed in the previous chapter, expressed in differential equation form (see Figure 4.2a):

$$LC\frac{d^2P_A}{dt^2} + RC\frac{dP_A}{dt} + P_A = P_{ao} \tag{5.1}$$

Suppose the input $P_{ao}$ were to assume the form of a sinusoidal waveform of amplitude $X_0$ and angular frequency $\omega$. Note that $\omega$ is related to absolute frequency, $f$, by the following relationship:

$$\omega = 2\pi f \tag{5.2}$$

To simplify the mathematics, we employ the generalized sinusoidal (or complex exponential) function instead of the sine or cosine:

$$P_{ao}(t) = X_0 e^{j\omega t} \tag{5.3}$$

where $X_0$ is a real constant. Solution of the inhomogenous differential equation, Equation (5.1), in which the right-hand side takes the particular form shown in Equation (5.3), yields a solution for $P_A$ that contains two parts, as discussed previously in Section 2.5. The complementary function represents the transient part of the response, while the particular solution characterizes the steady-state response. In this discussion, we will be concerned only with the steady-state response in $P_A$. With the input given by Equation (5.3), the only way for equality to hold for arbitrary values of $t$ (time) between the left-hand and right-hand sides of Equation (5.1) is for the particular solution of $P_A$ to contain the function $e^{j\omega t}$. Thus, we assume the following form for $P_A(t)$:

$$P_A(t) = Ze^{j\omega t} \tag{5.4}$$

which states that the output of the system defined by Equation (5.1) must also be sinusoidal with the same frequency as the input signal. In Equation (5.4), we allow the function $Z$ to be complex. Substituting Equation (5.4) into Equation (5.1), we obtain, after canceling $e^{j\omega t}$ from both sides of the equation and rearranging terms:

$$Z = H_o(\omega)X_0 \tag{5.5}$$

where

$$H_o(\omega) = \frac{1}{(1 - LC\omega^2) + jRC\omega} \tag{5.6}$$

$H_o(\omega)$ is a complex function of the frequency of the input and can be expressed in polar form as

$$H_o(\omega) = |H_o(\omega)|e^{j\phi_o(\omega)} \tag{5.7}$$

where the magnitude is

$$|H_o(\omega)| = \frac{1}{\sqrt{(1 - LC\omega^2)^2 + R^2C^2\omega^2}} \tag{5.8}$$

and the phase component is

$$\phi_o(\omega) = -\tan^{-1}\left(\frac{RC\omega}{1 - LC\omega^2}\right) \tag{5.9}$$

The complex function $H_o(\omega)$ represents the relationship between the sinusoidal input $P_{ao}(t)$ and the sinusoidal output $P_A(t)$. Substituting Equation (5.5) and (5.7) back into Equation (5.4), we obtain the following expression for $P_A(t)$:

$$P_A(t) = |H_o(\omega)|X_0e^{j(\omega t + \phi(\omega))} \tag{5.10}$$

Since $P_{ao}(t)$ is $X_0e^{j\omega t}$, Equation (5.10) imples that, although the output $P_A(t)$ remains sinusoidal at the same angular frequency $\omega$, its amplitude and phase are different from those of $P_{ao}(t)$. The ratio between the output and input amplitudes, or the *gain*, is given by $|H_o(\omega)|$, while the *phase difference* is represented by $\phi_o(\omega)$. It is important to note that both gain and phase difference are functions of the forcing frequency $\omega$. Figure 5.1 shows predictions of $P_A(t)$ produced by the lung mechanics model when $P_{ao}(t)$ assumed the form of sinusoidal waves of *unit amplitude* at absolute frequencies of 1, 4, and 8 Hz. The values of the parameters employed here were $R = 0.3\ \mathrm{cm\,H_2O\,s\,L^{-1}}$, $C = 0.1\ \mathrm{L\,cm\,H_2O^{-1}}$, and $L = 0.01\ \mathrm{cm\,H_2O\,s^2\,L^{-1}}$. At very low frequencies, $P_A$ oscillates virtually in synchrony with

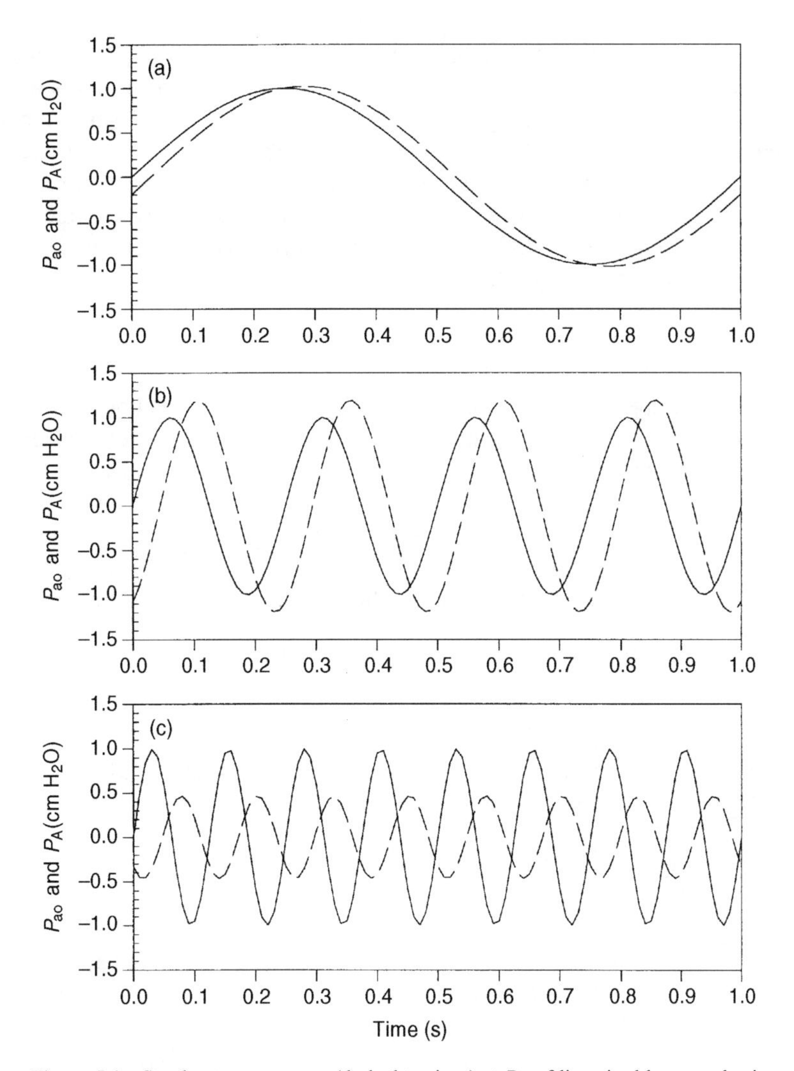


**Figure 5.1** Steady-state responses (dashed tracings) at $P_A$ of linearized lung mechanics model to sinusoidal excitation (solid tracings) at $P_{ao}$. Frequencies of sine waves are (a) 1 Hz, (b) 3 Hz and (c) 8 Hz.

$P_{ao}$ and is of the same amplitude. At very high frequencies, $P_A$ lags substantially behind $P_{ao}$ and is significantly attenuated. Using this set of parameters, however, there is a range of frequencies over which $P_A$ is amplified and becomes larger in amplitude than $P_{ao}$. The amplification is greatest at approximately 4 Hz. If one determined the impulse or step response for this model with these parameter values, one would find an underdamped response with an oscillation frequency of about 4 Hz. Thus, exciting the system with an external sinusoidal input at this frequency produces *resonance*, since the applied forcing acts to reinforce the natural vibrations of the system.

The complex function $H_o(\omega)$ contains all the information shown in Figure 5.1 and much more. It predicts how the lung mechanics model will respond to sinusoidal inputs of unit amplitude and all possible frequency values. As such, it is also called the *frequency response* of the system. Figure 5.2 illustrates one method of graphically representing the frequency response of the lung mechanics model. At each absolute frequency, $f$, we evaluate the gain and phase of $H_o(\omega)$. Two frequency responses are shown in this diagram. The first represents the underdamped system, with $R = 0.3\,\text{cm}\,H_2O\,\text{s}\,L^{-1}$; this frequency response encompasses the results shown in Figure 5.1. The second frequency response shown represents the overdamped system where $R = 1\,\text{cm}\,H_2O\,\text{s}\,L^{-1}$. In this case, it is clear that there is no resonance peak, so that the gain continually decreases with increasing frequency.

Note from Equation (4.4) that the transfer function corresponding to the lung mechanics model described by Equation (5.1) is

$$H_o(s) \equiv \frac{P_A(s)}{P_{ao}(s)} = \frac{1}{LCs^2 + RCs + 1} \tag{5.11}$$

An alternative approach to deriving the frequency response function $H_o(\omega)$ is by evaluating $H_o(s)$ along the imaginary axis on the $s$-plane, i.e., *by setting* $s = j\omega$. Substituting $j\omega$ for $s$ in Equation (5.11), we obtain

$$H_o(\omega) = \frac{1}{LC(j\omega)^2 + RCj\omega + 1} \tag{5.12}$$

Since $j^2 = -1$, rearranging terms in Equation (5.12) leads to Equation (5.6).

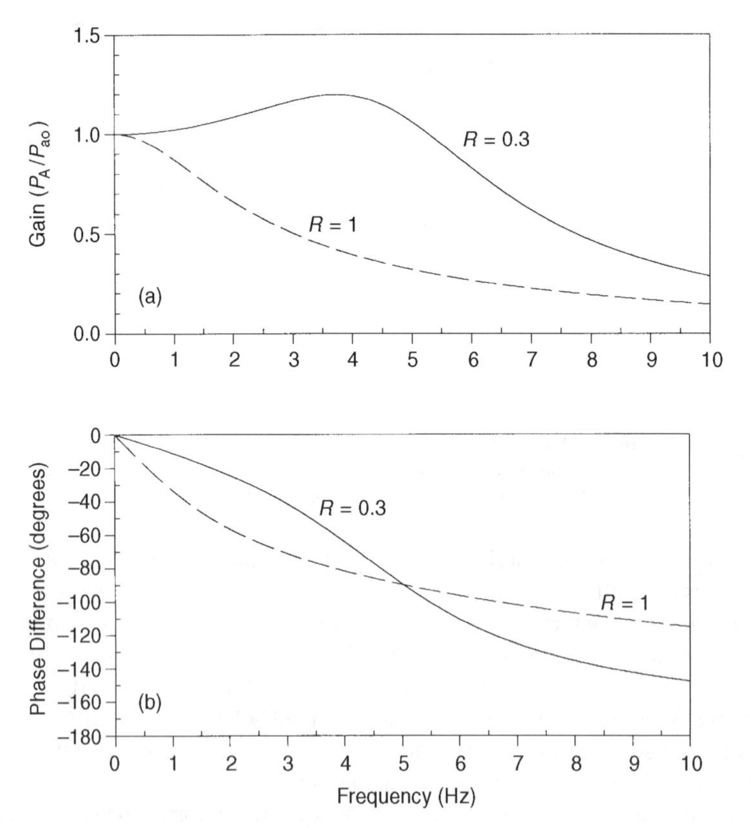


**Figure 5.2** Frequency responses of the linearized lung mechanics model in underdamped (solid curves) and overdamped (dashed curves) conditions. The values shown for $R$ are in cm $H_2O\,\text{s}\,L^{-1}$. Other parameter values are $C = 0.1$ L cm $H_2O^{-1}$ and $L = 0.01$ cm $H_2O$ $s^2$ $L^{-1}$.

### 5.1.2 Closed-Loop Frequency Response

Now, consider the closed-loop situation where there is proportional feedback of $P_A(t)$. The corresponding transfer function would be as given in Equation (4.5b). Evaluating the frequency response, we obtain

$$H_c(\omega) = \frac{1}{(1 + k - LC\omega^2) + jRC\omega} \tag{5.13}$$

Thus, for the closed-loop case, the magnitude and phase of the frequency response are given by

$$|H_c(\omega)| = \frac{1}{\sqrt{(1 + k - LC\omega^2)^2 + R^2C^2\omega^2}} \tag{5.14}$$

and

$$\phi_c(\omega) = -\tan^{-1}\left(\frac{RC\omega}{1 + k - LC\omega^2}\right) \tag{5.15}$$

The closed-loop frequency responses are shown together with the open-loop responses in Figure 5.3. Here, the feedback gain $k$ has been assumed to be unity. Closing the loop leads to a reduction of the steady-state gain from 1 to 0.5, i.e., $|H_c(\omega = 0)| = 0.5$ compared to $|H_o(\omega = 0)| = 1$. This is consistent with the results that were previously presented in Chapters 3 and 4. Closing the loop also shifts the location of the resonance peak to a substantially higher frequency (~6.5 Hz) versus ~4 Hz). This also is consistent with the impulse and step responses of the underdamped system in Chapter 4 where we found an increase in frequency but decrease in amplitude of the transient oscillations. The phase portion of the frequency response shows a general decrease in the phase lag introduced by the closed-loop system vis-à-vis the open-loop system. This is equivalent to our earlier finding in Chapter 4 that closing the loop produces an increase in the speed of response of the system.

### 5.1.3 Relationship between Transient and Frequency Responses

Since the transfer function of any linear system is the Laplace transform of its impulse response and the frequency response can be deduced by replacing $s$ in the Laplace transform by $j\omega$, it follows that the frequency response can be derived by taking the Fourier transform of the impulse response. This implies that one should be able to deduce various features of the transient response from the corresponding frequency response. Consider an idealized linear system with the following frequency response:

$$\begin{aligned} H(\omega) &= e^{-j\omega\tau}, & -\omega_c \le \omega \le \omega_c \\ &= 0, & |\omega| > \omega_c \end{aligned} \tag{5.16}$$

i.e., this system has a gain of unity at angular frequencies between $-\omega_c$ and $\omega_c$ and zero gain at all frequencies outside this range. The *bandwidth* of a linear system is defined as the range of frequencies over which the system gain exceeds $1/\sqrt{2}$ or 0.7071 (see sections to follow). Thus, in this case, $\omega_c$ represents the system bandwidth. The phase of this system is linear with

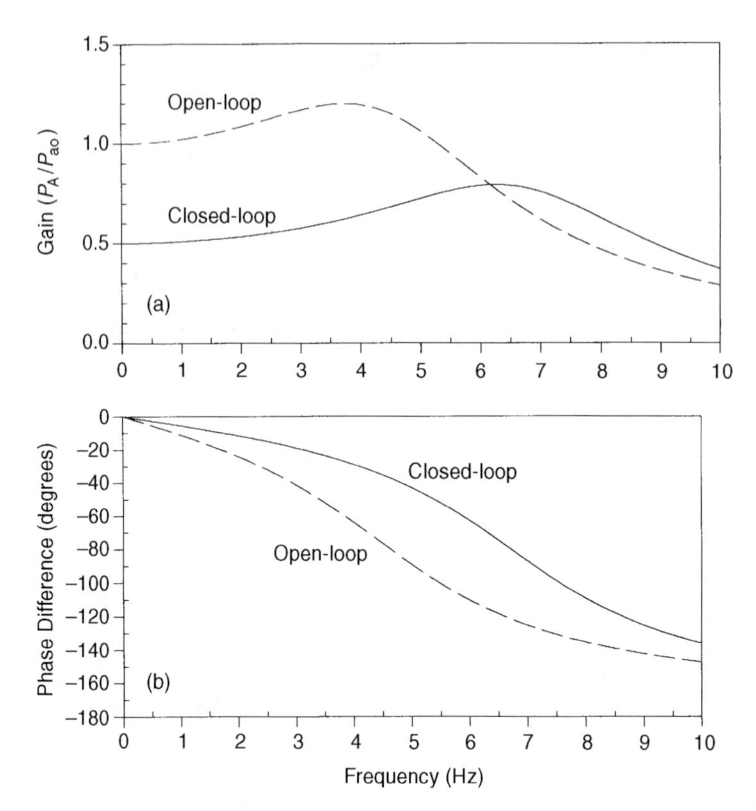


**Figure 5.3** Frequency responses of the linearized lung mechanics model in closed-loop (solid curves) and open-loop modes. The values of the parameters used are: $R = 0.3$ cm $H_2O$ $L^{-1}$, $C = 0.1$ L cm $H_2O^{-1}$, $L = 0.01$ cm $H_2O$ $s^2$ $L^{-1}$, and $k = 1$ (or equivalently, $\lambda = 2$).

frequency within the bandwidth; the slope of the phase curve (line) is $-\tau$, i.e., the output is delayed by $\tau$ relative to the input at all frequencies. This frequency response is displayed in Figures 5.4a and 5.4b.

To deduce the corresponding impulse response, $h(t)$, we take the inverse Fourier transform of Equation (5.16):

$$\begin{aligned} h(t) &= \frac{1}{2\pi}\int_{-\infty}^{\infty} H(\omega)e^{j\omega t}\, d\omega \\ &= \frac{1}{2\pi}\int_{-\omega_c}^{\omega_c} e^{-j\omega\tau}e^{j\omega t}\, d\omega \\ &= \frac{\omega_c}{\pi}\,\mathrm{sinc}[\omega_c(t-\tau)] \end{aligned} \tag{5.17}$$

where the function sinc($x$) represents $\sin(x)/x$. Figure 5.4c shows the form of $h(t)$. Note three important features in Equation (5.17). First, the impulse response peaks $\tau$ units of time after the input impulse has occurred, due to the delay inherent in $H(\omega)$. Secondly, the maximum value of the impulse response or the *peak amplitude* (see Section 4.4.2.1) is proportional to the bandwidth $\omega_c$. This makes sense since a larger bandwidth allows the impulse response to

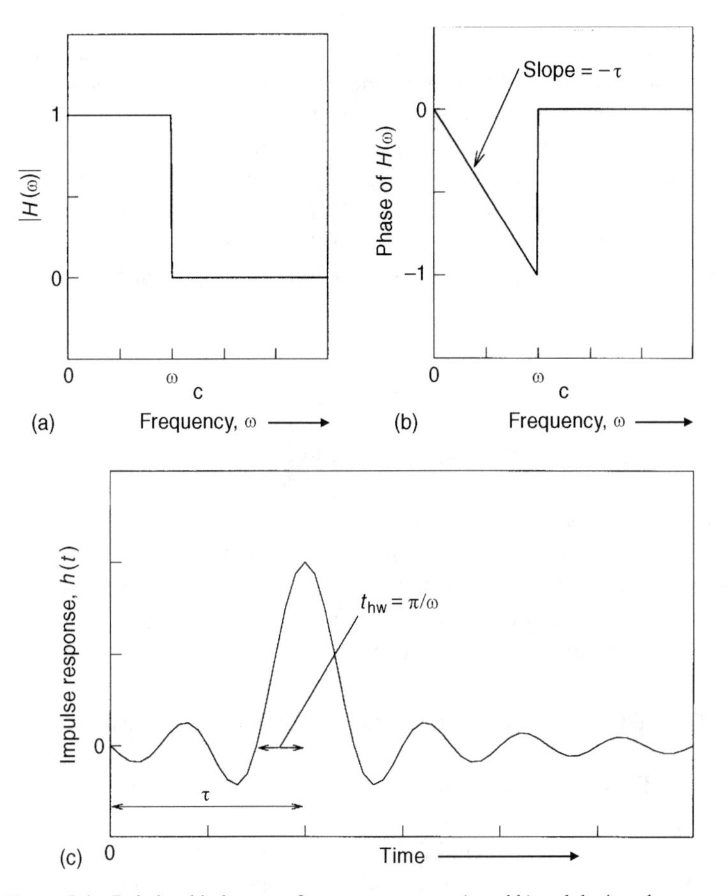


**Figure 5.4** Relationship between frequency response (a and b) and the impulse response (c) of a linear system.

be composed of a broader range of frequencies: in particular, the higher frequencies add sharpness or abruptness to the impulse response. In the limit, as bandwidth becomes infinite, the peak amplitude also becomes infinite as the impulse response approaches a delta function, i.e., the same form as the input itself. The third important feature relates to the *half-width*, $t_{hw}$, of the impulse response, i.e., the time taken for the main response to the impulse to fully develop. From Equation (5.17), $t_{hw}$ can be deduced from the interval between the impulse response peak and the preceding or subsequent zero-crossing (see Figure 5.4c):

$$t_{hw} = \frac{\tau}{\omega_c} \tag{5.18}$$

Since the step response is simply the integral of the impulse response, it also can be shown that the rise time, $T_r$, of the step response is proportional to $t_{hw}$ and therefore inversely proportional to $\omega_c$. In fact, for a second-order system such as our linearized lung mechanics model, the following approximate relationship holds:

$$T_r \approx \frac{2}{\omega_c} \tag{5.19}$$

As we discussed in Section 4.4.1, the *natural frequency* $\omega_n$ and *damping ratio* $\zeta$ are the two key parameters that characterize the impulse and step responses of the generalized second-order linear system. These parameters can easily be derived from the frequency response of the same system. Consider the linearized lung mechanics model in open-loop mode, the frequency response of which is given by Equations (5.8) and (5.9). Note from Equation (4.49) in Section 4.4.1 that $\omega_n = 1/(LC)^{1/2}$. Thus, $1 - LC\omega_n^2 = 0$, so that $\phi_o(\omega_n) = -\tan^{-1}(\infty) = -90°$. Therefore, by locating the frequency at which the phase plot attains a phase lag of 90°, we can deduce $\omega_n$. Evaluation of the frequency response magnitude at $\omega_n$ allows us to deduce $\zeta$, since

$$|H_o(\omega_n)| = \frac{1}{RC\omega_n} = R\sqrt{\frac{C}{L}} = \frac{1}{2\zeta} \tag{5.20}$$

where the last equality in Equation (5.20) is based on Equation (4.50) in Section 4.4.1.

For the (open-loop) case where $L = 0.01$, $C = 0.1$, and $R = 0.3$, note that $\omega_n = 31.62\,\text{rad s}^{-1}$, corresponding to a frequency of approximately 5 Hz. At this frequency, $|H_o(\omega_n)| \sim 1.05$. It is important to note that this does not correspond to the *resonant frequency*, which is located at $\sim$4 Hz ($\omega_r \sim 25$). Moreover, the peak value of $|H_o(\omega)|$, which occurs at the resonant frequency, is $\sim$1.2. In fact, resonance occurs at the natural frequency only when there is no damping in the system ($R = 0$).

## 5.2 GRAPHICAL REPRESENTATIONS OF FREQUENCY RESPONSE

### 5.2.1 Bode Plot Representation

In the field of control engineering, Bode plots represent one of the most accepted classical methods for displaying the frequency response of a linear system. These plots are similar to but differ from the graphs presented in Figures 5.2 and 5.3 in that the gain (or magnitude) and frequency scales are presented in logarithmic form, while the phase remains on a linear scale. In addition, the frequency scale is generally displayed in terms of the angular frequency, $\omega$, and therefore in units of radians per second.

The gain of the frequency response, $H(\omega)$, is expressed in units of *decibels* (dB), defined as follows:

$$\begin{aligned} G_{\text{dB}}(\omega) &= 10\ \log_{10}|H(\omega)|^2 \\ &= 20\ \log_{10}|H(\omega)| \end{aligned} \tag{5.21}$$

The first equality in Equation (5.21) is shown to emphasize the fact that for a linear system, $|H(\omega)|^2$ represents the ratio between the *power* of the output and the power of the input signal. The value of $\omega$ at which only half of the input signal power is transmitted is known as the *corner* or *cut-off* frequency, $\omega_c$. Thus, $|H(\omega_c)|^2 = 0.5$. Since $|H(\omega_c)| = 1/\sqrt{2}$, it turns out that the range of frequencies between 0 and $\omega_c$ is also the system bandwidth, as discussed previously. Using Equation (5.21), we find that $G_{\text{dB}}(\omega_c) = -3$ dB. This definition assumes only the case where the system takes the form of a low-pass filter, where the high frequencies are attenuated. For systems that are high-pass in nature, the corner frequency is defined as the frequency at which the input signal power is amplified and doubled at the output. In this case, $|H(\omega_c)|^2 = 2$, and therefore, $G_{\text{dB}}(\omega_c) = 3$ dB. Thus, in general, the corner frequency is the frequency at which the gain of the linear system is changed by 3 dB.

The logarithmic nature of $G_{\text{dB}}$ represents one of the major strengths of the Bode plot. In general, it is possible to factorize the numerator and denominator of any given frequency response function into a cascade of first-order systems. For instance, consider the following generalized frequency response function:

$$H(\omega) = \frac{G_{\text{SS}} \prod_{m=1}^{M} (1 + j\omega\tau_m)}{(j\omega)^N \prod_{i=1}^{P} (1 + j\omega\tau_i)} \tag{5.22}$$

Note that in Equation (5.22), $G_{\text{SS}}$ is the steady state gain for the case $N = 0$. If we express the magnitude of $H(\omega)$ in terms of logarithmic gain, then

$$G_{\text{dB}}(\omega) = 20 \log_{10} \left[ \frac{G_{\text{SS}} \prod_{m=1}^{M} \sqrt{1 + \omega^2\tau_m^2}}{\omega^N \prod_{i=1}^{P} \sqrt{1 + \omega^2\tau_i^2}} \right] \tag{5.23a}$$

Evaluating the logarithm of the expression in brackets in Equation (5.23a), we obtain

$$\begin{aligned} G_{\text{dB}}(\omega) = 20 \log_{10} G_{\text{SS}} &+ \sum_{m=1}^{M} 20 \log_{10}(1 + \omega^2\tau_m^2)^{1/2} \\ &+ 20 \log_{10} \omega^{-N} + \sum_{i=1}^{P} 20 \log_{10}(1 + \omega^2\tau_i^2)^{-1/2} \end{aligned} \tag{5.23b}$$

This representation converts the logarithm of the products of several factors into equivalent sums of the logarithms of these factors. As a result, the contribution of each term is additive, which makes it easy to determine how the individual factors contribute to the overall gain. The overall phase of $H(\omega)$ can also be decomposed into the sum of all its individual components:

$$\phi(\omega) = \sum_{m=1}^{M} \tan^{-1}(\omega\tau_m) - \frac{N\pi}{2} - \sum_{i=1}^{P} \tan^{-1}(\omega\tau_i) \tag{5.24}$$

From the above example, one can see that it is useful to consider the magnitude and phase contributions of the basic components of $H(\omega)$, which take the form of either $(j\omega)^{-1}$ or $(1 + j\omega\tau)^{\pm 1}$. Let us first consider the term $(j\omega)^{-1}$. Note that since $j\omega$ is purely imaginary and has no real part, its phase contribution is $\pi/2$ radians. Consequently, the phase contribution of $(j\omega)^{-N}$ is $-N\pi/2$ radians, the negative sign implying that the phase shift is a lag. The magnitude contribution of $(j\omega)^{-N}$ is infinite at zero frequency but zero at infinite frequency.

Now, consider the term $(1 + j\omega\tau)$. This factor adds a phase shift of $+\tan^{-1}(\omega\tau)$, i.e., a phase lead, to $H(\omega)$. At very low frequencies ($\omega \ll 1/\tau$), this phase lead would be close to zero. At very high frequencies, i.e., $\omega \gg 1/\tau$, this phase lead would approach $\pi/2$ radians or 90°. What about the gain contributions of this factor? At very low frequencies ($\omega \ll 1/\tau$), the gain would be $\sim 20 \log_{10}(1) = 0$ dB, i.e., this would appear as a straight line on the zero-decibel axis. At very high frequencies ($\omega \gg 1/\tau$), the gain would approximate $20 \log_{10}(\omega\tau)$ dB, and thus behave like a straight line (on the Bode plot) with a slope of 20 dB/decade. These two straight lines bound the actual gain plot and are known as the *low-frequency* and *high-frequency asymptotes*, respectively. Conversely, each $(1 + j\omega\tau)^{-1}$ factor would contribute a phase shift of $-\tan^{-1}(\omega\tau)$, i.e., a phase lag, to $H(\omega)$. As in the previous case, at very low frequencies the phase lag would approach zero while the low-frequency asymptote for the gain plot would coincide with the zero-decibel axis. At very high frequencies, the phase lag would approach 90°, and the high-frequency asymptote for the gain plot would be a straight line with slope −20 dB/decade. The Bode plots for these two basic functions are displayed in Figure 5.5, together with the low-frequency and high-frequency asymptotes.

The Bode plots of the frequency response of the linearized lung mechanics model discussed earlier are presented in Figure 5.6. It should be noted that these plots contain the same information as the linear frequency response plots displayed in Figure 5.3. One difference is that the frequency scale is expressed in terms of $\omega$, in units of radians per second. Another important feature is that the logarithmic scaling enhances the appearance of the resonance peaks in both the open-loop and closed-loop systems. In both cases, the bandwidth of the system can be readily determined as the frequency range over which the gain lies above −3 dB relative to the steady-state gain level.

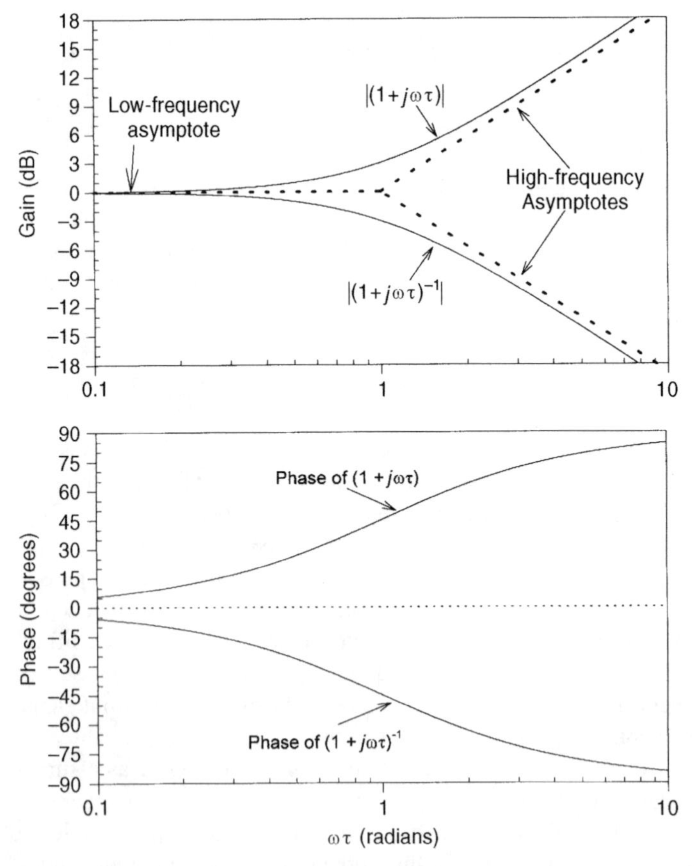


**Figure 5.5** Bode plots of the first-order frequency response functions $(1 + j\omega\tau)$ and $(1 + j\omega t)^{-1}$. Note, in this case, that the "frequency scale" has been normalized and presented in terms of the product "$\omega\tau$."

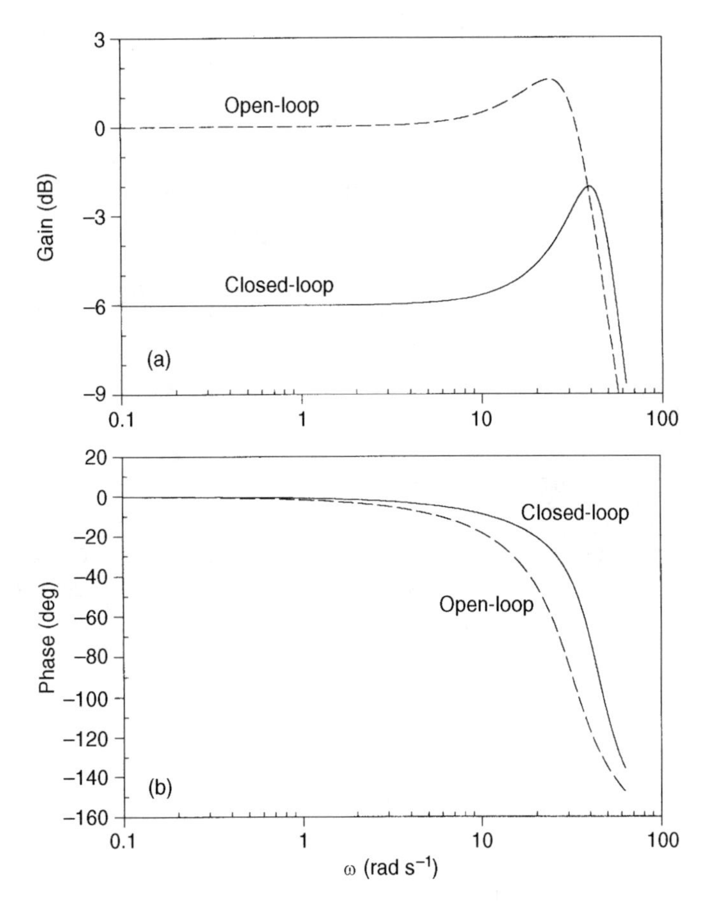


**Figure 5.6** Bode plots of the frequency response of the linearized lung mechanics model in open-loop and closed-loop modes.

### 5.2.2 Nichols Charts

Instead of presenting gain and phase in separate plots, an alternative approach is to plot the logarithmic magnitude in dB versus phase for a range of frequencies. These plots are known as *Nichols charts*. The log-magnitude vs. phase curves for the frequency responses $(1+j\omega\tau)$ and $(1+j\omega\tau)^{-1}$ are displayed in Figure 5.7. The solid circles on these plots correspond to the frequencies listed. In this case, the individual values of the product $\omega\tau$ are shown, so that the same plots would apply irrespective of the specific value of $\tau$ being employed. Figure 5.8 shows the Nichols charts for the linearized lung mechanics model in open-loop and closed-loop modes. The values placed next to the closed circles represent the corresponding angular frequencies, $\omega$, in radians per second. These curves convey the same information that was contained in the Bode plots of Figure 5.6 and the linear frequency response plots of Figure 5.3. However, as we will see later, the shapes of these curves at the points where gain approaches 0 dB and the phase approaches $-180°$ can yield useful information about system stability.

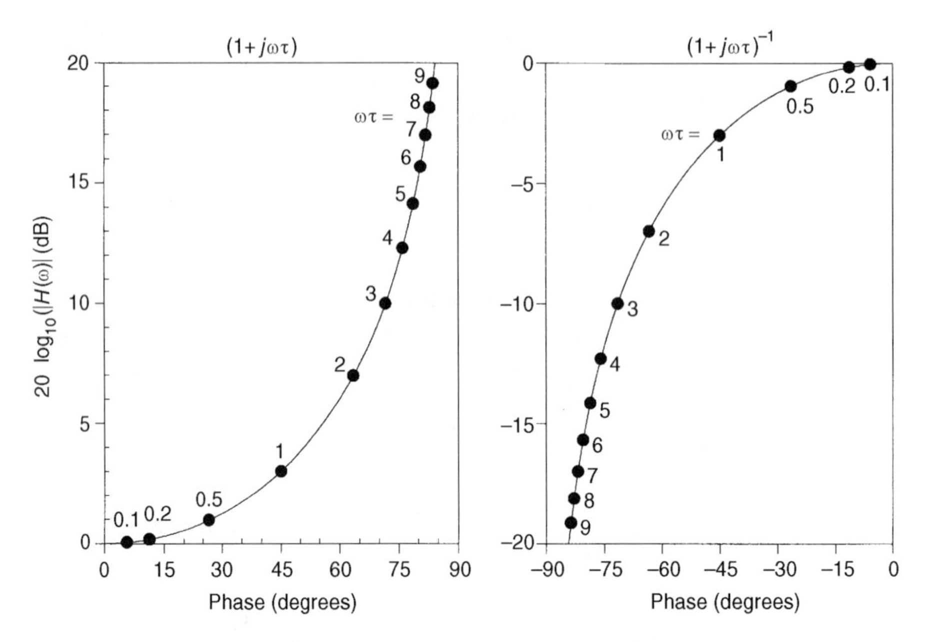


**Figure 5.7** Nichols charts for the frequency response functions $(1 + j\omega\tau)$ and $(1 + j\omega\tau)^{-1}$.

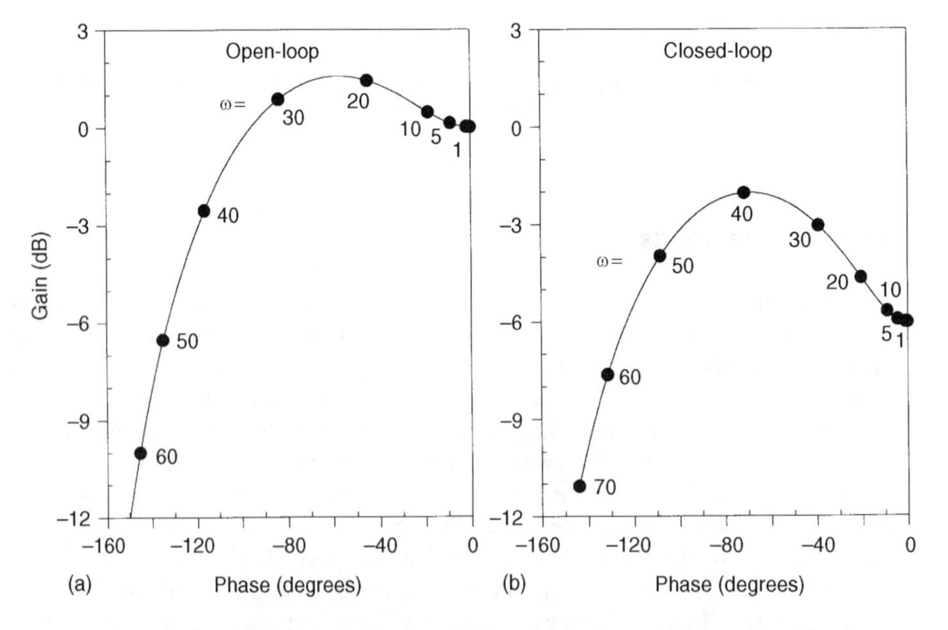


**Figure 5.8** Nichols charts for the linearized lung mechanics model in open-loop (a) and closed-loop (b) modes. The parameter values assumed are the same as the values employed in Figures 5.3 and 5.5.

### 5.2.3 Nyquist Plots

Nyquist plots are sometimes also called *polar plots*. Here, the frequency response $H(\omega)$ is plotted on a plane in which the horizontal axis reflects the magnitude of the real part of $H(\omega)$ while the vertical axis reflects the imaginary part. Thus, at any frequency $\omega$, $H(\omega)$ is represented by a vector linking the origin to the point in question, and the length of the vector represents the magnitude of $H(\omega)$. As illustrated in the inset in Figure 5.9, the angle subtended by this line and the positive real axis represents $\phi$, the phase of $H(\omega)$. The sign convention generally adopted is that *anticlockwise* rotations of the vector $H(\omega)$ from the positive real axis yield *positive values* for $\phi$.

Nyquist plots corresponding to the basic frequency response functions $1/j\omega\tau$, $(1+j\omega\tau)$ and $(1+j\omega\tau)^{-1}$ are shown in Figure 5.9. The plot for $1/j\omega\tau$ coincides with the negative portion of the imaginary axis: when $\omega = 0$, $1/j\omega\tau$ is at $-j\infty$, but as $\omega$ becomes large, $1/j\omega\tau$ approaches the origin along the imaginary axis. The locus of $(1+j\omega\tau)$ begins with a gain of unity on the real axis. As $\omega$ increases, this frequency response function moves vertically upward, tracing a path that is parallel with the positive imaginary axis. By contrast, the locus traced by $(1+j\omega\tau)^{-1}$ is a semi-circular arc that begins at 1 on the real axis when $\omega = 0$ and ends at the origin when $\omega = \infty$.

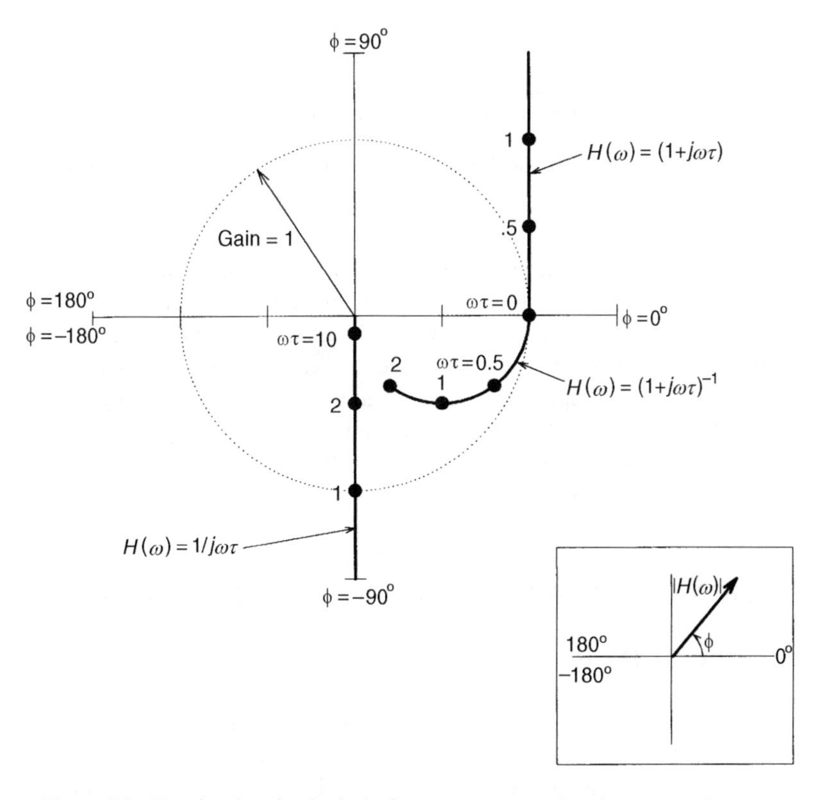


**Figure 5.9** Nyquist plots for the basic frequency response functions $1/j\omega\tau$, $(1+j\omega\tau)$, and $(1+j\omega\tau)^{-1}$. Selected values of $\omega\tau$ are shown as solid circles. The dotted circle represents the locus of points where gain equals unity. Inset shows definitions of $|H(\omega)|$ and $\phi$—anticlockwise rotations of vector $H(\omega)$ from the positive real axis yield positive values of $\phi$, and vice versa.

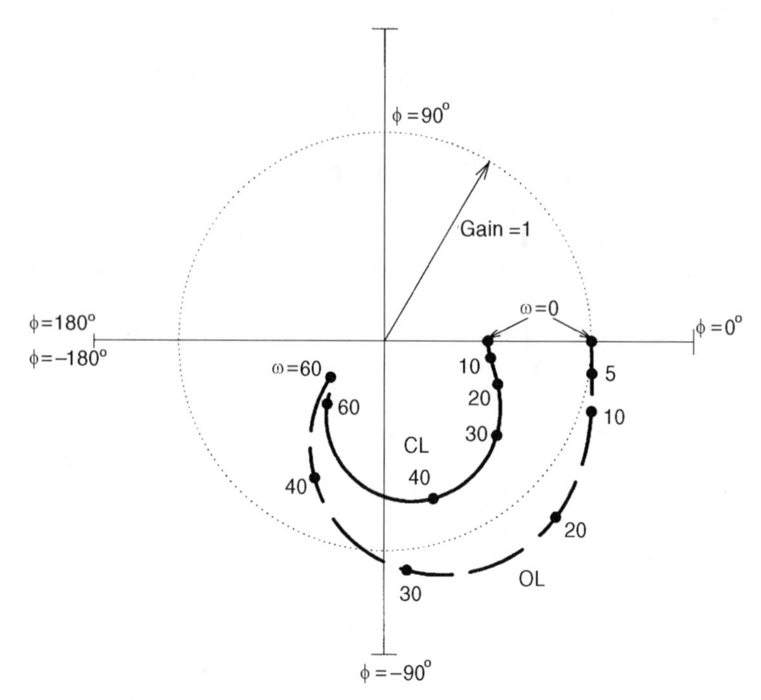


**Figure 5.10** Nyquist plots for the frequency responses of the linearized lung mechanics model in open-loop (OL, broken curve) and closed-loop (CL, solid curve) modes. The dotted circle represents the locus of points at which the gain equals 1.

The loop-like forms of the Nyquist plots presented in Figure 5.10 are more representative of the frequency responses of physiological systems. The particular plots shown characterize the frequency responses of the linearized lung mechanics model in both open-loop and closed-loop modes. These frequency responses represented here are exactly the same as those shown in Figure 5.3 where magnitude and phase were separately plotted against frequency. However, it is clear from the Nyquist plots that the points at which these curves intersect the imaginary axis (i.e., when $\phi = -90^\circ$ and $\omega = \omega_n$) do not correspond to the points of resonance at which the gain is maximum (and $\omega = \omega_r$).

## 5.3 FREQUENCY-DOMAIN ANALYSIS USING MATLAB AND SIMULINK

### 5.3.1 Using MATLAB

To demonstrate the utility of employing MATLAB to examine the frequency response of a known system, we turn again to our linearized model of lung mechanics. We will use the closed-loop transfer function expression given in Equation (4.5b), since this can be converted to the open-loop expression by simply setting $k$ equal to zero. Assuming that the values of $L$, $R$, $C$, and $k$ have been preassigned, the following command lines set up the transfer function, `Hs`, of the model and produces the frequency vector `w` which contains the range of frequencies (in rad $s^{-1}$) to be examined:

```
>> num = [1];
>> den = [L*C  R*C  (1 + k)];
>> Hs = tf(num, den);
>> f = 0:0.1:10
>> w = 2*pi*f;
```

The MATLAB Control System Toolbox function `freqresp` is used next to compute the frequency response, `Hw`, of `Hs` over the frequency range of 0 to 10 Hz. Since `Hw` is a complex mutlidimensional array, the `squeeze` function is used to collapse it into a complex vector, which is subsequently decomposed into magnitude and phase components using the `abs` and `angle` functions, respectively. Finally, the magnitude and phase components are plotted against `w`.

```
>> Hw = freqresp(Hs, w);
>> Hwmag = abs(squeeze(Hw))';
>> Hwpha = 180* angle (squeeze (Hw}))/pi;
>> subplot(2,1,1); plot(w,Hwmag);
>> ylabel('Freq Resp Magnitude'); grid on;
>> subplot(2,1,2); plot(w,Hwpha);
>> xlabel('Frequency (rad/s)');
>> ylabel('Freq Resp Phase (deg)'); grid on;
```

The above command lines will produce linearly-scaled frequency response plots of the type shown in Figures 5.2 and 5.3.

To produce Bode plots, the following commands can be used:

```
>> bode(Hs,w);
>> [Hwmag, Hwpha] = bode(Hs, w};
```

The first command line will lead to the automatic generation of Bode gain and phase diagrams. The second command line will not produce the plots but will save the results in the variables "`Hwmag`" and "`Hwpha`."

In similar fashion, the Nichols chart can be generated using the following lines:

```
>> nichols (Hs,w);
>> [Hwmag, Hwpha] = nichols(Hs, w);
```

Again, the second line will only compute the results but will not produce the plots.

Finally, the Nyquist plot can be produced as follows:

```
>> nyquist(Hs,w);
>> [Hwreal, Hwimag] = nyquist(Hs, w);
```

In this case, however, the second command line yields the real and imaginary parts of the frequency response, Hw, and not the magnitude and phase.

All of the above command lines are contained in a script file called "fda_llm.m," which has been included in the library of MATLAB and SIMULINK files that accompany this book.

### 5.3.2 Using SIMULINK

The MATLAB functions described in the previous section are extremely useful when the exact transfer function of the system being analyzed is known. However, with more complicated models where there may exist several subsystems connected through forward and feedback loops, deriving a closed form for the overall frequency response can be very laborious. In such situations, an alternative approach would be to perturb the model with a known input, monitor the resulting output, and use both input and output to deduce the frequency response of the system. This is a basic *system identification* technique; systems identification and parameter estimation will be discussed in greater detail in Chapter 7.

Figure 5.11 provides an illustration of how the frequency response of our linearized lung mechanics model can be "measured." Here, we assume the following values for the model parameters (see Equation (4.5b)): $L = 0.01\,\mathrm{cm\,H_2O\,s^2\,L^{-1}}$, $R = 0.3\,\mathrm{cm\,H_2O\,s\,L^{-1}}$, $C = 0.1\,\mathrm{L\,cm\,H_2O^{-1}}$, and $k = 1$ (i.e., closed-loop mode), so that the particular transfer function employed here is given by

$$H(s) = \frac{1}{0.001s^2 + 0.03s + 2} \tag{5.25}$$

This model is represented in Figure 5.11 by the LTI system block, labeled Hs, found in the SIMULINK Controls Toolbox library. Although Hs is represented by the simple form shown in this example, in general it could be composed of several interconnected subsystems. The point of relevance here is that one has to identify the input and output that relate to the overall transfer function of the model. White noise is fed into the input of Hs, and both input and output are fed into a block known as Spectrum Analyzer, found in the Extra Sinks sublibrary of the Simulink Extras blockset. The Spectrum Analyzer produces the graphical results shown in Figure 5.12, where the top panel displays the input and output time-courses, and the lower two panels show the frequency response magnitude and phase. Note that since the results displayed are computed from datasets of finite duration, it is inevitable that "noise" will appear in the estimated frequency response plots. The model

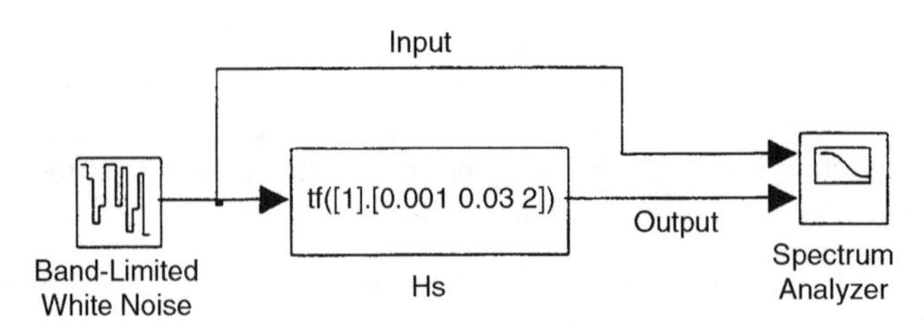


**Figure 5.11** SIMULINK model used for determining the frequency response of the linearized lung mechanics (closed-loop) transfer function.

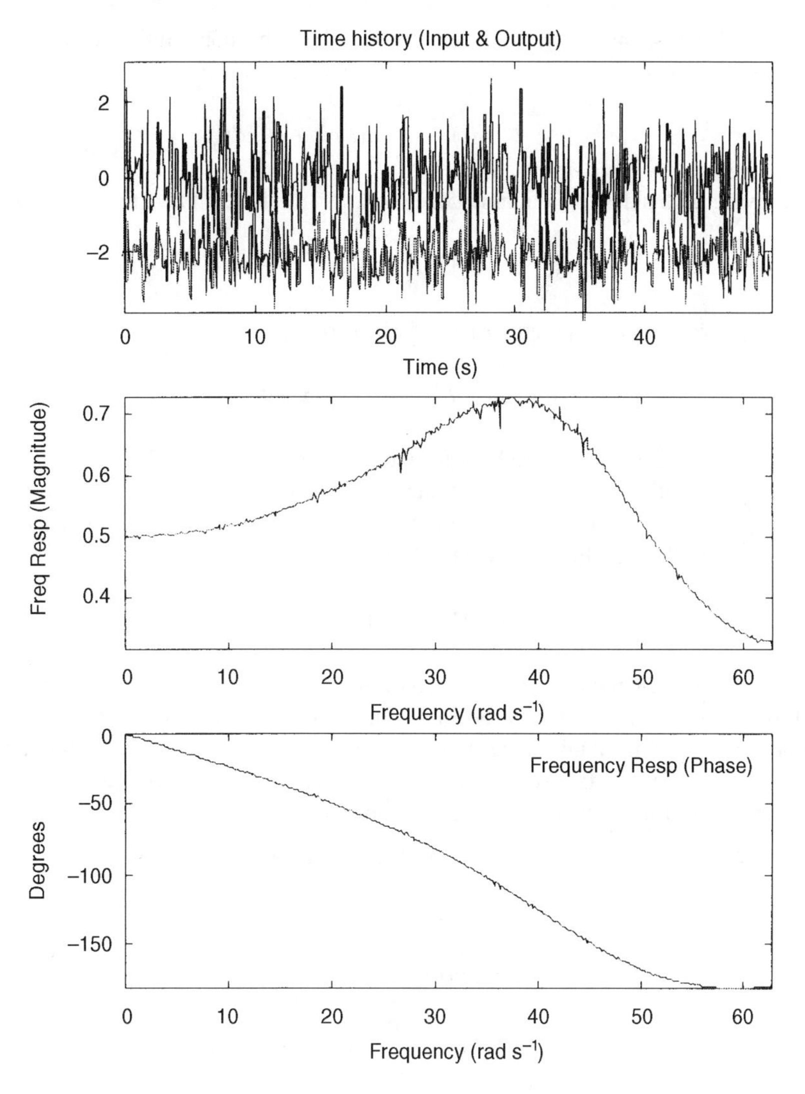


**Figure 5.12** Determination of frequency response of linearized lung mechanics model using the SIMULINK program shown in Figure 5.11. Top panel shows time-courses of the input to and output from the model. Note that the actual mean level of the output signal was zero: the displayed time-course was shifted vertically to enhance clarity of presentation.

shown in Figure 5.11 has been saved in the SIMULINK model file "`fdallm2.mdl`," which may be found in the library of MATLAB/SIMULINK files that accompany this book.

The algorithm embedded in `Spectrum Analyzer` is based on the following principle that is valid for all linear systems: The frequency response of the system under study can be derived by dividing the cross-spectrum of the input and output by the spectrum of the input. This principle is derived from the basic linear properties of convolution and superposition. Assuming $x(t)$ and $y(t)$ to represent the input and output of the linear system with impulse response $h(t)$, we begin by recalling the convolution equation displayed in Equation (2.45):

$$y(t) = \int_0^\infty h(t')x(t - t')\, dt' \tag{5.26}$$

Multiplying both sides of Equation (5.26) by $x(t - \tau)$ and then taking expectations, we obtain

$$E[x(t - \tau)y(t)] = \int_0^\infty h(t')E[x(t - \tau)x(t - t')]\, dt' \tag{5.27a}$$

where the *expectations* operator $E[\cdot]$ is defined by

$$E[z] = \int_{-\infty}^\infty zp(z)\, dz \tag{5.28}$$

and $p(z)$ is the probability distribution function of the variable $z$. However, by definition, the left-hand side of Equation (5.27a) yields the *cross-correlation* function between $x$ and $y$, $R_{xy}(\tau)$, while the expectation term on the right-hand side is equal to the *autocorrelation* function of $x$, $R_{xx}(\tau)$. Thus, we replace Equation (5.27a) with

$$R_{xy}(\tau) = \int_0^\infty h(t')R_{xx}(\tau - t')\, dt' \tag{5.27b}$$

It can be shown that *Fourier transformation* of $R_{xy}$ and $R_{xx}$ yields the cross-spectrum $S_{xy}$ and autospectrum $S_{xx}$, respectively; this equivalence principle is also known as the *Wiener–Khinchine* theorem. And since the frequency response $H(\omega)$ is obtained by Fourier transforming $h(t)$, the time-convolution on the right-hand side of Equation (5.27b) can be converted into a product in the frequency domain:

$$S_{xy}(\omega) = H(\omega)S_{xx}(\omega) \tag{5.29a}$$

where

$$S_{xy}(\omega) = \int_{-\infty}^\infty R_{xy}(\tau)e^{-j\omega\tau}\, d\tau, \tag{5.30a}$$

$$S_{xx}(\omega) = \int_{-\infty}^\infty R_{xx}(\tau)e^{-j\omega\tau}\, d\tau \tag{5.30b}$$

$$H(\omega) = \int_{-\infty}^\infty h(\tau)e^{-j\omega\tau}\, d\tau \tag{5.30c}$$

Thus, the frequency response can be estimated from

$$H(\omega) = \frac{S_{xy}(\omega)}{S_{xx}(\omega)} \tag{5.29b}$$

Note that for good estimates of $H(\omega)$ to be derived from Equation (5.29b), it is important for $S_{xx}$ to be positive over the range of interest for $\omega$. Ideally, the input or stimulus signal should have a spectrum that is relatively flat over a wide bandwidth, i.e., the input should be *broadband*. In fact, the optimal type of input, from the viewpoint of estimation, is white noise (see Chapter 7).

An important detail that the user should note is that, when setting up `Spectrum Analyzer`, a value is required for the "`Sampling Interval`." This value allows the block to assign a time-scale to the input and output data. The sampling interval also determines the frequency range over which the estimated frequency response is plotted. In our example, we were concerned with the frequency range of 0 to 10 Hz, corresponding to a range in $\omega$ of 0 to $\sim$63 rad s$^{-1}$. Thus, we chose a minimum sampling frequency of 20 Hz, which translated to a sampling interval of 0.05 s.

## 5.4 FREQUENCY RESPONSE OF A MODEL OF CIRCULATORY CONTROL

The regulation of heart rate and systemic blood pressure is achieved in the short-term primarily through the feedback control via the arterial baroreflexes. However, both cardiovascular variables are continually perturbed by respiration. Breathing can affect heart rate and arterial blood pressure through a number of mechanisms. First, respiratory-induced intrathoracic pressure changes exert a direct effect on arterial pressure which, in turn, affects heart rate through the baroreflexes. Secondly, the present evidence suggests a direct coupling between the respiratory pattern generator in the medulla and the autonomic centers that influence heart rate. Thirdly, vagal feedback from the pulmonary stretch receptors during breathing has been shown to reflexively affect heart rate. And, finally, changes in heart rate can lead to changes in cardiac output which, in turn, produce arterial blood pressure fluctuations that alter heart rate through the baroreflexes. The overall effect of respiration on heart rate, commonly referred to as the *respiratory sinus arrhythmia*, can be quantified in terms of a frequency response function. Changes in phase and/or magnitude of this frequency response function would suggest changes in one of the factors that influence autonomic control of heart rate.

### 5.4.1 The Model

The model of circulatory control that we will examine was developed by Saul and coworkers (1991) from the Harvard Medical School and Massachusetts Institute of Technology. The SIMULINK implementation of this model (filename: "`rsa.mdl`") is shown in Figure 5.13. Respiration, measured in the form of lung volume change, $V$, is assumed to directly affect the autonomic inputs to the sinoatrial node: inspiration leads to decreases in both vagal and sympathetic efferent activity (note signs in summing blocks). The model does not distinguish between respiratory input from the pulmonary stretch receptors from the central drive that originates in the medullary centers. Feedback from the baroreceptors also directly influences the autonomic inputs to the heart: a rise in arterial blood pressure, `abp`, produces a decrease in sympathetic activity and an increase in parasympathetic activity. During inspiration, the decrease in vagal efferent activity acts on the sinoatrial node to increase heart rate, `hr`. The transfer function that models the dynamics of this relationship is a simple low-pass filter with a cutoff frequency (fp) that is on the order of 0.2 Hz and a negative gain, $-K_p$. In contrast, the response of the sinoatrial node to sympathetic stimulation is considerably slower. In addition to a latency of 1–2 s, the transfer function that characterizes the dynamics of sympathetic activity to heart rate conversion has a cutoff frequency, fs, of 0.015 Hz. In this case, the gain is positive.

Changes in heart rate are assumed to affect arterial blood pressure after a delay of 0.42 s. For simplicity, the transfer function representing the properties of the arterial vasculature is assumed static with a gain of 0.01 (mm Hg) min bt$^{-1}$. As well, since the

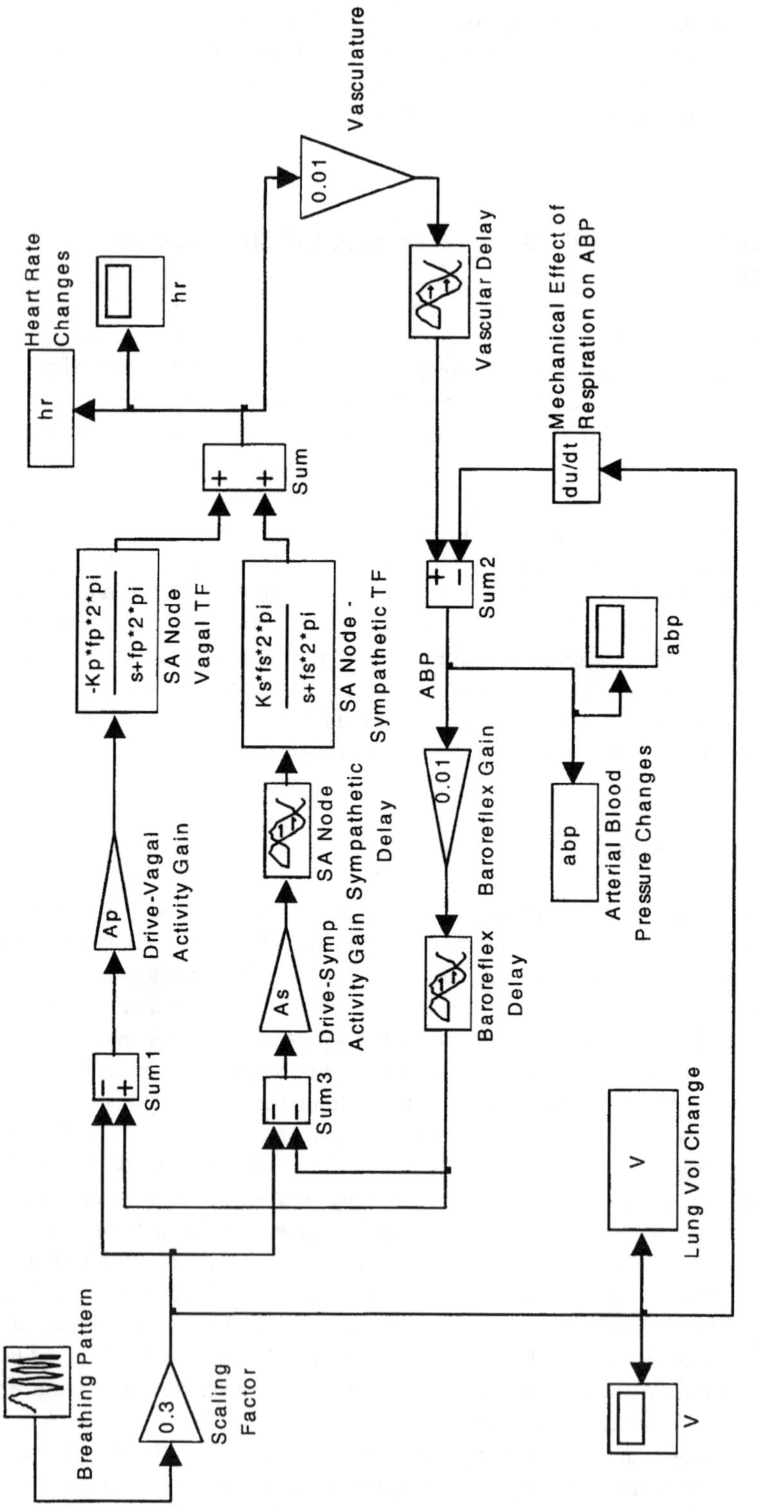


**Figure 5.13** SIMULINK model of circulatory control that accounts for the effect of respiration on heart rate and arterial blood pressure.

transduction of abp into baroreceptor output occurs with very rapid dynamics, we assume that the baroreflex can be adequately represented by a static gain (equal to 0.01) in series with a fixed delay of 0.3 s. Finally, the direct mechanical effects of respiration on abp are modeled as a negative differentiator, i.e., inspiration tends to decrease abp while expiration tends to increase it. Thus, the model simulates respiratory sinus arrhythmia by allowing the direct autonomic stimulation of heart rate. As well, the resulting changes in heart rate and the direct mechanical effects of respiration produce fluctuations in abp, which subsequently affect hr via the baroreflexes.

### 5.4.2 Simulations with the Model

To determine the frequency response of the circulatory control model, we employ a source block that produces a chirp signal. This is a sine wave, the frequency of which increases linearly with time. In our case, we set the parameters of the chirp block such that we start off with a frequency of 0.005 Hz and end with a frequency of 0.5 Hz after a duration of 300 s (simulation time). Since the amplitude of the chirp signal is not adjustable, a gain block of 0.3 is included between the source block and the rest of the model. This limits the peak-to-peak amplitude of the "respiration signal" to 0.6 liter. Before starting the simulation, the m-file "rsa_var.m" has to be executed in order to assign values to the adjustable parameters of the model. The following nominal parameter values represent the normal subject in supine posture: SA node vagal transfer function gain, $K_p = 6$; SA node sympathetic transfer function gain, $K_s = 18$; SA node vagal transfer function cutoff frequency, $f_p = 0.2$ Hz; SA node sympathetic transfer function cutoff frequency, $f_s = 0.015$ Hz. The relative weight factors for the conversion of respiratory drive or baroreflex drive to efferent neural activity are: $A_p$ (for the vagal branch) = 2.5 and $A_s$ (for the sympathetic branch) = 0.4.

Figure 5.14 displays the results obtained from one simulation run; for the sake of clarity, only 100 s of the simulated "data" are shown. The top panel shows the chirp signal (respiratory input) used to stimulate the model. The corresponding changes in heart rate predicted by the model are displayed in the middle panel. Note that at low frequencies, heart rate fluctuates almost in synchrony with lung volume change; however, at the higher frequencies, it tends to lag respiration. Also, the amplitude of the heart rate signal decreases with increasing frequency, underscoring the low-pass nature of the overall frequency response. The predicted behavior of arterial blood pressure is somewhat different: as frequency increases, the respiratory-induced changes in abp become larger. This results from the growing influence of the direct mechanical effects of breathing on blood pressure as frequency increases.

### 5.4.3 Frequency Response of the Model

Using the method described in Section 5.3.2, the frequency response of the model can be deduced from the input and simulated output. Instead of inserting the Spectrum Analyzer block, the reader can also save the input (V) and output (hr or abp) variables to the Workspace, and use the following MATLAB code (saved as "rsa_tf.m") to deduce the frequency response:

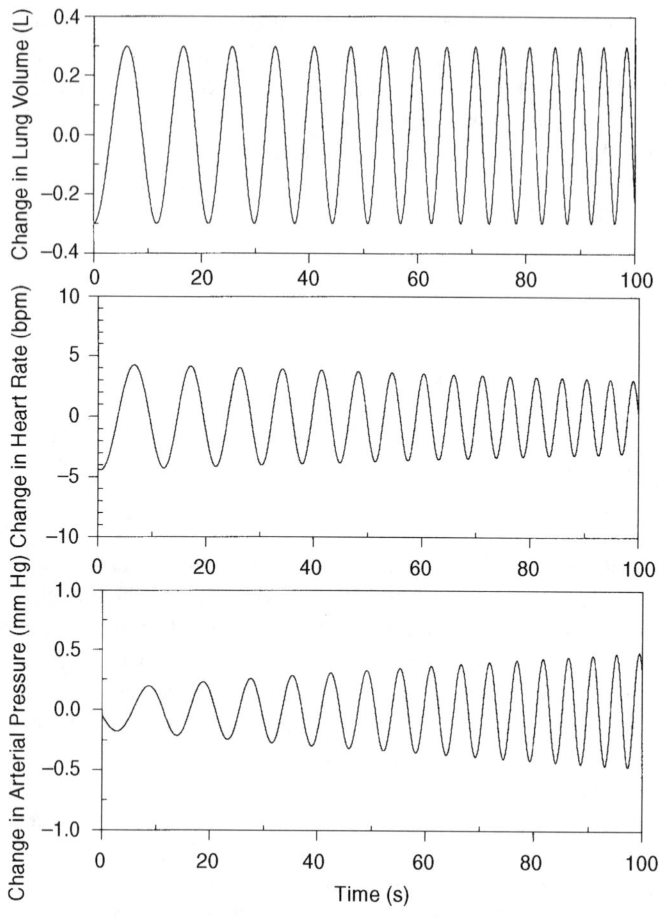


**Figure 5.14** Responses in heart rate and arterial blood pressure to a controlled breathing pattern (slow-to-high frequency), as predicted by the SIMULINK model of circulatory control ("normal" conditions).

```
% We assume the sampling interval is 0.1 s so that N = 3000
% for a total simulation time of 300 s
>>freq = [0:1/300:5]';
% compute Power spectrum of V and Cross-spectrum between
% V and hr
>> Pv = psd(V, N, 10);
>> Pvhr = csd(V, hr, N,10);
% compute Frequency Response magnitude and phase
>> Hvhr = Pvhr./Pv;
>> Hvhrmag =abs(Hvhr);
>> Hvhrpha = angle(Hvhr)*180/pi;
```

The chirp signal is useful as an input waveform since it produces a reasonably broad spectrum over the frequency range of interest: 0 to 0.4 Hz. Figure 5.15 displays the magnitude (top panel) and phase (lower panel) components of the frequency response between respiration and heart rate estimated for the simulated supine normal subject (solid curves). The low-pass nature of the magnitude response is clearly evident; however, the frequency response values toward the low (0 Hz) and high (0.4 Hz) ends of the range displayed cannot be regarded as accurate since most of the spectral power of the chirp input is contained in the frequencies in the middle of this range.

The results of two other simulation cases are also presented in Figure 5.15. The first simulates how the frequency response of the respiratory sinus arrhythmia would change if the "subject" were given a dose of atropine ("+Atropine", dashed curves) that produces complete *parasympathetic blockade*. In addition, the model parameters are also modified to simulate the subject in a standing posture, when the sympathetic influence on heart rate is enhanced. Under such conditions, heart rate control would be modulated predominantly by the sympathetic nervous system. Not surprisingly, the resulting frequency response magnitude curve shows a substantial increase at frequencies below 0.03 Hz and a large decrease at frequencies higher than 0.1 Hz. The phase curve shows a much steeper slope, indicating an

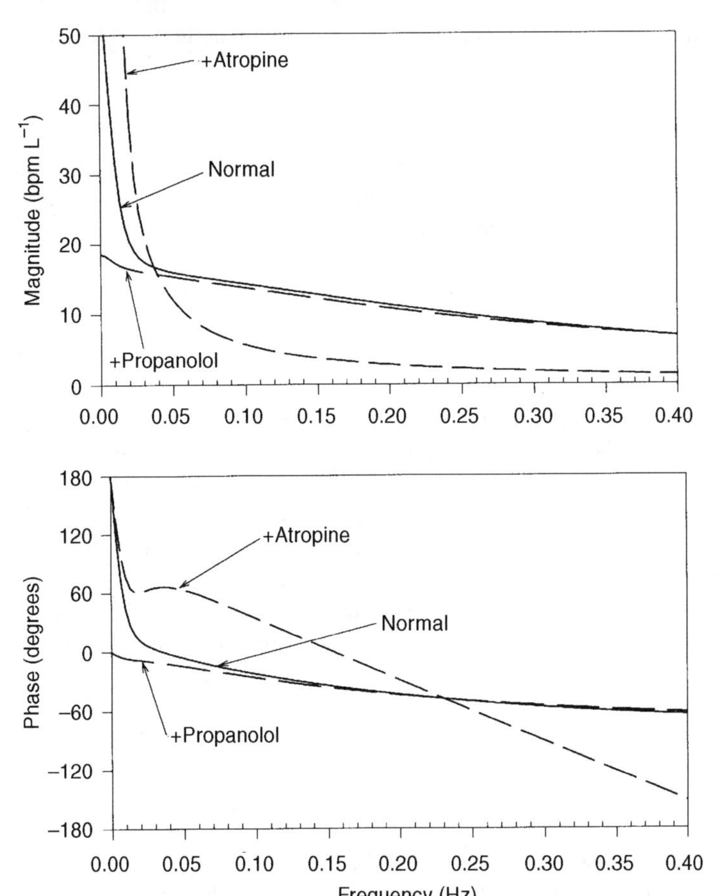


**Figure 5.15** Frequency responses of the circulatory control model under conditions that simulate normal heart rate control, complete $\beta$-adrenergic blockade ("+Atropine"), and complete parasympathetic blockade ("+Propanolol").

increase in the lags inherent in the system. The values of the model parameters employed here are: $A_p = 0.1$, $K_p = 1$, $f_p = 0.07$ Hz, $A_s = 4.0$, $K_s = 9$ and $f_s = 0.015$ Hz.

In the other simulation case, the "subject" is given a dose of propanolol, which produces *β-adrenergic blockade*. Furthermore, we assume a supine posture, thus making vagal modulation the predominant mode of control. The frequency response curves corresponding to this condition are labeled "+Propanolol". Compared to the control case, there is little change in the frequency response above 0.05 Hz. However, loss of sympathetic modulation leads to a significant decrease in frequency response magnitude and phase at the very low frequencies. Under this "purely vagal" state, the phase difference between respiration and heart rate is relatively small over the 0 to 0.4 Hz range, indicating that the respiratory-induced changes in heart rate occur rapidly. The parameter values used to represent this state are: $A_p = 2.5$, $K_p = 6$, $f_p = 0.2$ Hz, $A_s = 0.1$, $K_s = 1$, $f_s = 0.015$ Hz.

## 5.5 FREQUENCY RESPONSE OF GLUCOSE–INSULIN REGULATION

In Section 3.6, we examined the model of glucose and insulin regulation proposed by Stolwijk and Hardy but limited the scope of our analysis to the steady state. Here, we present the complete, dynamic version of this model as well as its SIMULINK implementation.

### 5.5.1 The Model

Employing the same notation as that presented in Section 3.6, Equations (3.40a) and (3.40b), which characterize the mass balance of glucose in the blood plasma, may be extended to incorporate dynamics in the following way:

$$C_G \frac{dx}{dt} = U(t) + Q_L - \lambda x - \nu xy, \qquad x \le \theta \tag{5.31a}$$

$$C_G \frac{dx}{dt} = U(t) + Q_L - \lambda x - \nu xy - \mu(x - \theta), \qquad x > \theta \tag{5.31b}$$

In the above equations, $C_G$ represents the glucose capacitance in the extracellular space and $U(t)$ represents the time-course with which external glucose is infused into the bloodstream, as part of the "glucose tolerance test." Basically, Equations (5.31a) and (5.31b) state that the net difference between the rate at which glucose is added to the blood and the rate at which it is eliminated equals the rate at which the glucose concentration, $x$, will increase (or decrease). It is important to note that the cross-product term between $x$ and $y$, the insulin concentration, makes the above equations nonlinear (or strictly speaking, bilinear). The corresponding dynamic mass balance for insulin is simply a straightforward extension of Equations (3.43a) and (3.43b):

$$C_I \frac{dy}{dt} = -\alpha y, \qquad x \le \phi \tag{5.32a}$$

$$C_I \frac{dy}{dt} = -\alpha y + \beta(x - \phi), \qquad x > \phi \tag{5.32b}$$

where $C_I$ is the insulin capacitance of the extracellular space.

The SIMULINK implementation of this model (filename "`glucose.mdl`") is displayed in Figure 5.16b. The top half of the interconnected block structures represents Equations (5.31a) and (5.31b), characterizing the dynamics of blood glucose build-up and elimination, while the bottom half models insulin dynamics, as described by Equations

(5.32a) and (5.32b). Saturation blocks are employed to function as thresholding operators, with the lower limit set equal to zero and the upper limit set equal to a very large number (so that there is effectively no saturation at the high end). One of these saturation blocks allows for the disappearance of the last term in Equation (5.31b) when $x \leq \theta$, where $\theta$ is the threshold concentration below which all glucose is reabsorbed in the kidneys, which results in no glucose loss in urine. The other saturation block in the insulin portion of the model allows for the disappearance of the last term in Equation (5.32b) when $x \leq \phi$, where $\phi$ is the threshold glucose concentration for insulin secretion.

The glucose–insulin regulation model is encapsulated into a `Subsystem` block, which is shown in relation to the source and sink blocks in Figure 5.16a. The `Subsystem` block is created by dragging its icon from the `Connections` block library to the model window. By double-clicking on the `Subsystem` block, a subsystem window will appear and the user can proceed to create the model in question within this window. `Inport` blocks are used for all signals entering the subsystem while `Outport` blocks form the terminal points for all signals leaving the subsystem.

### 5.5.2 Simulations with the Model

Glucose infusion into the model is simulated through the use of a `pulse generator` block, which produces a rectangular wave. The period of the output waveform is set equal to the simulation time of 5 hours, with time step being 0.01 hour. The duty cycle is set to 5% so that glucose infusion occurs over a duration of 0.25 hour or 15 minutes, starting at $t = 0.5$ hours. The glucose infusion rate (amplitude of the rectangular wave) is set equal to $100\,000\,\mathrm{mg\,h^{-1}}$.

Examples of two simulation runs are displayed in Figure 5.17. The input waveform is shown in the top panel of this figure. The resulting time-courses in glucose concentration and insulin concentration are shown in the middle and lower panels, respectively. Two classes of subjects are examined here: the normal adult (solid curves) and the Type-2 diabetic (dashed curves). Note that in the diabetic, the steady-state levels for glucose and insulin are both higher than corresponding levels in the normal, which confirms what we found by a graphical method of solution in Section 3.6. As well, in the diabetic, the decay of glucose and insulin concentrations toward steady-state levels following the infusion is noticeably slower compared to the normal subject. Furthermore, the glucose time-course does not show the slight undershoot exhibited by the corresponding time-course in the normal.

### 5.5.3 Frequency Responses of the Model

Using the method outlined in Section 5.3.2 (see MATLAB m-file "`gireg.m`"), estimates of the frequency responses of this model are deduced and shown in Figure 5.18. The top panel displays the frequency response magnitude curves ($H_{ux}$) relating the glucose infusion to resulting glucose concentration for both normal and Type-2 diabetic. The slower response of the diabetic is reflected in the higher values for the frequency response magnitude at frequencies below $0.4\,\mathrm{cycles\,h^{-1}}$. The frequency response magnitude curve for insulin ($H_{uy}$) shows a very similar shape (middle panel of Figure 5.18). Finally, we also display the frequency response relating glucose concentration as input to insulin concentration as output

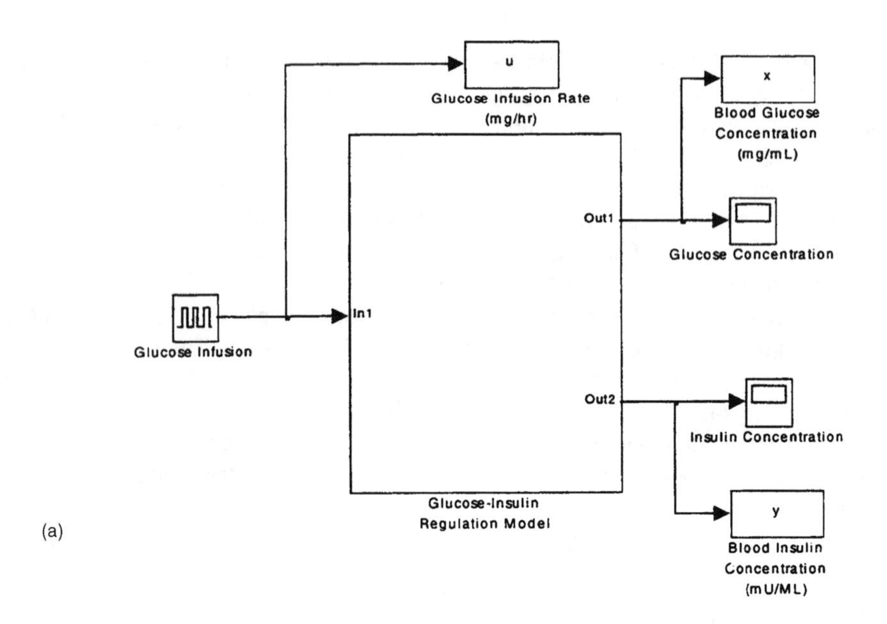


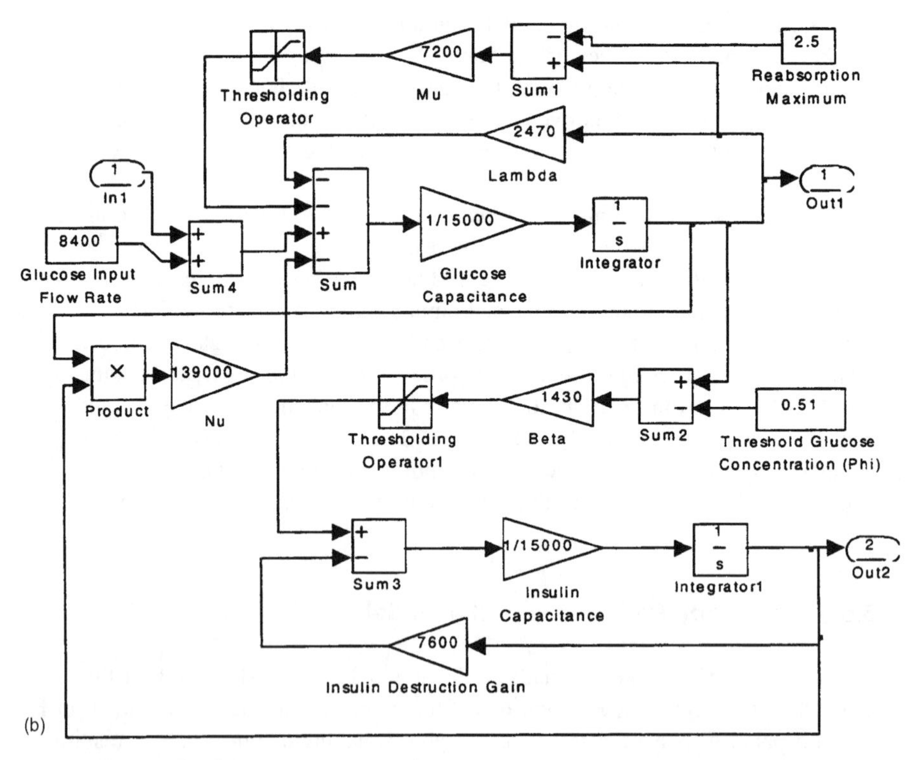


**Figure 5.16** SIMULINK model of blood glucose–insulin regulation. (a) The input to and outputs from the model. (b) Details of the dynamic structure.

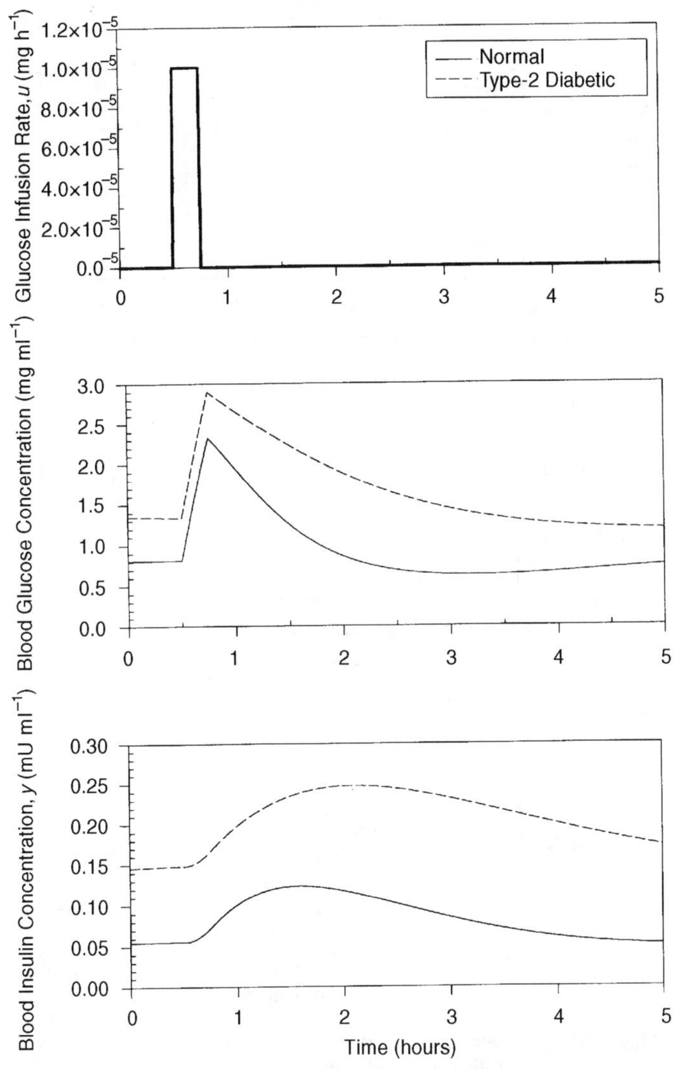


**Figure 5.17** Response of the glucose–insulin regulation model to a rapid (15 min) infusion of 25 g of glucose in a simulated normal human (solid lines) and a Type-2 diabetic (dashed lines).

($H_{xy}$). In this case, the transfer magnitude for the diabetic is higher than the normal over a much wider range of frequencies (0 to 1 cycle h$^{-1}$).

An important point to note in this example is that there are a number of nonlinearities embedded in the model: the thresholds $\theta$ and $\phi$, as well as the multiplicative interaction between $x$ and $y$. These are clearly exhibited in the steady-state curves representing glucose–insulin regulation in Figure 3.13. Therefore, strictly speaking, the application of frequency response analysis, a linear technique, is not valid. On the other hand, it is also clear, from the graphs presented in Figure 3.13, that the degree of nonlinearity is not so high as to preclude replacing segments of the hyperbolic relationship with piecewise linear approximations. Thus, assuming that the fluctuations in $x$ and $y$ are reasonably limited in range, particularly under physiological conditions, frequency response analysis remains useful in providing some insight into the dynamic behavior of this system.

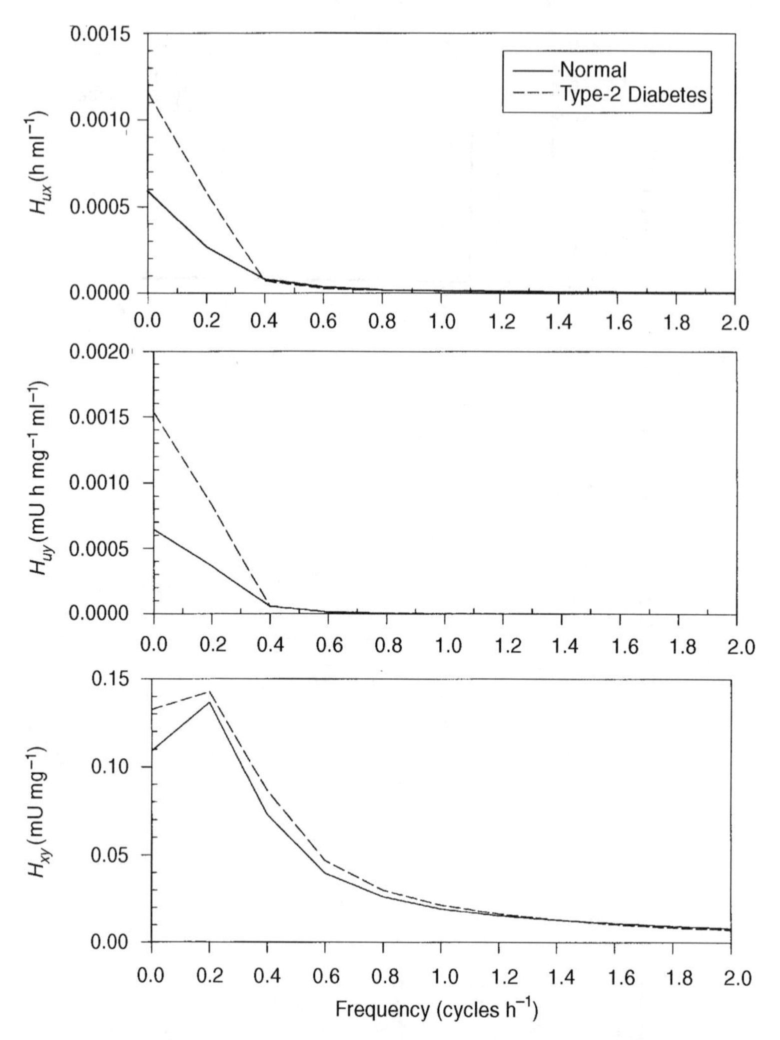


**Figure 5.18** Frequency responses of glucose–insulin regulation model in conditions simulating a typical normal human (solid lines) and a Type-2 diabetic (dashed lines).

## BIBLIOGRAPHY

Dorf, R.C., and R.H. Bishop. *Modern Control Systems*, 7th ed. Addison-Wesley, Reading, MA, 1995.

Kuo, B.C. *Automatic Control Systems*, 4th ed. Prentice-Hall, Englewood Cliffs, NJ, 1994.

Milhorn, H.T. *The Application of Control Theory to Physiological Systems*. W.B. Saunders, Philadelphia, 1966.

Milsum, J.H. *Biological Control Systems Analysis*. McGraw-Hill, New York, 1966.

Morris, N.M. *Control Engineering*, 4th ed. McGraw-Hill (UK), London, 1991.

Saul, J.P., R.D. Berger, P. Albrecht, S.P. Stein, M.H. Chen, and R.J. Cohen. Transfer function analysis of the circulation: unique insights into cardiovascular regulation. *Am. J. Physiol.* **261** (*Heart Circ. Physiol.* **30**): H1231–H1245, 1991.

Soechting, J.F., P.A. Stewart, R.H. Hawley, P.R. Paslay, and J. Duffy. Evaluation of neuromuscular parameters describing human reflex motion. *Trans. ASME, Series G* **93**: 221–226, 1971.

Stolwijk, J.E., and J.D. Hardy. Regulation and control in physiology. In: *Medical Physiology*, 13th ed. (edited by V.B. Mountcastle). C.V. Mosby, St. Louis, 1974; pp. 1343–1358.

Strum, R.D., and D.E. Kirk. *Contemporary Linear Systems using MATLAB*. PWS Publishing Co., Boston, MA, 1994.

## PROBLEMS

**P5.1.** Consider the simplified model of eye-movement control displayed in Figure P4.1. Assuming that $G/J = 14{,}400\,\text{rad}^2\,\text{s}^{-2}$, $B/J = 24\,\text{rad}\,\text{s}^{-1}$ and $k_v = 0.01$, compute the frequency response for this model. Display the magnitude and phase components of the frequency response in the form of:
(a) linear-scale frequency response plots (e.g., Figures 5.2 and 5.3);
(b) Bode plots;
(c) Nichols charts;
(d) Nyquist plots.

**P5.2.** In the eye-movement control model considered above, determine how the frequency response of the model would change if:
(a) there is no velocity feedback ($k_v = 0$);
(b) the velocity feedback gain is negative ($k_v < 0$).

Employ the parameter values given in Problem P5.1. Compare the control case ($k_v = 0.01$) to cases (a) and (b) using linear-scale frequency response plots.

**P5.3.** Compute and plot the frequency responses of the system shown in Figure P4.2 when:
(a) the feedback loop is open; and
(b) the feedback loop is closed.

In both cases, assume the time delay, $T$, to be 1 second.

**P5.4.** Determine the frequency responses of the ventilatory control model shown in Figure P4.3, assuming (a) $\alpha = 0$ (no rate sensitivity); (b) $\alpha = \frac{1}{2}$ (lag–lead feedback); and (c) $\alpha = 2$ (lead–lag feedback).

**P5.5.** Derive a closed-form expression for the frequency response of the circulatory control model shown in Figure 5.13, with respiration as the input and heart rate as the output. Using the parameter values given in the SIMULINK implementation of the model (“`rsa.mdl`”) and Section 5.4.3, deduce and plot the frequency responses of this model for the three cases shown in Figure 5.15: (a) normal supine subject; (b) following atropine administration; and (c) following propanolol infusion. Compare these plots with those presented in Figure 5.15.

**P5.6.** The model of Figure 5.13 can be used to investigate the dynamics of blood pressure regulation by the baroreflexes in the following way. In the SIMULINK model, “`rsa.mdl`,” remove the respiratory input from the model and add an external source that imposes a random excitation directly on arterial blood pressure. This can be achieved experimentally in approximate fashion by imposing positive and negative pressure changes on the neck, thereby changing carotid sinus pressure. Assuming the applied pressure time-course to be the input and the resulting heart rate changes to be the output, use the method outlined in Section 5.3.2 to deduce the frequency response of the closed-loop baroreflex control system.

**P5.7.** Derive an expression for the closed-loop frequency response of the neuromuscular reflex model displayed in Figure 4.11, assuming the external moment, $M_x$, to be the input and angular displacement of the forearm, $\theta$, to be the output. Using the parameter values given in Section 4.6.2, display the magnitude and phase plots of the frequency response.

**P5.8.** Using the SIMULINK implementation of the neuromuscular reflex model (“`nmrflx.mdl`”), displayed in Figure 4.12, estimate the frequency response of the closed-loop system by using the method discussed in Section 5.3.2. Employ a random noise source as the driving input, $M_x$. Check your results against the magnitude and phase plots of frequency response deduced analytically in Problem P5.7.

# 6

# Stability Analysis: Linear Approaches

## 6.1 STABILITY AND TRANSIENT RESPONSE

The concept of stability is best explained by referring to the illustration in Figure 6.1. The four plots in this figure represent the impulse responses from a linear system under different conditions. In case (a), the response is similar to that exhibited by an overdamped or critically damped system. In case (b), the system exhibits an underdamped response, i.e., there is some oscillation in the response but it is eventually damped out. In both cases, the system is said to be *stable*. However, in case (d), the impulsive stimulus produces a response that is oscillatory with growing amplitude; as a consequence, the system output never returns to its original operating point prior to stimulation. This is the hallmark of an *unstable* system. With these examples in mind, we introduce the following definition of stability: *a stable dynamic system is one that will respond to a bounded input with a bounded response*. Apart from the clear-cut cases of "stable" and "unstable" systems, there are *conditionally or marginally stable* systems that exhibit undamped oscillations (e.g. case (c)). Here, the response is bounded but never returns to the steady operating level prior to perturbation; however, such systems can exhibit unbounded responses if stimulated with certain bounded inputs, such as sinusoidal waves with frequencies that match the characteristic frequencies of the system.

We examined in considerable detail in Chapter 4 how first-order and second-order systems, operating in either open-loop or closed-loop modes, respond to impulsive or step inputs. Recall that the dynamics of the system responses, i.e., the "transient responses," were determined by the roots of the denominator, or the *poles*, of the model transfer function. For the closed-loop ($k > 0$) generalized second-order system, this was shown in Equation (4.51) to be:

$$\frac{Y(s)}{X(s)} = \frac{G_{SS}\omega_n^2}{s^2 + 2\zeta\omega_n s + (1 + kG_{SS})\omega_n^2} \tag{6.1}$$

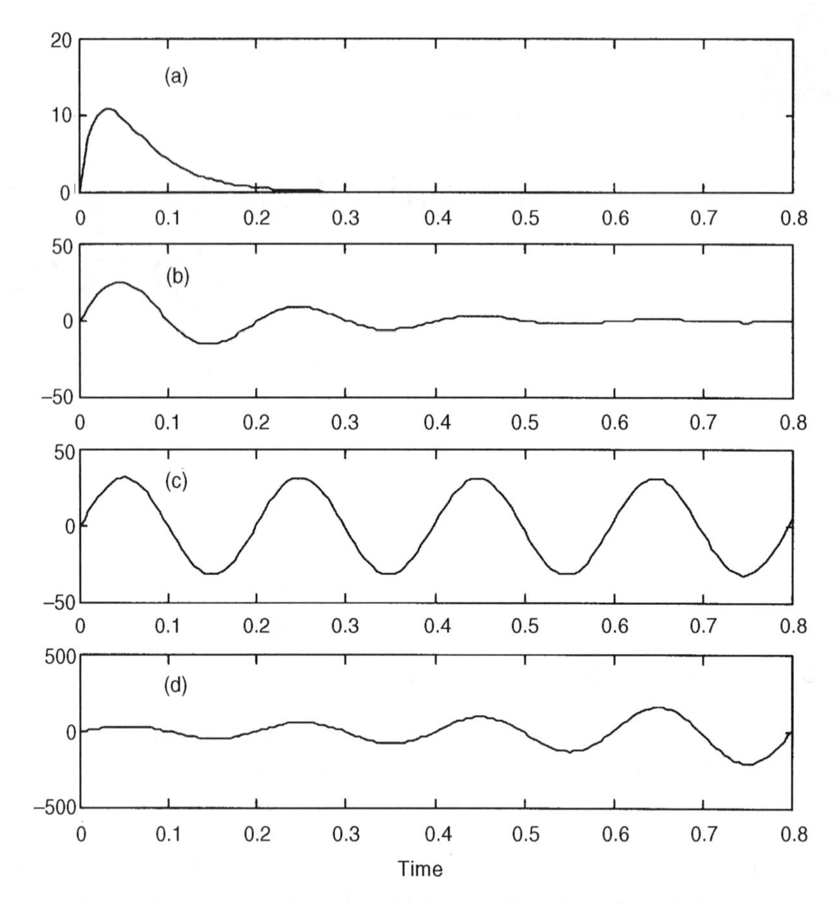


**Figure 6.1** Responses of a system that is (a) stable and overdamped; (b) stable and underdamped; (c) marginally stable or oscillatory; (d) unstable.

so that the system poles are given by the roots ($\alpha_1$ and $\alpha_2$) of

$$s^2 + 2\zeta\omega_n s + (1 + kG_{SS})\omega_n^2 = 0 \tag{6.2}$$

i.e.,

$$\alpha_{1,2} = -\zeta\omega_n \pm \omega_n\sqrt{\zeta^2 - (1 + kG_{SS})} \tag{6.3}$$

The poles can be real, imaginary or complex, depending on the size of $\zeta^2$ relative to the term $1 + kG_{SS}$, as shown by Equation (6.3). Thus, the impulse response corresponding to Equation (6.1) is given by

$$h(t) = \frac{G_{SS}\omega_n^2}{\alpha_2 - \alpha_1}(e^{\alpha_2 t} - e^{\alpha_1 t}) \tag{6.4}$$

For positive values of $\zeta$, $\omega_n$, $k$ and $G_{SS}$, Equation (6.4) shows that the transient response will take on one of the forms represented in cases (a), (b), and (c) in Figure 6.1, but not case (d). This arises from the fact that in cases (a), (b), and (c), the real parts of the roots are either negative or zero, so that the terms within the parentheses in Equation (6.4) represent exponential decaying, exponentially damped, or simply sinusoidal dynamics. The exponentially growing behavior of the *unstable* system would only occur *if the real parts of* $\alpha_1$ *and/or* $\alpha_2$ *were positive*. This could be so if $\zeta$ or $\omega_n$ were negative; however, this would not be physically feasible. The only possible way in which the closed-loop model represented by Equation (6.1) could realistically be made unstable would be by making $k$ negative, i.e. with positive feedback. If $k$ were to be negative and to assume a magnitude larger than $1/G_{SS}$, one of the roots in Equation (6.3) would start to become positive real, and the resulting impulse response (Equation (6.4)) would increase exponentially with time.

We can summarize the conclusions from the above discussion on stability by extending the results to more generalized systems of the forms shown in Figure 6.2. In Figure 6.2a, the dynamics of the forward and feedback components are characterized by transfer functions $P(s)$ and $Q(s)$, respectively. The gain of the forward path is controlled by the static factor $K$, which can be varied between zero and infinity. Figure 6.2b shows a similar configuration except that, here, the feedback gain can be varied by varying $K$ between zero and infinity. For the kind of system represented in Figure 6.2a, the closed-loop transfer function is

$$H_A(s) = \frac{Y(s)}{X(s)} = \frac{KP(s)}{1 + KP(s)Q(s)} \tag{6.5}$$

For the type of system shown in Figure 6.2b, the closed-loop transfer function is

$$H_B(s) = \frac{Y(s)}{X(s)} = \frac{P(s)}{1 + KP(s)Q(s)} \tag{6.6}$$

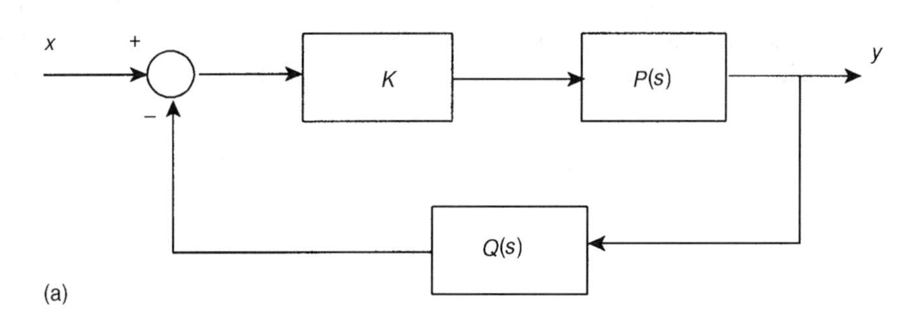


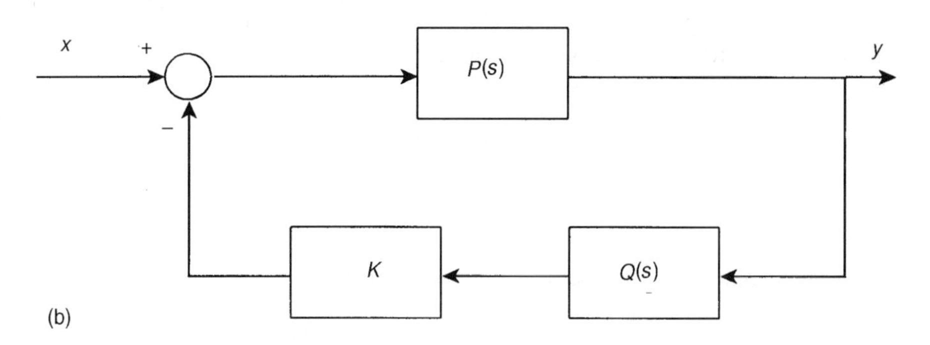


**Figure 6.2** Closed-loop systems with variable forward (a) and feedback (b) gains.

Note that, in both cases, the denominator of the closed-loop transfer function is the same. Thus, for given forms of $P(s)$ and $Q(s)$ and gain $K$, both types of systems will have the same transient response. As in the example of the lung mechanics model, the transient response and stability of both these systems are determined by the poles of their closed-loop transfer functions, i.e., the roots of the following characteristic equation:

$$1 + KP(s)Q(s) = 0 \tag{6.7}$$

Note that the product $KP(s)Q(s)$ yields the *loop transfer function* $\{LG(s)\}$ of both closed-loop systems. (Recall from Section 3.2 that the magnitude of $LG(s)$ is the *loop gain* of the feedback control system.) Thus, Equation (6.7) can be generalized further to

$$1 + LG(s) = 0 \tag{6.8}$$

Extending the result arrived at earlier, we can conclude that, in each of the closed-loop systems shown in Figure 6.2, the transient response will be unstable if the *real part of any root* of its characteristic equation (Equation (6.7)) becomes *positive*.

## 6.2 ROOT LOCUS PLOTS

The root locus method is a classical procedure used to determine how the poles of the closed-loop transfer function would change as a function of a system parameter (generally, some gain constant), given the location of the open-loop poles and zeros. In fact, the "root locus" is the path on the complex $s$-plane traced by the closed-loop poles when the system parameter in question varies over a range of values. To illustrate how this method is applied, we turn once again to our simple lung mechanics model. Referring back to Figure 4.2b, we find that the lung mechanics model is merely a special case of the closed-loop form shown in Figure 6.2, where

$$P(s) = \frac{1}{LCs^2 + RCs + 1} \tag{6.9}$$

$$Q(s) = 1 \tag{6.10}$$

and

$$K = k \tag{6.11}$$

For given values of the lung mechanics parameters ($L$, $C$, and $R$), the root locus will show us how the dynamic behavior of the model changes as the feedback gain takes on a range of values. Applying Equation (6.7), the task then is to solve for the roots of the following characteristic equation as $k$ varies:

$$1 + \frac{k}{LCs^2 + RCs + 1} = 0 \tag{6.12}$$

This is equivalent to solving

$$LCs^2 + RCs + 1 + k = 0 \tag{6.13}$$

The general solution for the roots of Equation (6.13) is

$$\alpha_{1,2} = \frac{-R \pm \sqrt{R^2 - 4L(1+k)/C}}{2L} \tag{6.14}$$

To solve Equation (6.14), we will assume, as in Section 4.3, that $L = 0.01$ $\mathrm{cm\,H_2O\,s^2\,L^{-1}}$, $C = 0.1\,\mathrm{L\,cm\,H_2O^{-1}}$, and $R = 1\,\mathrm{cm\,H_2O\,s\,L^{-1}}$. It is generally useful to determine the locations of the closed-loop poles when $k$ assumes its two most extreme values. First, note that when $k = 0$, solving Equation (6.13) becomes equivalent to determining the locations of the poles of the open-loop system, $P(s)$. Substituting the above values of $L$, $C$, $R$, and $k$ into Equation (6.14), we obtain the following solutions: $\alpha_1 = -88.73$ and $\alpha_2 = -11.27$. For these parameter values, both poles are real and negative. At the other extreme, when $k$ becomes infinitely large, then Equation (6.14) yields the solutions: $\alpha_1 = -50 + j\infty$ and $\alpha_2 = -50 - j\infty$. Thus, for very large $k$, the poles are complex. Finally, the value of $k$ at which the two real poles become complex can be found by solving for the value of $k$ where the expression inside the square-root operation in Equation (6.14) goes to zero, i.e.,

$$k = \frac{R^2 C}{4L} - 1 \tag{6.15}$$

Substituting in values for $L$, $R$, and $C$, we obtain $k = 1.5$.

The complete root locus plot of the system in question can be obtained easily with the use of the MATLAB Control System Toolbox function "`rlocus`." This function assumes that the product $P(s)Q(s)$ yields a transfer function $H$ that takes the form of the ratio of two polynomial functions of $s$:

$$P(s)Q(s) = H(s) = \frac{N(s)}{D(s)} \tag{6.16}$$

Then, Equation (6.7) can be recast into the following form:

$$D(s) + KN(s) = 0 \tag{6.17}$$

"`rlocus`" finds the solution to Equation (6.17) for all values of $K$ between 0 and infinity. For our specific example, the following MATLAB command lines can be used to plot the corresponding root locus:

```
>> Ns = [1];
>> Ds =[L*C R*C} 1];
>> Hs = tf(Ns, Ds);
>> rlocus(Hs);
```

The resulting root locus plot is displayed in Figure 6.3. Note that the locations of the closed-loop poles when $k$ equals zero, 1.5, and infinity are exactly as we had deduced earlier. Also, since the poles always lie on the left-hand side of the $s$-plane (i.e., real parts of poles are always negative), the closed-loop system is stable for all positive values of feedback gain, $k$. Note that the root locus gives us a good global picture of the transient response characteristics of the system but tells us little about its frequency response.

As a further example, we consider the linear lung mechanics model when integral feedback control is employed instead of proportional feedback, i.e., in this case, the fluctuations in alveolar pressure are integrated before being fed back to the comparator.

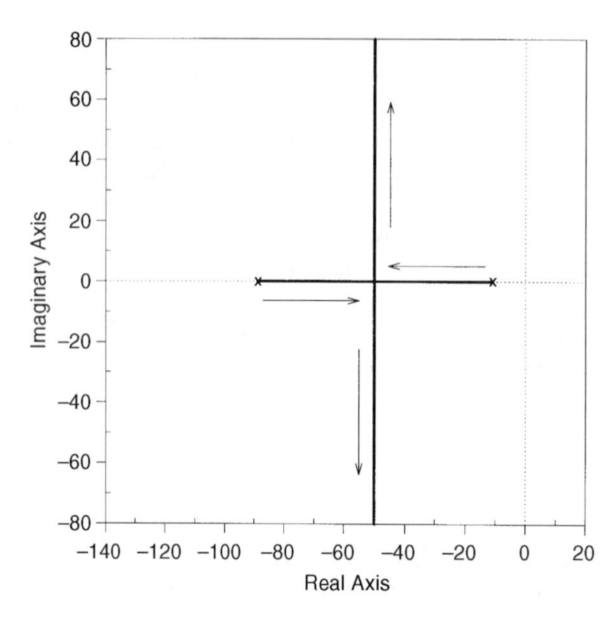


**Figure 6.3** Root locus plot for the lung mechanics model with proportional feedback. Locations marked " × " indicate positions of poles when $k = 0$. Arrows indicate direction in which the poles move as $k$ is increased from zero to infinity. Dotted horizontal and vertical lines represent the real and imaginary axes, respectively.

This system is displayed in Figure 6.4. Referring to Figure 6.2b, the forward transfer function $P(s)$ remains the same while the feedback transfer function $Q(s)$ is now given by

$$Q(s) = \frac{1}{s} \tag{6.18}$$

The characteristic equation (Equation (6.17)) now assumes the specific form

$$LCs^3 + RCs^2 + s + k = 0 \tag{6.19}$$

The above equation can be solved easily using MATLAB by simply inserting an extra term into the row vector Ds that represents $D(s)$:

```
>> Ns = [1];
>> Ds = [L*C R*C} 1 0];
>> Hs = tf(Ns, Ds);
>> rlocus(Hs);
```

The corresponding root locus plot, displayed in Figure 6.5, shows a form that differs significantly from the plot in Figure 6.3. Because the characteristic equation is now third-order, there are three poles instead of two. When $k = 0$, all three poles are located on the real axis, one of which is situated at the origin ($s = 0$). As $k$ increases, the most negative pole becomes progressively more negative while remaining real. However, when $k$ increases beyond 2.64, the other two poles become complex, i.e., the transient response becomes a damped oscillation that becomes less and less damped, and the frequency of which progressively increases, as $k$ continues to increase. Finally, when $k$ exceeds 100, the system becomes unstable as these two poles move into the right-hand side of the $s$-plane, i.e., the impulse response assumes the form of a growing oscillation. These results

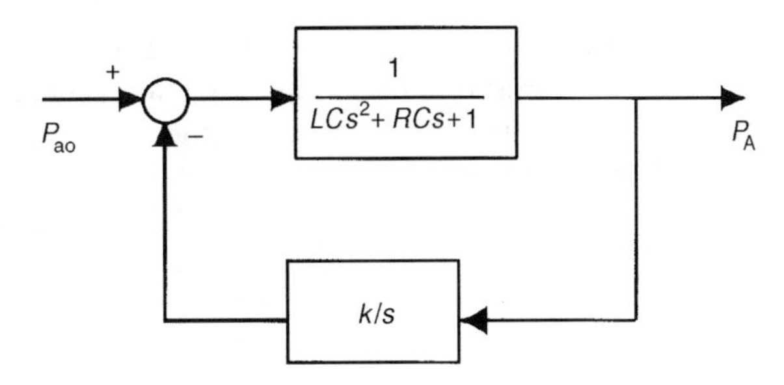


**Figure 6.4** Linear lung mechanics model with integral feedback.

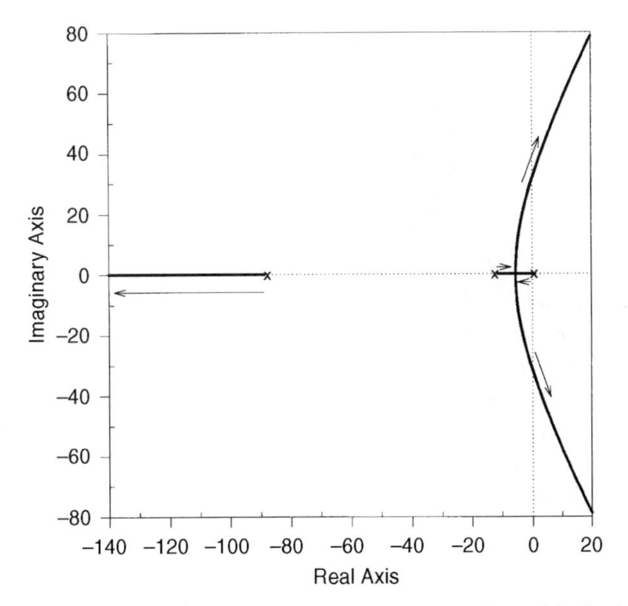


**Figure 6.5** Root locus plot for the lung mechanics model with integral feedback. Poles at $k = 0$ are marked "×". Arrows indicate direction in which the three poles move as $k$ is increased toward infinity.

demonstrate that the integral feedback system depicted here exhibits dynamics that are less stable than the proportional feedback case, although it gives a faster response.

## 6.3 ROUTH–HURWITZ STABILITY CRITERION

The root locus method requires the evaluation of all roots of the characteristic equation in order to determine whether a given system is stable or unstable. The Routh–Hurwitz technique is a classical stability test that enables such a determination without the need to actually evaluate the roots. With the computational tools that are available nowadays, this test has become somewhat obsolete. Nonetheless, we will describe it here for the sake of completeness.

We assume the following general form for the characteristic equation of the closed-loop system in question:

$$a_n s^n + a_{n-1} s^{n-1} + \cdots + a_1 s + a_0 = 0 \tag{6.20}$$

The Routh–Hurwitz technique requires the computation of values based on the coefficients of the characteristic equation and the arrangement of these values into an array of the following construction (with each row corresponding to a power of $s$, as indicated in the margin to the left of the array):

$$\begin{array}{c|cccc} s^n & a_n & a_{n-2} & a_{n-4} & \cdot \\ s^{n-1} & a_{n-1} & a_{n-3} & a_{n-5} & \cdot \\ s^{n-2} & b_1 & b_2 & b_3 & \cdot \\ s^{n-3} & c_1 & c_2 & c_3 & \cdot \\ \cdot & \cdot & \cdot & \cdot & \cdot \\ s^1 & p_1 & & & \\ s^0 & q_1 & & & \end{array} \tag{6.21}$$

where

$$b_1 = \frac{a_{n-1}a_{n-2} - a_n a_{n-3}}{a_{n-1}}, \qquad b_2 = \frac{a_{n-1}a_{n-4} - a_n a_{n-5}}{a_{n-1}}. \qquad c_1 = \frac{b_1 a_{n-3} - b_2 a_{n-1}}{b_1} \tag{6.22}$$

and so on. The Routh–Hurwitz criterion states that the number of closed-loop poles located on the right-hand side of the $s$-plane is given by the number of changes in sign in the first column of the constructed array in Equation (6.21). Thus, for a *stable* system, there should be *no changes in sign in the first column of the array.*

To illustrate the use of the Routh–Hurwitz test, we will apply it to the two examples discussed in Section 6.2. Consider first the linear lung mechanics model with proportional feedback. The characteristic equation here is given by Equation (6.13). Using Equation (6.21), the Routh array is

$$\begin{array}{c|cc} s^2 & LC & 1+k \\ s & RC & 0 \\ s^0 & 1+k & 0 \end{array} \tag{6.23}$$

For positive values of $L$, $C$, and $R$ and for $k \geq 0$, there are no changes in sign in the terms of the *first column* of the array. Therefore, by the Routh–Hurwitz criterion, the system defined by Equation (6.13) will always be stable.

In the lung mechanics model with integral feedback, the characteristic equation is given by Equation (6.19). In this case, the first two columns of the Routh array are

$$\begin{array}{c|cc} s^3 & LC & 1 \\ s^2 & RC & k \\ s & 1 - \dfrac{Lk}{R} & 0 \\ s^0 & k & 0 \end{array} \tag{6.24}$$

For positive values of $L$, $C$, $R$, and $k$, note that all terms in the first column of Equation (6.24) will be positive, except for the third term which can become negative if

$$k > \frac{R}{L} \tag{6.25}$$

For $R = 1\ \mathrm{cm\,H_2O\,s\,L^{-1}}$ and $L = 0.01\ \mathrm{cm\,H_2O\,s^2\,L^{-1}}$, Equation (6.25) predicts that there will be one change of sign in Equation (6.24) when $k$ exceeds the value of 100. This is exactly the result obtained in Section 6.2 when we employed the root locus method.

## 6.4 NYQUIST CRITERION FOR STABILITY

One primary disadvantage of the Routh–Hurwitz test is that it becomes difficult to apply when the characteristic equation cannot be simply expressed as a polynomial function of $s$. Pure time delays are found abundantly in physiological systems. Although these can be approximated by rational polynomial expressions (see Problem P4.3), the resulting characteristic equation can become extremely unwieldy. For these kinds of problems, it is generally more convenient to employ the Nyquist stability test.

The formal mathematical development of the Nyquist stability criterion will not be presented here, as it involves a fair bit of complex variable theory; the interested reader can find this in most engineering texts on control systems. Instead, we will employ a more intuitive approach by illustrating the basic notions underlying this criterion. Consider the very simple negative feedback system shown in Figure 6.6. The input, $u$, represents a disturbance to the system that is nonzero for only a brief period of time. The system output, $x$, is fed back through a static gain $-K$ and added to the input before being fed forward through the time-delay ($T_D$) block. Note that the negativity in feedback control is implemented through the assignment of a negative value to the feedback gain and the addition (not subtraction) of the resulting feedback signal ($z$) at the level of the summing junction.

Figure 6.7 shows three of the many possibilities with which the system responds following the imposition of a transient disturbance, $u$. In all these cases, we have assumed that $u$ takes the form of a half-sine wave. Consider case (a) in which we assume $K = 0.5$ and $T_D$ is half the duration of $u$. Here, the initial passage of $u$ through the forward and feedback blocks produces an inverted and attenuated (by 50%) half-sine wave at $z$, labeled 1 in Figure 6.7a (upper panel). Propagation of this signal around the loop a second time would produce a response at $z$ of the form labeled as 2. Similarly, the third and fourth traversals around the entire loop would produce 3 and 4, respectively. This process would continue until the "echoes" of the initial $u$ are finally damped out totally. Each time there is a complete traversal around the loop, there is a change in sign, reflecting the negative nature of the feedback. Superimposition of all these individual effects on the initial disturbance leads to the complete response at $x$ shown in the lower panel of Figure 6.7a. Thus, the system response to $u$ is a rapidly damped oscillation, i.e., the system in this case is stable. In Figure 6.7b, $K$ is assigned the value of 0.5 again, but $T_D$ is now increased to a value equal to the duration of $u$. This produces the responses in $z$ labeled as 1 after the first traversal around the loop and 2 after the second traversal (Figure 6.7b, upper panel). The net result again is a damped oscillation (Figure 6.7b, lower panel). However, in this case, the oscillation appears to be more slowly damped and the frequency of the oscillation is also lower. Note also, that the corrective actions (1 in response to $u$ and 2 in response to 1) have been delayed so much that they occur

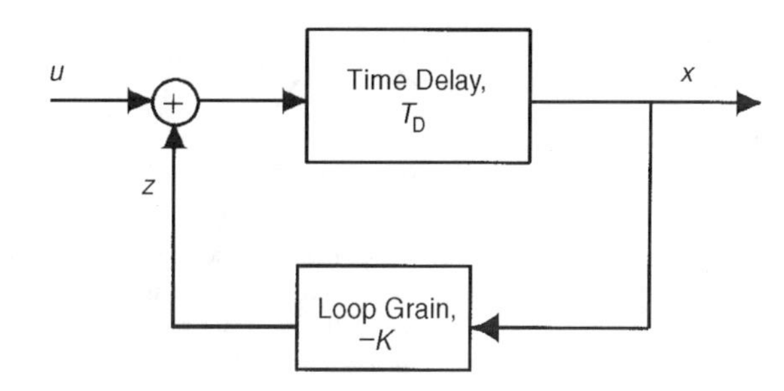


**Figure 6.6** Simple negative feedback system with delay. Note that, in this example, the negativity in feedback is embedded in the "loop gain" block.

out of phase (i.e., at 180°) with the preceding fluctuations. Finally, in case (c), the time delay is kept the same as that for case (b), but the loop gain magnitude is increased to 1. As in case (b), due to the increased delay, the feedback signal tends to reinforce the effect of the initial disturbance rather than to cancel it out. However, in this case, since there is no attenuation in the feedback loop, the net result is a sustained oscillation of period equal to twice the length of $T_D$. It is easy to see that if the loop gain were to be increased further to a value exceeding 1, the system response would be an oscillation with growing amplitude.

From the examples shown in Figure 6.7, it is clear that a closed-loop system can become unstable if the total phase lag imposed by all system components around the loop equals 180° *and* the loop gain magnitude is at least unity. This is the basic notion underlying the Nyquist criterion. Thus, in order to determine whether a specific closed-loop system is stable or unstable, we would first deduce the *loop transfer function* (i.e., the product of all component transfer functions around the closed loop). The Nyquist plot of the loop transfer function is generated. Note that a loop gain of unit magnitude and phase lag of 180° is represented by the point $(-1 + j0)$ on the complex plane. *The system is stable if the* $(-1 + j0)$ *point lies to the left of the Nyquist plot as the locus is traversed in the direction of increasing*

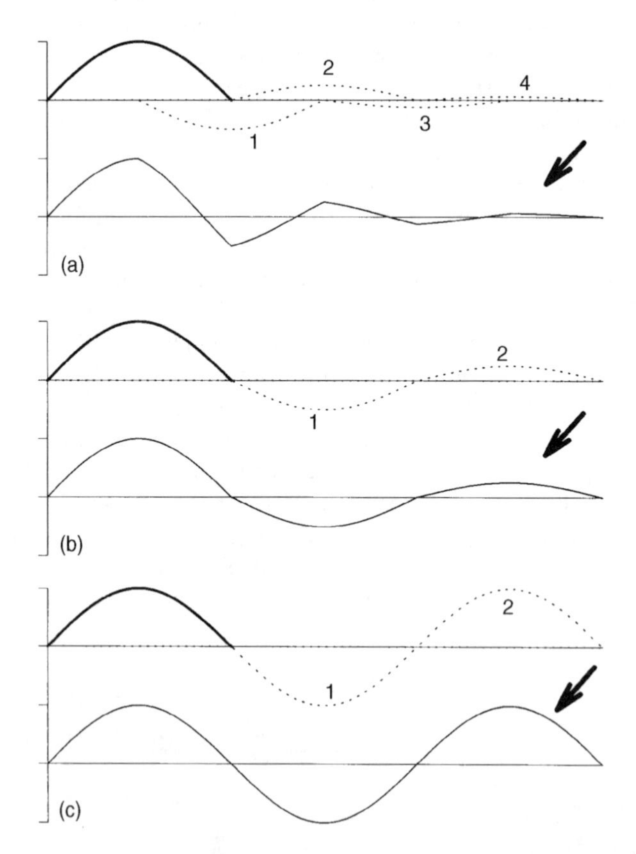


**Figure 6.7** Response of the negative feedback system with delay to an initial disturbance when: (a) loop gain magnitude $< 1$ and phase lag $< 180°$; (b) loop gain magnitude $< 1$ and phase lag $= 180°$; and (c) loop gain magnitude $= 1$ and phase lag $= 180°$.

*frequency.* Another way of stating this criterion is that in the stable system the Nyquist plot *will not encircle* the $(-1+j0)$ point. In order to make this determination, it is generally necessary to evaluate the loop transfer function from zero frequency to infinity, or at least over a wide band of frequencies. This criterion, as stated above, is valid as long as the loop transfer function does not contain any poles with positive real parts. If this condition does not apply, one has to employ a different version of the Nyquist criterion; more details of the method under such circumstances can be found elsewhere (e.g., Dorf and Bishop, 1995).

The reader may recall that the Nyquist representation was previously discussed in Section 5.2.3. However, one should be cautioned that, in Chapter 5, the examples shown were those in which we characterized the frequency responses corresponding to the open-loop and closed-loop transfer functions of the systems in question. For a determination of stability, we need to evaluate the *loop transfer function*, which will yield a Nyquist plot quite different from the Nyquist plots that correspond to the open-loop or closed-loop transfer functions of the same system.

To illustrate the application of the Nyquist stability criterion, we turn once again to the two examples that have been discussed earlier. For the linear lung mechanics model with proportional feedback, the loop transfer function, $H_{L1}(s)$, is given by

$$H_{L1}(s) = \frac{k}{LCs^2 + RCs + 1} \tag{6.26}$$

The frequency response corresponding to Equation (6.26) is obtained by substituting $j\omega$ for $s$:

$$H_{L1}(\omega) = \frac{k}{(1 - LC\omega^2) + jRC\omega} \tag{6.27}$$

The Nyquist plots corresponding to Equation (6.27) are shown in Figure 6.8 for three values of feedback gain ($k = 1$, 10, and 100). Note that at $\omega = 0$, $H_{L1}(\omega) = k$. However, the zero frequency values for $H_{L1}(\omega)$ when $k = 10$ and $k = 100$ lie outside the range displayed. In Equation (6.27), also notice that when $\omega \to \infty$, $H_{L1}(\omega) \to 0$. Thus, for each plot, the Nyquist locus begins at the point $(k + j0)$ at zero frequency and ends at the origin at infinite frequency. The direction of traversal of each locus with increasing values of frequency is indicated by the arrows (Figure 6.8). Except for the hypothetical case when $k$ becomes infinite, it can be seen that none of the Nyquist loci touch or encircle the $(-1+j0)$ point (represented as the filled circle in Figure 6.8). Thus, this system is stable for all finite values of feedback gain.

The loop transfer function for the linear lung mechanics model with integral feedback, $H_{L2}(s)$, is given by

$$H_{L2}(s) = \frac{k}{LCs^3 + RCs^2 + s} \tag{6.28}$$

The frequency response corresponding to the above transfer function is

$$H_{L2}(\omega) = \frac{k}{-RC\omega^2 + j\omega(1 - LC\omega^2)} \tag{6.29}$$

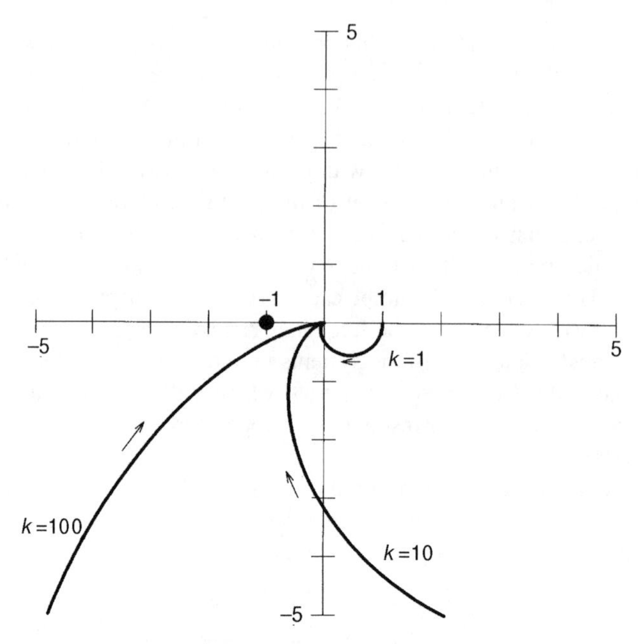


**Figure 6.8** Nyquist plots for the linear lung mechanics model with proportional feedback. Feedback gains shown are $k = 1$, 10 and 100. For the latter two cases, only portions of the Nyquist plots lie outside of the scale shown. Arrows indicate direction of Nyquist trajectories as frequency increases from 0 to infinity. Filled circle represents location of the $-1 + j0$ point.

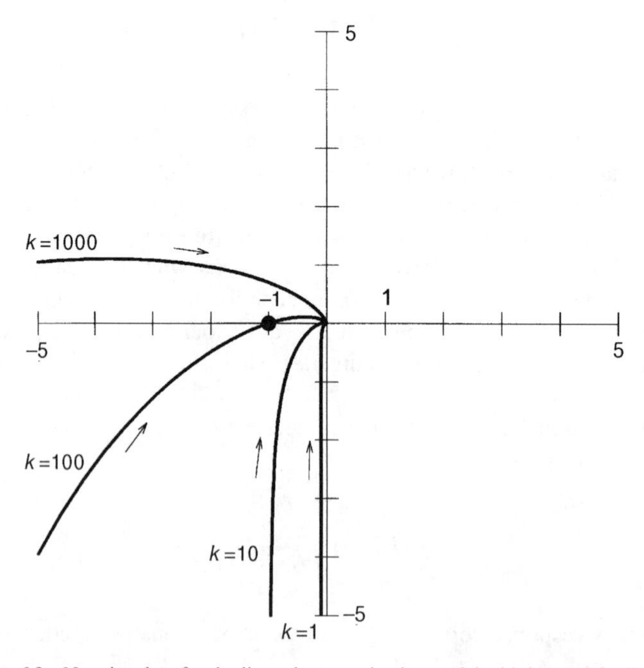


**Figure 6.9** Nyquist plots for the linear lung mechanics model with integral feedback. In all cases ($k = 1$, 10, 100, 1000), portions of the Nyquist plot lie outside of the scale shown. Arrows indicate the direction of Nyquist trajectories as frequency increases from 0 to infinity. Filled circle represents location of the $-1 + j0$ point.

In this case, note that when $\omega \to 0$, $H_{L2}(\omega) \to -j\infty$. When $\omega$ is very large, the term in $\omega^3$ will become much more important than the other terms in the denominator in Equation (6.29). Thus, when $\omega \to \infty$, $H_{L2}(\omega) \to 0$, with the Nyquist locus at high values of $\omega$ tending to approach the origin along the positive imaginary axis. Figure 6.9 shows the behavior of Nyquist loci at four different values of feedback gain ($k = 1, 10, 100$, and 1000). All loci start off from $-j\infty$ and curve in toward the origin as frequency increases toward infinity. For values of $k$ below 100, the Nyquist loci remain to the right of the $(-1 + j0)$ point, so the system remains stable. When $k = 100$, the Nyquist locus passes through the $(-1 + j0)$ point, indicating that the system becomes conditionally stable in this condition. Then, when $k > 100$, the system becomes unstable with the Nyquist plot encircling the $(-1 + j0)$ point.

## 6.5 RELATIVE STABILITY

While it is useful to be able to determine under what conditions a closed-loop system would become marginally stable or unstable, it is equally useful to have a means of assessing how "far" the point of instability is from the current state of system stability. Consider two stable systems. Suppose the effect of an impulsive disturbance elicits a rapidly damped, nonoscillatory response from the first system while the same disturbance produces an underdamped oscillatory response from the second system. Although both systems are "stable," it would be reasonable to conclude that the first may be considered "more stable" than the second. This is the basis underlying the notion of *relative stability*.

The relative stability of a given system can be quantified in terms of either the *gain margin* or the *phase margin*. Both provide measures of the $(-1 + j0)$ point from specific points on the locus of the loop transfer function. The *gain margin* refers to the *factor by which the loop gain corresponding to a phase of* $-180°$ *has to be increased before it attains the value of unity*. The frequency at which the phase of the loop transfer function becomes $-180°$ is known as the *phase crossover frequency*. To illustrate this point, consider a special case of the linear lung mechanics model with integral feedback that has the following specific loop transfer function:

$$H_{L2}(s) = \frac{1}{s^3 + 3s^2 + 2s} \tag{6.30}$$

The Bode magnitude and phase plots corresponding to Equation (6.30) are displayed in Figure 6.10. Note that at the phase crossover frequency, $\omega_{pc}$, when the phase of $H_{L2}(\omega_{pc})$ becomes equal to $-180°$, the gain of $H_{L2}(\omega_{pc})$ remains less than unity, implying that this system is stable. By definition, the gain margin ($GM$) in this case is given by

$$|H_{L2}(\omega_{pc})| \times GM = 1 \tag{6.31}$$

Therefore, the gain margin, expressed in decibels, is

$$GM_{dB} = 20 \log_{10}\left(\frac{1}{|H_{L2}(\omega_{pc})|}\right) = -20 \log_{10} |H_{L2}(\omega_{pc})| \tag{6.32}$$

As shown in the top panel of Figure 6.10, $GM_{dB}$ is given by the vertical distance between the 0 dB axis and the point on the gain plot at which the phase crossover frequency is attained.

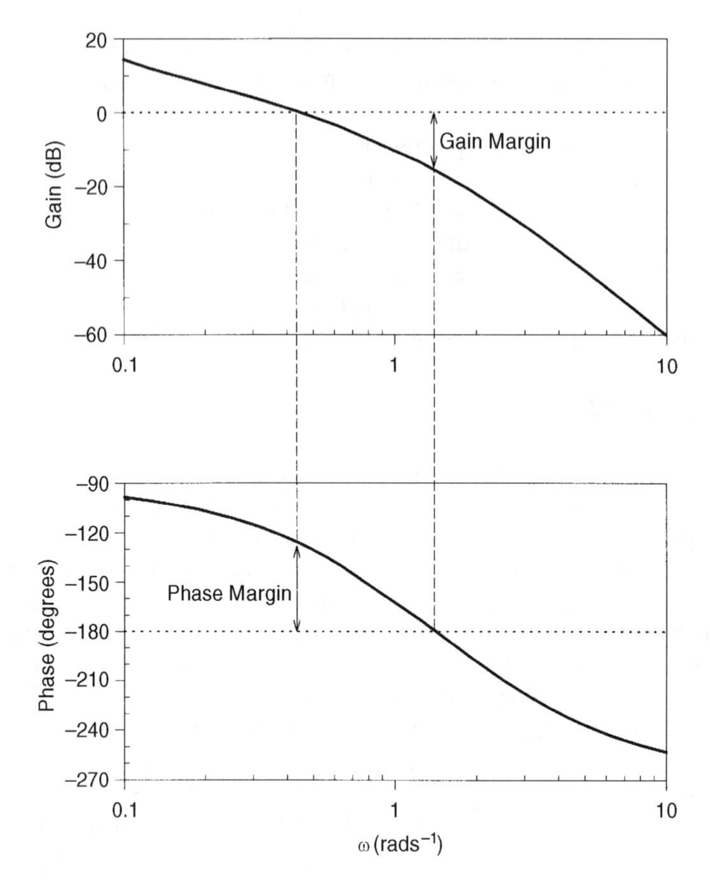


**Figure 6.10** Derivation of gain and phase margins from the Bode plots of the loop transfer function.

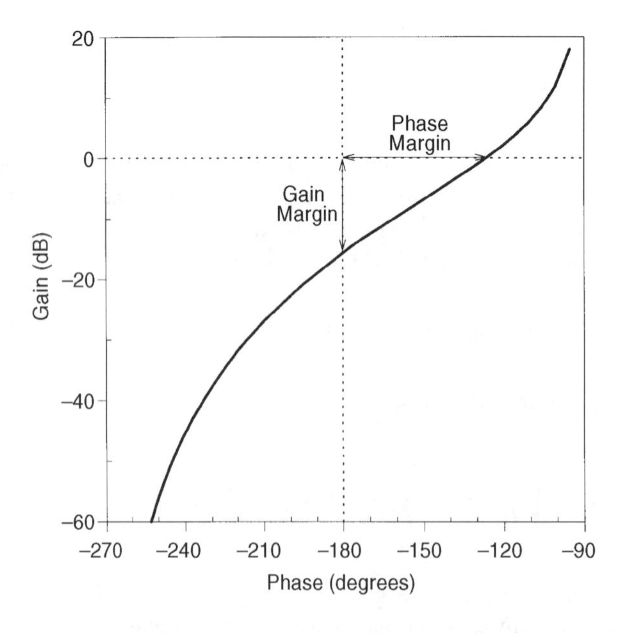


**Figure 6.11** Derivation of gain and phase margins from the Nichols chart of the loop transfer function.

The other measure of relative stability, the *phase margin*, is defined as the *shift in phase at unit loop gain necessary to produce a phase lag of* 180°. The frequency at which the loop gain becomes equal to unity is known as the *gain crossover frequency* ($\omega_{gc}$). Referring to the lower panel of Figure 6.10, note that the phase margin is given by the vertical distance between the −180° line and the point on the Bode phase plot at which the gain crossover frequency is attained.

The gain and phase margins can also be readily deduced from the Nichols chart (Figure 6.11) and the Nyquist plot (Figure 6.12). In the case of the Nyquist plot, the intersection between the Nyquist locus and the negative horizontal axis yields $|H_{L2}(\omega_{pc})|$ which, by Equation (6.31), gives the reciprocal of $GM$. Note, however, that here $GM$ is expressed as a ratio and not in terms of decibels. It can be appreciated from Figures 6.10 through 6.12 that larger positive values for the gain or phase margins imply greater relative stability. On the other hand, negative values for either gain or phase margins would imply that the system in question is already unstable.

Numerical evaluation of the gain and phase margins can be performed easily using the function `margin` in the MATLAB Control System Toolbox. For instance, to evaluate the gain and phase margins for the loop transfer function defined by Equation (6.30), the following MATLAB command lines (contained in script file "`gpmargin.m`") may be applied:

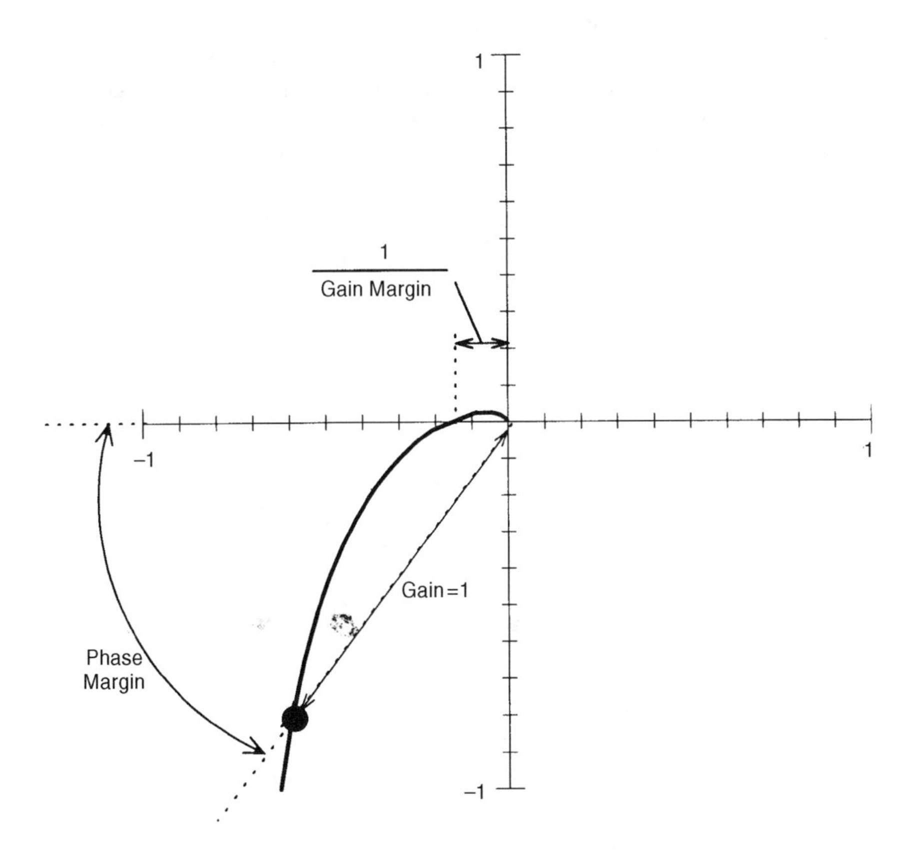


**Figure 6.12** Derivation of gain and phase margins from the Nyquist plot of the loop transfer function.

```
>>% First construct loop transfer function and vector of frequencies
>> num = [1];
>> den = [1 3  2  0];
>> Hs = tf(num, en);
>> f = 0.01:0.01:10;
>> w = 2*pi*f;

>>% Compute magnitude (mag) and phase (pha)of loop transfer function
>> [mag, pha] = bode(Hs, w)

>>% Compute gain margin (GM), phase margin (PM), gain crossover
>>% frequency (wcG) and phase crossover frequency (wcP)
>> [GM,PM, wcG, wcP] = margin (mag, pha, w);

>>% Plot Bode diagram showing gain and phase margins
>> margin(mag,pha,w);
```

## 6.6 STABILITY ANALYSIS OF THE PUPILLARY LIGHT REFLEX

The pupillary light reflex has been studied extensively using control system analysis, most notably by bioengineering pioneers Lawrence Stark and Manfred Clynes. The purpose of this reflex is to regulate the total light flux reaching the retina, although the same pupil control system is also used to alter the effective lens aperture so as to reduce optical aberrations and increase depth of focus. The reflex follows the basic scheme shown in Figure 6.13a. An

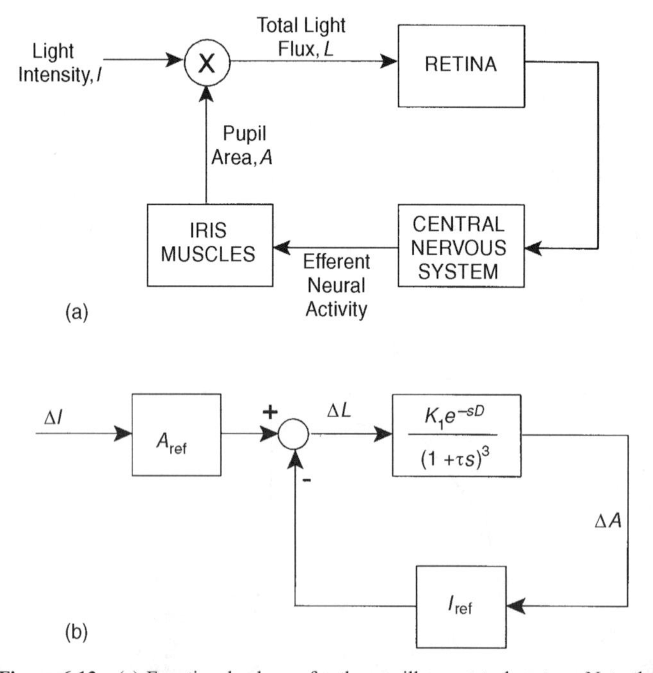


**Figure 6.13** (a) Functional scheme for the pupillary control system. Note that the total light flux, $L$, is given by the product of light intensity, $I$, and pupil area, $A$. (b) Linearized (small-signal) model of the pupillary light reflex.

increase in the intensity ($I$) of ambient light elevates the total light flux ($L$) received by the retina, which converts the light into neural signals. The afferent neural information is sent via the optic nerve to the lateral geniculate body and then to the pretectal nucleus. Subsequently, the Edinger–Westphal nucleus sends efferent neural signals back toward the periphery to the iris sphincter and dilator muscles which, respectively, contract and relax to reduce the pupil area ($A$).

Not surprisingly, quantitative investigations into this feedback control scheme have revealed significant nonlinearities in each of the system components. However, Stark came up with a linear characterization that provides a reasonably good approximation of the underlying dynamics when the changes involved are relatively small. This linearized model is schematized in Figure 6.13b. Using an ingenious experimental design (see Figure 7.13), he was able to functionally "open the loop" of this reflex and measure the dynamics of this system. (This technique is discussed further in Section 7.4.5.) He found that the dynamics could be modeled by a third-order transfer function with time constant $\tau$, in series with a pure time delay $D$. In Figure 6.13b, $\Delta I$ represents a small change in light intensity from the reference intensity level, $I_{\text{ref}}$, while $\Delta A$ represents the corresponding change in pupil area from the reference value, $A_{\text{ref}}$, and $\Delta L$ is the change in total light flux reaching the retina. Based on the model, the closed-loop transfer function of the pupillary reflex can be deduced as

$$\frac{\Delta A}{\Delta I} = \frac{\dfrac{A_{\text{ref}} K_1 e^{-sD}}{(1+\tau s)^3}}{1 + \dfrac{I_{\text{ref}} K_1 e^{-sD}}{(1+\tau s)^3}} \tag{6.33}$$

By inspection of Figure 6.13b and Equation (6.33) above, one can readily infer that the *loop transfer function* of this model is given by

$$H_{\text{L}}(s) = \frac{K e^{-sD}}{(1+\tau s)^3} \tag{6.34}$$

where $K = I_{\text{ref}} K_1$. Therefore, the characteristic equation for the closed-loop model is

$$1 + \frac{K e^{-sD}}{(1+\tau s)^3} = 0 \tag{6.35}$$

From his measurements on normal humans, Stark found the following values for the model parameters: $K = 0.16$, $D = 0.18$ s and $\tau = 0.1$ s. In the next two subsections, we will assume the above values of $D$ and $\tau$ in our analyses and determine the critical value of $K$ above which the model becomes unstable.

### 6.6.1 Routh–Hurwitz Analysis

In Section 6.3, when the Routh–Hurwitz stability criterion was first discussed, application of the test was simple since the examples considered had characteristic equations that could be expressed as polynomials in $s$. In Equation (6.35) however, the presence of the time delay complicates matters a little. Llaurado and Sun (1964) suggested that this problem could be circumvented by expanding $e^{-sD}$ as a power series in $s$:

$$e^{-sD} \approx 1 - Ds + \frac{D^2 s^2}{2} - \frac{D^3 s^3}{6} \tag{6.36}$$

Substituting the above approximation for $e^{-sD}$ into Equation (6.35) and collecting terms for each power of $s$, we obtain the following third-order polynomial expression in $s$:

$$\left(\tau^3 - \frac{KD^3}{6}\right)s^3 + \left(3\tau^3 + \frac{KD^2}{2}\right)s^2 + (3\tau - KD)s + (1 + K) = 0 \tag{6.37a}$$

Inserting the numerical values for $D$ and $\tau$ into Equation (6.37a), we obtain

$$(0.001 - 0.000972K)s^3 + (0.03 + 0.0162K)s^2 + (0.3 - 0.18K)s + (1 + K) = 0 \tag{6.37b}$$

The Routh array corresponding to Equation (6.37b) is

$$\begin{array}{c|ccc} s^3 & 0.001 - 0.000972K & 0.3 - 0.18K & 0 \\ s^2 & 0.03 + 0.0162K & 1 + K & 0 \\ s & \dfrac{890 - 431.72K - 19.44K^2}{300 + 162K} & 0 & 0 \\ s^0 & 1 + K & 0 & 0 \end{array} \tag{6.38}$$

The requirement for stability is that all terms in the first column of the above array should have the same sign. Since the term corresponding to $s^0$ is $1 + K$, and $K$ must be positive, then for the system to be stable, all terms in the first column of the Routh array must also be positive. Thus, the following inequalities have to be satisfied simultaneously:

$$0.001 - 0.000972K > 0 \tag{6.39a}$$

$$0.03 + 0.0162K > 0 \tag{6.39b}$$

$$\frac{890 - 431.72K - 19.44K^2}{300 + 162K} > 0 \tag{6.39d}$$

$$1 + K > 0 \tag{6.39d}$$

From the first inequality (Equation (6.39a)), we find that $K < 10.3$. The second and fourth inequalities are satisfied for all values of $K$ that are greater than zero. In the third inequality, the quadratic expression in the numerator of the left-hand side has to be factorized first. From this, it can be deduced that the inequality is satisfied if $-24.196 < K < 1.996$. Thus, combining this result with that from the first inequality, we conclude that for the closed-loop system to be stable, $K$ must be less than 1.996. Since the average value of $K$ measured by Stark was 0.16, we can conclude from Routh–Hurwitz analysis that the normal pupillary reflex is a highly stable negative feedback system.

### 6.6.2 Nyquist Analysis

The frequency response, $H_L(\omega)$, corresponding to the loop transfer function can be obtained by substituting $j\omega$ for $s$ in Equation (6.34):

$$H_L(\omega) = \frac{Ke^{-j\omega D}}{(1 + j\omega\tau)^3} = \frac{Ke^{-j\omega D}}{(1 - 3\omega^2\tau^2) + j\omega\tau(3 - \omega^2\tau^2)} \tag{6.40}$$

The problem of evaluating the time-delay transfer function in Equation (6.40a) can be approached in a number of ways. One way is to apply the power series expansion approach employed in the previous section. Another possibility is to employ a Padé approximation to the delay, as we had illustrated in Problem P4.2. The first of these methods converts the time-delay transfer function into a polynomial function of $s$ while the second approximates it with a transfer function that consists of the ratio of two polynomials in $s$. However, a third approach is to express $H_L(\omega)$ in polar form and recognize that the delay will only affect the phase component of the transfer function. MATLAB offers a convenient means of performing this computation over a given range of frequencies. An illustration of this procedure is given in the MATLAB command lines below.

```
>>% Construct undelayed transfer function & evaluate frequency response
>>num =[K]; den = [tau^3 3*tau^2 3*tau 1];
>> Hs} = tf(num, den);
>> [R, I] = nyquist(Hs); I = squeeze(I); R = squeeze(R);

>>% Add delay to results
>> Rdel = real((R + j*I).*exp(-j*w*D));
>> Idel= imag((R + j*I).*exp(-j*w*D));

>>% Plot final Nyquist diagram
>> plot(Rdel,Idel);
```

The complete script file (named "`pupil.m`") for evaluating Equation (6.40) and plotting the Nyquist diagrams is included in the library of MATLAB/SIMULINK files accompanying this book.

Figure 6.14a displays the Nyquist plot for the normal pupil control system with $K = 0.16$, generated using the above MATLAB code and assuming the parameter values, $D = 0.18$ s and $\tau = 0.1$ s. The Nyquist plots corresponding to increased values of $K$, 1.6 (dotted curve) and 2 (unbroken curve), are shown in Figure 6.14b. Note that the scale of the axes in Figure 6.14b has been increased to cover a substantially larger range of values. It is clear that the pupillary reflex model is unstable when $K = 2$ but stable when $K = 1.6$. Further computations show that the critical value of $K$ for the development of self-sustained oscillations is 1.85. This is slightly smaller than the 1.996 value deduced in Section 6.6.1. However, the critical value arrived at here is probably the more accurate prediction since, in the previous section, an approximation had to be assumed in order to represent the pure-delay transfer function as a power series.

Examination of all the Nyquist plots in Figure 6.14 shows that the intersection of each locus with the negative real axis (i.e., phase $= -180°$) occurs at the same critical frequency,

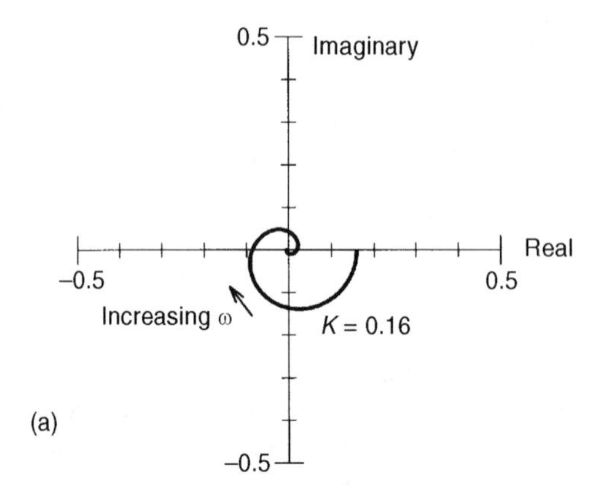


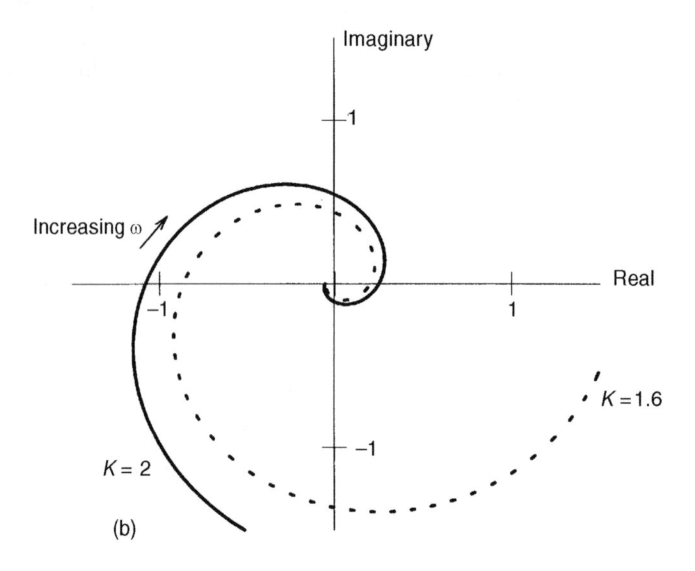


**Figure 6.14** Nyquist plots of the linearized pupillary reflex model at (a) normal loop gain factor ($K = 0.16$), (b) elevated loop gain factors ($K = 1.6$ and $K = 2$).

$\omega_c$, regardless of the value of $K$. This can be confirmed by analysis of the phase of $H_L(\omega)$. At the critical frequency, since the phase of $H_L(\omega)$ must equal $-\pi$ radians (or $-180^\circ$), we have

$$-\pi = -\omega_c D - \tan^{-1}\left(\frac{\omega_c \tau(1 - \omega_c^2 \tau^2)}{(1 - 3\omega_c^2 \tau^2)}\right) \tag{6.41}$$

Note that $\omega_c$ can be deduced by solving Equation (6.41). However, since none of the terms depends on $K$, $\omega_c$ must also be independent of $K$. $\omega_c$ can be found from solution of Equation (6.41) or from inspection of the Nyquist plots to be equal to 7.1 rad s$^{-1}$, which corresponds to an absolute frequency of 1.1 Hz. This predicted frequency is close to the frequency of continuous oscillations of the pupil, known as "hippus," which have been observed under certain pathological conditions. Stark was able to artificially induce hippus in normal subjects using a clever experimental design. He focused a thin beam of light at the edge of the pupil.

This stimulation of the reflex led to constriction of the pupil, which produced a large decrease in retinal illumination (since much of it was now blocked by the iris). This, in turn, acted through the reflex to dilate the pupil, restoring the effect of the applied retinal illumination. This experimental design was tantamount to elevating the loop gain of the closed-loop system tremendously, thereby setting the stage for a self-sustained oscillation to occur. The frequency of this oscillation was found to be close to that predicted by the model. The good agreement between model prediction and experimental observation supports the approximate validity of the linear assumption in this case.

## 6.7 MODEL OF CHEYNE–STOKES BREATHING

The term *periodic breathing* refers to the cyclic modulation of respiration that occurs over the time-scale of several breaths. The resulting ventilatory pattern may or may not include periods of apnea, in which breathing ceases altogether. Periodic breathing does not commonly occur in normals during wakefulness; however, its frequency of incidence increases dramatically during ascent to altitude as well as during sleep onset. An exaggerated form of periodic breathing, known as *Cheyne–Stokes breathing*, is frequently observed in patients with congestive heart failure. A large body of evidence suggests that periodic breathing results from an instability in the feedback control system that regulates ventilation and arterial blood gases. In this section, we will demonstrate that this is a reasonable hypothesis by applying stability analysis to a linearized model of chemoreflex regulation of ventilation. Recall that a steady-state nonlinear model for arterial $CO_2$ and $O_2$ regulation was discussed previously in Section 3.7. In the present model, we assume that the system is operating under normoxic conditions, so that the chemoreflex response to hypoxia can be ignored. However, the various components of the model are assigned dynamic properties. Since the response of the "central" chemoreceptors located in the ventral medulla is much more sluggish than that of the "peripheral" (carotid body) chemoreceptors, it is convenient to assume that there are functionally two feedback loops in this system: one representing the central chemoreflex and the other representing the peripheral chemoreflex. We also incorporate into the model the delays taken to transport blood from the lungs and the chemoreceptors. A simplified schematic diagram of this dynamic model is shown in Figure 6.15.

### 6.7.1 $CO_2$ Exchange in the Lungs

The dynamic equivalent of the gas exchange equation given in Equation (3.48) is

$$V_{\text{lung}} \frac{dP_{ACO_2}}{dt} = (\dot{V}_E - \dot{V}_D)(P_{ICO_2} - P_{ACO_2}) + 863Q(C_{vCO_2} - C_{aCO_2}) \tag{6.42}$$

where $Q$ represents pulmonary blood flow, $V_{\text{lung}}$ is the effective $CO_2$ storage capacity of the lungs, and $C_{aCO_2}$ and $C_{vCO_2}$ are the $CO_2$ concentrations in arterial and mixed venous blood, respectively. Other symbols are as defined previously in Section 3.7. It should also be noted that, in the steady state, the last term in Equation (6.42) would equal $863\dot{V}_{CO_2}$, where $\dot{V}_{CO_2}$ is the metabolic production rate of $CO_2$.

Suppose that small perturbations are imposed on $\dot{V}_E$ ($\Delta\dot{V}_E$) and that these lead to small perturbations in $P_{ACO_2}$ ($\Delta P_{ACO_2}$) and $C_{aCO_2}$ ($\Delta C_{aCO_2}$). If we ignore the effect of the arterial blood gas fluctuations on mixed venous $CO_2$ concentration (since the body tissues represent a very large buffer of $CO_2$ changes), assume that dead space ventilation remains constant, and

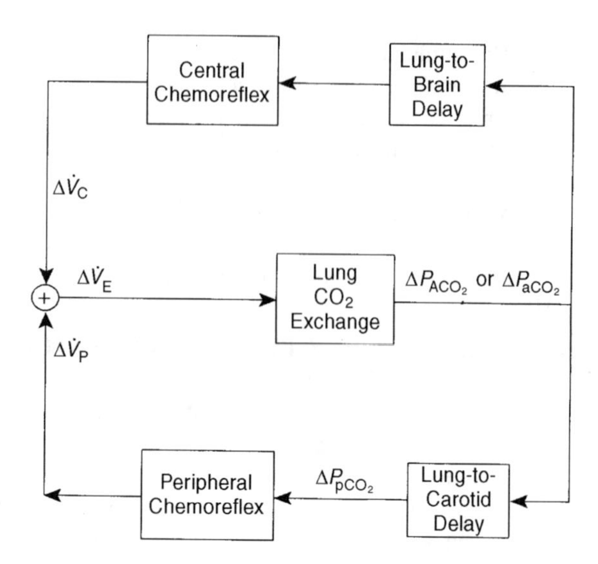


**Figure 6.15** Linearized dynamic model of the chemoreflex control of ventilation.

ignore terms involving the product $\Delta\dot{V}_E\,\Delta P_{ACO_2}$, we can derive the following small-signal expression from Equation (6.42):

$$V_{lung}\frac{d(\Delta P_{ACO_2})}{dt} = -(\dot{V}_E - \dot{V}_D)\,\Delta P_{ACO_2} + (P_{ICO_2} - P_{ACO_2})\,\Delta\dot{V}_E - 863Q\,\Delta C_{aCO_2} \quad (6.43a)$$

If we approximate the blood $CO_2$ dissociation curve with a straight line relating $C_{aCO_2}$ to $P_{aCO_2}$ with slope $K_{CO_2}$, and impose the assumption of alveolar–arterial $P_{CO_2}$ equilibration, we obtain from Equation (6.43a) the following result:

$$V_{lung}\frac{d(\Delta P_{aCO_2})}{dt} + (\dot{V}_E - \dot{V}_D + 863QK_{CO_2})\,\Delta P_{aCO_2} = (P_{ICO_2} - P_{aCO_2})\,\Delta\dot{V}_E \quad (6.43b)$$

Note that in Equations (6.43a) and (6.43b), $\dot{V}_E$ and $P_{aCO_2}$ represent the steady-state operating levels of minute ventilation and arterial $CO_2$ tension, respectively. If we take the Laplace transform of Equation (6.43b) and rearrange terms, we can obtain the following expression for the transfer function, $H_{lung}(s)$, of the lungs:

$$H_{lung}(s) \equiv \frac{\Delta P_{aCO_2}}{\Delta\dot{V}_E} = \frac{-G_{lung}}{\tau_{lung}s + 1} \quad (6.44)$$

where

$$\tau_{lung} = \frac{V_{lung}}{\dot{V}_E - \dot{V}_D + 863QK_{CO_2}} \quad (6.45)$$

and

$$G_{lung} = \frac{P_{aCO_2} - P_{ICO_2}}{\dot{V}_E - \dot{V}_D + 863QK_{CO_2}} \quad (6.46)$$

Thus, Equations (6.44) through (6.46) indicate that, under small-signal conditions, the dynamics of $CO_2$ exchange in the lungs may be modeled approximately as a simple first-order system with time constant $\tau_{lung}$ and gain $G_{lung}$. Note, however, that $\tau_{lung}$ and $G_{lung}$ will vary, depending on the steady-state operating levels of $\dot{V}_E$ and $P_{aCO_2}$. This reflects the fundamentally *nonlinear* nature of the gas exchange process. Another important detail is that the negative value for $H_{lung}(s)$ in Equation (6.44) merely implies that the negative feedback in this closed-loop system is embedded in the $CO_2$ exchange process (i.e., when ventilation increases, $P_{aCO_2}$ decreases). $G_{lung}$ will always be positive since $P_{aCO_2}$ must be greater than $P_{ICO_2}$ (for positive metabolic $CO_2$ production rates).

### 6.7.2 Transport Delays

We assume that pulmonary end-capillary blood returning to the heart will take some time ($T_p$) to arrive at the peripheral chemoreceptors (carotid bodies) and a slightly longer time ($T_c > T_p$) to first appear at the site of the central (medullary) chemoreceptors. Thus,

$$\Delta P_{pCO_2}(t) = \Delta P_{aCO_2}(t - T_p) \tag{6.47a}$$

$$\Delta P_{cCO_2}(t) = \Delta P_{aCO_2}(t - T_c) \tag{6.47b}$$

All mixing effects in the vasculature during the convective process are ignored. The Laplace transforms corresponding to Equations (6.47a) and (6.47b) are

$$\Delta P_{pCO_2}(s) = e^{-sT_p} \Delta P_{aCO_2}(s) \tag{6.48a}$$

$$\Delta P_{cCO_2}(s) = e^{-sT_c} \Delta P_{aCO_2}(s) \tag{6.48b}$$

### 6.7.3 Controller Responses

Following Bellville et al. (1979), we assume the following dynamic relations for the peripheral and central chemoreflex responses:

$$\tau_p \frac{d\dot{V}_p}{dt} + \dot{V}_p = G_p[P_{pCO_2} - I_p] \tag{6.49a}$$

$$\tau_c \frac{d\dot{V}_c}{dt} + \dot{V}_c = G_c[P_{cCO_2} - I_c] \tag{6.49b}$$

and

$$\dot{V}_c + \dot{V}_p = \dot{V}_E \tag{6.49c}$$

In the above controller equations, $\tau_p$ and $\tau_c$ represent the characteristic response times of the peripheral and central chemoreflexes, respectively; it is assumed that $\tau_c \gg \tau_p$. $G_p$ and $G_c$ represent the steady-state gains for the peripheral and central controllers, respectively. The square brackets on the right-hand side of Equations (6.49a) and (6.49b) are used to imply a thresholding operation: i.e., these terms will be set equal to zero if the quantities within the brackets become negative. Thus, $I_p$ and $I_c$ represent the corresponding apneic thresholds for the peripheral and central chemoreceptors, respectively.

Assuming that the "set-point" of operation is nowhere in the vicinity of the apneic thresholds, Equations (6.49a) and (6.49b) can be linearized using small-signal analysis. The result of this analysis following Laplace transformation yields:

$$\Delta \dot{V}_p(s) = \frac{G_p}{\tau_p s + 1} \Delta P_{pCO_2}(s) \tag{6.50a}$$

$$\Delta \dot{V}_c(s) = \frac{G_c}{\tau_c s + 1} \Delta P_{cCO_2}(s) \tag{6.50b}$$

### 6.7.4 Loop Transfer Functions

Corresponding to the two feedback loops in this model are two loop transfer functions: one for the peripheral chemoreflex loop ($H_{Lp}(s)$) and the other for the central chemoreflex loop ($H_{Lc}(s)$). These are derived by combining Equations (6.44), (6.48a), and (6.50a):

$$H_{Lp}(s) \equiv \frac{\Delta \dot{V}_p(s)}{\Delta \dot{V}_E(s)} = \frac{G_{lung} G_p e^{-sT_p}}{(\tau_{lung} s + 1)(\tau_p s + 1)} \tag{6.51a}$$

$$H_{Lc}(s) \equiv \frac{\Delta \dot{V}_c(s)}{\Delta \dot{V}_E(s)} = \frac{G_{lung} G_c e^{-sT_c}}{(\tau_{lung} s + 1)(\tau_c s + 1)} \tag{6.51b}$$

Since the overall frequency response of the loop transfer function is defined as

$$H_L(\omega) \equiv \frac{\Delta \dot{V}_p(\omega) + \Delta \dot{V}_c(\omega)}{\Delta \dot{V}_E(\omega)} \tag{6.52}$$

if we combine Equations (6.51a) and (6.51b) and introduce the substitution $s = j\omega$, we obtain the following expression for the overall frequency response of the two loops:

$$H_L(\omega) = \frac{G_{lung}}{(1 + j\omega\tau_{lung})} \left( \frac{G_p e^{-j\omega T_p}}{(1 + j\omega\tau_p)} + \frac{G_c e^{-j\omega T_c}}{(1 + j\omega\tau_c)} \right) \tag{6.53}$$

Thus, the stability of the respiratory control model for a given set of parameters ($V_{lung}$, $Q$, $K_{CO_2}$, $G_p$, $\tau_p$, $G_c$, $\tau_c$) and conditions ($\dot{V}_E$, $P_{aCO_2}$, $P_{ICO_2}$) can be tested by applying the Nyquist criterion to Equation (6.53).

### 6.7.5 Nyquist Stability Analysis Using MATLAB

The generation of the Nyquist diagram from Equation (6.53) can be carried out relatively easily using MATLAB. The following shows sample lines of MATLAB code that can be used for this purpose.

```
>>% Construct loop transfer functions
>> num1 = [Glung*Gp]; den1 = [taulung*taup  (taulung + taup) 1];
>> Hs1 = tf(num1, den1);
>> num2 = [Glung*Gc]; den2 = [(taulung*tauc)  taulung +tauc) 1;
>> Hs2 = tf(num2, den2);

>>% Compute Nyquist results, excluding effect of pure delays
>> [R1, I1] = nyquist(Hs1, w); R1 = squeeze(R1); I1 = squeeze(I1);
>> [R2,I2] = nyquist(Hs2, w);R2 = squeeze(R2); I2 = squeeze(I2);

>>% Add delay to results
>> R1del = real((R1 + j*I1).*exp(-j*w*Tp));
>>I1del = imag((R1 + j*I1).*exp(-j*w*Tp));
>> R2del = real((R2 + j*I2).*exp(-j*w*Tc));
>> I2del = imag((R2 + j*I2).*exp(-j*w*Tc));
>> Rdel = R1del + R2del;
>> Idel = I1del + I2del;

>>% Plot Nyquist diagram of overall frequency response
>> axis square; plot(Rdel,Idel); grid;
```

Figure 6.16 shows Nyquist plots representing the overall frequency responses in the case of the typical normal subject (N) and the patient with congestive heart failure (C). We have assumed the following parameter values to represent *both* types of subjects: $V_{lung} = 2.5\,L$, $K_{CO_2} = 0.0065\,mm\,Hg^{-1}$, $G_p = 0.02\,L\,s^{-1}\,mm\,Hg^{-1}$, $G_c = 0.04\,L\,s^{-1}$

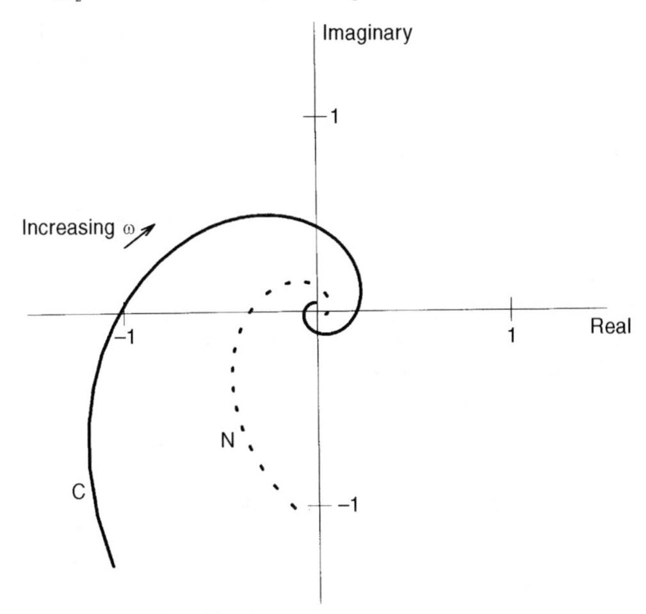


**Figure 6.16** Nyquist plots representing the frequency responses of the linearized ventilatory control model in the normal subject (N, dotted curve) and patient with congestive heart failure (C, continuous curve). Frequencies represented in plots range from 0.01 to 0.1 Hz.

mm $Hg^{-1}$, $\tau_p = 20$ s, $\tau_c = 120$ s, $\dot{V}_E = 0.12\,L\,s^{-1}$, $\dot{V}_D = 0.03\,L\,s^{-1}$, $P_{ICO_2} = 0$, and $P_{aCO_2} = P_{ACO_2} = 40$ mm Hg. In the normal subject, the following circulatory parameter values were assigned: $T_p = 6.1$ s, $T_c = 7.1$ s, and $Q = 0.1\,L\,s^{-1}$. In the patient with congestive heart failure, we assumed a halving of cardiac output and a doubling of the circulatory delays: $Q = 0.05\,L\,s^{-1}$, $T_p = 12.2$ s, and $T_c = 14.2$ s. The MATLAB script file ("nyq_resp.m") used to generate the Nyquist plots shown is included with the set of MATLAB/SIMULINK files that accompany this book.

The Nyquist plots in Figure 6.16 represent a bandwidth of frequencies that range from 0.01 to 0.1 Hz. These correspond to interbreath periodicities of cycle durations 10 to 100 s. In the normal subject (N), the Nyquist plot shows a stable system with a critical loop gain (i.e., at $-180°$) of 0.34. This critical point occurs at $f = 0.0295$ Hz or the equivalent of a periodicity of 34 s. Thus, when transient oscillations in ventilation appear in the normal subject, we would expect these oscillations to have a cycle duration of 34 s. On the other hand, in the subject with congestive heart failure, the halving of $Q$ and doubling of circulatory delays lead to a rotation and stretching of the Nyquist locus. Since the locus encircles the $(-1 + j0)$ point, we can conclude that under the assigned conditions, this system is unstable. The loop gain at $-180°$ is now 1.02, with the critical frequency occurring at $f = 0.0165$ Hz, which is equivalent to a periodicity of approximately 61 s. These cycle durations are consistent with the oscillation periods that have been observed in normals and heart failure subjects who exhibit Cheyne–Stokes respiration. For more complicated models and analyses, the reader is referred to journal reports such as those published by Khoo et al. (1982), Carley and Shannon (1988), and Nugent and Findley (1987).

## BIBLIOGRAPHY


Bellville, J.W., B.J. Whipp, R.D. Kaufman, G.D. Swanson, K.A. Aqleh, and D.M. Wiberg. Central and peripheral chemoreflex loop gain in normal and carotid body-resected subjects. *J. Appl. Physiol.* **46**: 843–853, 1979.

Carley, D.W., and D.C. Shannon. A minimal mathematical model of human periodic breathing. *J. Appl. Physiol.* **65**: 1400–1409, 1988.

Clynes, M. Computer dynamic analysis of the pupil light reflex. *Proc. 3d Int. Conf. Med. Electron.* International Federation of Medical Electronics and Biomedical Engineering, and Institute of Electrical Engineers (London), 1960, pp. 356–358.

Dorf, R.C., and R.H. Bishop. *Modern Control Systems*, 7th ed. Addison-Wesley, Reading, MA, 1995.

Khoo, M.C.K., R.E. Kronauer, K.P. Strohl, and A.S. Slutsky. Factors inducing periodic breathing in humans: a general model. *J. Appl. Physiol.* **53**: 644–659, 1982.

Llaurado, J.G., and H.H. Sun. Modified methods for studying stability in biological feedback control systems with transportation delays. *Med. Electron. Biol. Eng.* **2**: 179–184, 1964.

Nugent, S.T., and J.P. Finley. Periodic breathing in infants: a model study. *IEEE Trans. Biomed. Eng.* **34**: 482–485, 1987.

Stark, L. Stability, oscillations, and noise in the human pupil servomechanism. *Proc. I.R.E.* **47**: 1925–1939, 1959.


## PROBLEMS

**P6.1.** Figure P6.1 shows a simple negative feedback control system with a variable gain, $K$, in the feedback element. Determine the smallest value of $K$ that would render this closed-loop system unstable, using (a) the Routh–Hurwitz test and (b) the Nyquist stability criterion.

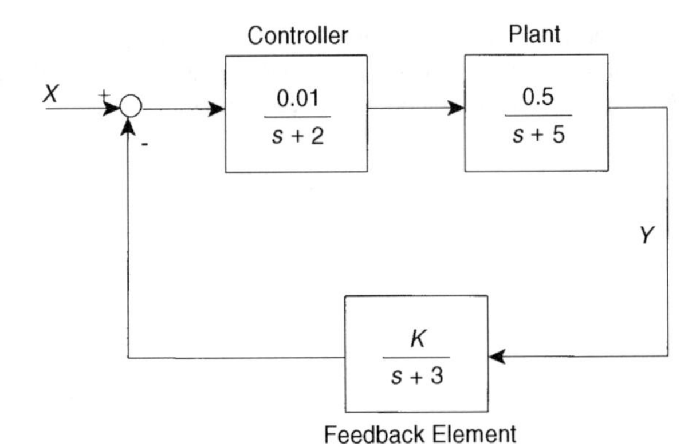


**Figure P6.1** Simple control system with variable feedback gain.

**P6.2.** Consider the simple model of eye-movement control shown schematically in Figure P4.1. Assume that $G/J = 14\,400\ \text{rad}^2\,\text{s}^{-2}$, $B/J = 24\ \text{rad}\,\text{s}^{-1}$.
(a) If $k_v = 0.01$, deduce the gain and phase margins of this closed-loop system.
(b) Using the Routh–Hurwitz and Nyquist stability tests, determine the value of $k_v$ at which you might expect the model to exhibit self-sustained oscillations in $\theta$.

**P6.3.** In the model of ventilatory control with feedback from the intrapulmonary $CO_2$ receptors shown in Figure P4.3, determine how rate sensitivity is expected to affect relative stability. Compute the gain and phase margins of this system, and display the corresponding Bode diagrams, when the rate sensitivity factor, $\alpha$, assumes the following values: (a) $\alpha = 0$; (b) $\alpha = \frac{1}{2}$; and (c) $\alpha = 2$.

**P6.4.** In the analysis of the pupillary light reflex model discussed in Section 6.6, the transfer function representing the pure delay, $D$, was approximated as a power series in the Laplace variable $s$. Repeat the Routh–Hurwitz and Nyquist stability analyses, assuming a first-order Padé approximation to the delay, i.e., assume that

$$e^{-sD} = \frac{1 - \dfrac{Ds}{2}}{1 + \dfrac{Ds}{2}}$$

In each case, find the value of the steady-state loop gain, $K$, that would lead to the production of self-sustained oscillations in pupil diameter.

**P6.5.** Consider the neuromuscular reflex model of Figure 4.11. Develop a MATLAB program that would enable you to assess the relative stability of this model as the feedback gain, $\beta$, is changed. Assume the following values for the rest of the model parameters: $J = 0.1\ \text{kg}\,\text{m}^2$, $k = 50\ \text{N}\,\text{m}$, $B = 2\ \text{N}\,\text{m}\,\text{s}$, $\tau = 1/300\ \text{s}$, $T_d = 0.02\ \text{s}$, and $\eta = 5$. Determine whether your prediction of the critical value of $\beta$ for instability to occur is compatible with simulation results using the SIMULINK program "`nmreflex.mdl`."

**P6.6.** It is known that hyperoxia, induced by breathing a gas mixture with high $O_2$ content, can substantially attenuate the $CO_2$ sensitivity of the peripheral chemoreceptors. As a first approximation, we can assume that this sets the parameter $G_p$ in the chemoreflex model of Section 6.7 equal to zero. This effectively reduces the model to only one feedback loop—that involving the central chemoreflex. Employing Routh–Hurwitz and Nyquist stability analyses, show that administration of inhaled $O_2$ would eliminate Cheyne–Stokes breathing in the patient with congestive heart failure. Use the parameter values given in Section 6.7.5.

**P6.7.** Develop a SIMULINK representation of the chemoreflex model described in Section 6.7, using the parameter values pertinent to the normal subject. Investigate the stability of the SIMULINK model by introducing impulsive perturbations into the closed-loop system.

Determine how changes in the following model parameters may promote or inhibit the occurrence of periodic breathing: (a) $V_{lung}$, (b) $P_{ICO_2}$, (c) $G_c$, and (d) $G_p$. Verify your conclusions using the Nyquist analysis technique illustrated in Section 6.7.

# 7

# Identification of Physiological Control Systems

## 7.1 BASIC PROBLEMS IN PHYSIOLOGICAL SYSTEM ANALYSIS

In the past several chapters, we have examined a variety of techniques for analyzing the steady-state and dynamic characteristics of feedback control systems. A common thread among all these different methods has been the use of the *systems approach*: the physiological process under study is decomposed into a number of interconnected "systems" or "subsystems," under the assumption that each of these components can be characterized functionally by a set of differential equations or their Laplace equivalents. This is displayed schematically in Figure 7.1a. Up to this point, we have always assumed that the equations in each of the "boxes" are known or can be derived by applying physical principles and physiological insight to the process in question. Thus, knowing the form of the independent variable (the "input"), the equations (representing the "system") can be solved to deduce the form of the dependent variable (the "output"). This type of analysis is known as the *forward problem* or *prediction problem*. Predictions allow us to determine whether the model postulated provides an accurate characterization of the process under study. A somewhat greater challenge is posed by the *inverse problem*. Here, a model of the process in question is available and the output is measured; however, the input is not observable and therefore has to be deduced. This is known as the *diagnosis problem* and often involves the need for *deconvolution* of the model impulse response with the output in order to deduce the input.

The third type of problem, that of *system identification*, is the most pervasive in physiological system analysis. There are two basic approaches to system identification. It is often the case, when dealing with physiological processes, that we have only very limited insight into the underlying mechanisms or that the complexity of the processes involved is just too overwhelming. Under such conditions, it would be difficult to begin with physical principles to derive the differential equations that appropriately characterize the system under study. At this level of knowledge (or lack of it), it probably would be more useful to probe the system in question with known stimuli and to record the system's response to these inputs. This is the *black-box* or *nonparametric* approach to system identification, where little is

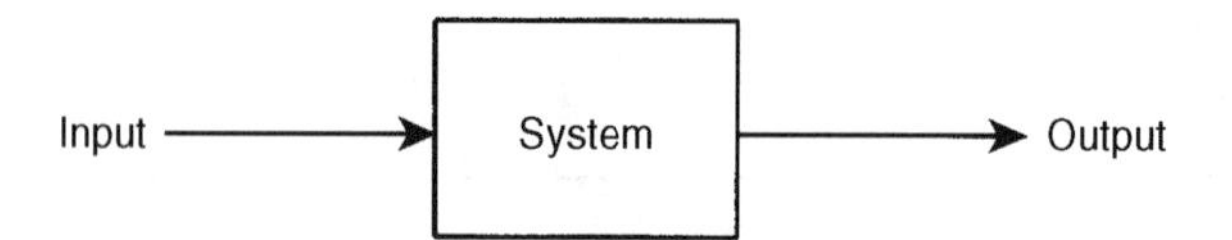


| Problem | Input | System | Output |
|---|---|---|---|
| Prediction | Known | Known | Unknown |
| Diagnosis (Deconvolution) | Unknown | Known | Known |
| Identification | Known | Unknown | Known |

(a)

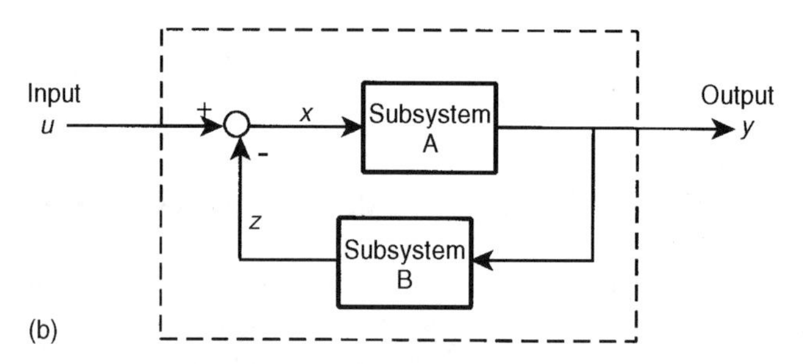


(b)

**Figure 7.1** (a) Three fundamental problems in system analysis. (b) Identification of a closed-loop system.

assumed about the system except, perhaps, whether we expect it to be linear or nonlinear. Ideally, we would be able to deduce, from the measured input and output, the system impulse response (if it is linear) or kernels (if it is nonlinear; see Marmarelis and Marmarelis, 1978) and use these to catalog the behavior of the unknown process. The result would be a purely *empirical* model of the system under study.

In the case of systems for which *some* knowledge regarding mechanisms is available, it is generally possible to put this knowledge to use by coming up with a mathematical description (which could consist of a set of differential or difference equations or their frequency-domain equivalents). This characterizes the second approach to system identification, in which a *structural* or *gray-box* model is constructed. While the "structure" of such a model is derived from what we know about the physiological process being studied, there remain unknown model coefficients or *parameters* that have to be determined. Thus, the next stage in system identification in such cases, following model-building, is the problem of *parameter estimation*. For this reason, structural models fall into the category referred to as *parametric models*. In the case of the linear lung mechanics model that we analyzed in Chapters 4 through 6, the unknown parameters ($R$, $L$ and $C$) each bear a one-to-one correspondence to a physiological entity—airway resistance, fluid inertance, and lung compliance. But this is not a requirement of all parametric models. *Functional models* are models that contain only parameters that can be estimated from input–output data. Frequently, some of these parameters may be related to the underlying physiological entitites but a one-to-one correspondence may not exist. In many models of pharmacokinetics, for example, there are often assumed "compartments" that may be used to account for effects arising from many different sources, but not one single definable physiological entity. In some other functional models, a negative delay may have to be postulated; such a parameter clearly has no physiological meaning but may be needed in order to fully characterize the observed system behavior.

The control engineering literature is replete with countless methods of system identification, particularly for linear systems. In this chapter, we will discuss the few basic techniques that have been most commonly applied in physiological system analysis. While there is a large body of literature on the theory of system identification in simple single-input–single-output systems, there has been relatively much less work published on the identification of closed-loop systems. The fact that most physiological systems are closed-loop can introduce some complications into the process of system identification or parameter estimation. Referring to the example illustrated in Figure 7.1b, if we could only measure the input ($u$) and the output ($y$) of the overall closed-loop system, but the internal variables $x$ and $z$ were unobservable, it would be impossible for us to know (just based on the measurements of $u$ and $y$) that this is in fact a closed-loop system. On the other hand, if we could measure $x$ in addition to $u$ and $y$, we would in principle be able to identify subsystem A; then, having identified the overall closed-loop system, we would be able to determine subsystem B. Similarly, if we could measure the feedback variable $z$, it would be possible (at least in theory) to determine subsystem B; then, from knowledge of the overall closed-loop model, we would be able to deduce subsystem A. This is the fundamental basis of *closed-loop estimation*. In some cases, it may be possible to "open the loop," through surgical, physiological, or pharmacological interventions. Indeed, some of the biggest advances in physiology have resulted from clever experimental designs that allowed the researchers to "open the loop" in one or more of these ways. We will review some examples of these in the sections to follow.

## 7.2 NONPARAMETRIC AND PARAMETRIC IDENTIFICATION METHODS

We begin by reviewing some of the basic computational techniques commonly employed in the identification of single-input–single-output, open-loop systems. While most physiological models have been developed assuming a continuous-time base, in practice physiological measurements are generally obtained as *discrete-time* samples of the signals under study. Some of these time series contain measurements obtained at a fixed sampling rate, e.g., arterial blood pressure and the electroencephalograph. However, some measurements that involve pulsatile or cyclic quantities, e.g., heart rate and respiratory rate, are sampled at a fixed phase of each cycle; these intervals are generally not equally spaced in time. Since the process of system identification requires the use of real data, the vast majority of identification techniques that have been developed assume a discrete-time base. Thus, for the most part, system identification problems are solved by numerical methods and do not have closed-form analytical solutions. An important assumption that we will make, however, is that the sampling interval has been selected to be small enough that the time series obtained adequately capture the fastest dynamics present in the observed signals. Sampling the input and output signals at rates that are lower than one-half of the highest frequency present in the signals can lead to the problem of *aliasing*, in which the sampled data may appear to contain dynamic components that were really not contained in the original signals. Use of these aliased input and output time series would definitely lead to erroneous estimates of the system impulse responses or transfer functions.

### 7.2.1 Numerical Deconvolution

The most direct nonparametric techniques for linear system identification have been discussed earlier in Chapter 4. The response to the step input has been one of the most commonly used methods for characterizing physiological system dynamics, provided the stimulus can indeed be made to follow a time-course that closely approximates a step. Having found the step response, the impulse response can be deduced by differentiating the former with respect to time. In general, impulsive inputs cannot easily be implemented in physiological applications. If the step input is also not a convenient option, one might resort to stimulating the system under study with a bolus type of input. Then, in order to estimate the impulse response from the bolus response and the input, one can employ the method of numerical deconvolution.

Assuming that the data samples are obtained at a uniformly spaced time interval, $T$, the convolution expression relating input ($u$) and impulse response ($h$) to output ($y$) is represented in discrete time as:

$$y(nT) = \sum_{k=0}^{n} h(nT - kT)u(kT)T \tag{7.1}$$

where the current time $t = nT$. In the special cases where $n = 0$, 1, and 2, Equation (7.1) becomes

$$y(0) = h(0)u(0)T \tag{7.1a}$$

$$y(T) = [h(T)u(0) + h(0)y(T)]T \tag{7.1b}$$

Thus, from

$$y(2T) = [h(2T)u(0) + h(T)u(T) + h(0)u(2T)]T \tag{7.1c}$$

Equation (7.1a), assuming $u(0) \neq 0$, we find that

$$h(0) = \frac{y(0)/T}{u(0)} \tag{7.2a}$$

Similarly, rearranging Equation (7.1b), we obtain

$$h(T) = \left\{\frac{y(T)}{T} - u(T)h(0)\right\} / u(0) \tag{7.2b}$$

so that, once $h(0)$ has been deduced from Equation (7.2a), the next point in the impulse response function, $h(T)$, can be determined from Equation (7.2b). Subsequently, from Equation (7.1c), we get

$$h(2T) = \left\{\frac{y(2T)}{T} - h(T)u(T) - h(0)u(2T)\right\} / u(0) \tag{7.2c}$$

where $h(2T)$ can be determined, since $h(0)$ and $h(T)$ are now known. This estimation procedure is continued for all subsequent values of $h(t)$. Thus, the general deconvolution formula is

$$h(nT) = \left\{\frac{y(nT)}{T} - \sum_{k=1}^{n} h(nT - kT)u(kT)\right\} / u(0) \tag{7.2d}$$

While Equation (7.2d) is valid in principle, in practice it is hardly used. The reason is that small values for $u(0)$ can amplify errors enormously, and errors made in each sequential estimate of $h$ tend to accumulate.

### 7.2.2 Least Squares Estimation

A key problem of numerical deconvolution is that the estimated impulse response function is "forced" to satisfy Equation (7.1) even when it is clear that the output measurements, $y$, will contain noise. The effect of this noise accumulates with each step in the deconvolution process. One way to get around this problem is to build some averaging into the estimation procedure. To do this, we restate the problem in the following way: Given $N$ pairs of input–output measurements, estimate the impulse response function (consisting of $p$ points, where $p \ll N$) that would allow Equation (7.1) to be satisfied *on average*. To develop this mathematically, we recast the relationship between input and output measurements in the following form:

$$y(nT) = \sum_{k=0}^{p-1} h(kT)u(nT - kT)T + e(nT), \qquad n = 0, 1, \ldots, N-1 \tag{7.3}$$

where $e(nT)$ represents the error between the measured (noisy) value and the "best estimate" of the response at time $t = nT$. The "best estimate" of the response is obtained by selecting the impulse response function $\{h(kT),\ k = 0,\ 1, \ldots,\ p-1\}$ that would minimize the sum of the squares of all the errors, $\{e(nT),\ n = 0,\ 1, \ldots,\ N-1\}$. Thus, this method is analogous to the fitting of a straight line to a given set of data-points, except that the "line" in this case is a multidimensional surface.

To find the least-squares estimate of $h(t)$, we proceed by defining the following matrix and vector quantities:

$$\underline{y} = [y(0)\ y(T) \quad \ldots \quad y((N-1)T)]' \tag{7.4}$$

$$\underline{h} = [h(0)\ h(T) \quad \ldots \quad h((N-1)T)]' \tag{7.5}$$

$$\underline{e} = [e(0)\ e(T) \quad \ldots \quad e((N-1)T)]' \tag{7.6}$$

and

$$\mathbf{U} = \begin{bmatrix} u(0) & 0 & \ldots & 0 \\ u(T) & u(0) & \ldots & 0 \\ \vdots & \vdots & \vdots & \vdots \\ u((N-1)T) & u((N-2)T) & \ldots & u((N-p)T) \end{bmatrix} T \tag{7.7}$$

In Equations (7.4) through (7.6), $\underline{y}$, $\underline{h}$ and $\underline{e}$ are column vectors (and the prime $'$ represents the transpose operation). Then, the system of equations represented by Equation (7.3) can be compactly rewritten in matrix notation as

$$\underline{y} = \mathbf{U}\underline{h} + \underline{e} \tag{7.8}$$

Let $J$ represent the sum of squares of the errors. Then,

$$J = \sum_{n=0}^{N-1} e(nT)^2 = \underline{e}'\underline{e} \tag{7.9}$$

Combining Equation (7.8) and Equation (7.9), we get

$$J = (\underline{y} - \mathbf{U}\underline{h})'(\underline{y} - \mathbf{U}\underline{h}) \tag{7.10}$$

To find the minimum $J$, we differentiate Equation (7.10) with respect to the vector $\underline{h}$, and equate all elements in the resulting vector to zero:

$$\frac{\partial J}{\partial \underline{h}} = -2\mathbf{U}'\underline{y} + 2\mathbf{U}'\mathbf{U}\underline{h} = \mathbf{0} \tag{7.11}$$

Rearranging Equation (7.11), we find that the least squares solution for the impulse response function is

$$\underline{h} = (\mathbf{U}'\mathbf{U})^{-1}\mathbf{U}'\underline{y} \tag{7.12}$$

It can be shown further that a lower bound to the estimate of the variance associated with the estimated elements of $\underline{h}$ is given by

$$\mathrm{var}(\underline{h}) = (\mathbf{U}'\mathbf{U})^{-1}\sigma_{\mathrm{e}}^2 \tag{7.13}$$

where $\sigma_{\mathrm{e}}^2$ is the variance of the residual errors $\{e(nT),\ n = 0,\ 1, \ldots,\ N-1\}$, i.e.,

$$\sigma_{\mathrm{e}}^2 = \frac{1}{N-1}\sum_{n=0}^{N-1} e(nT)^2 \tag{7.14}$$

This method produces much better results for $h(t)$ than numerical deconvolution, since we are using $N$ pieces of information to deduce estimates of $p$ unknowns, where $p$ should be substantially smaller than $N$. How small the ratio $p/N$ should be depends on the relative magnitude of noise in the data. As a rough rule of thumb, $p/N$ should be smaller than 1/3. Another requirement for obtaining good estimates of $h(t)$ is that the matrix $\mathbf{U}'\mathbf{U}$ must not be ill-conditioned since it has to be inverted: as one can see from Equation (7.13), the variance of $h(t)$ becomes infinite if $\mathbf{U}'\mathbf{U}$ is singular. Since $\mathbf{U}$ consists of all the input measurements, the conditioning of $\mathbf{U}'\mathbf{U}$ depends on the time-course of the stimulus sequence. This problem will be discussed further in Section 7.3.

The practical implementation of this method is relatively straightforward in MATLAB. An example of the MATLAB code that can be used to estimate the elements of vector $\underline{h}$ and their associated standard errors (in column vector `hse`) is given in the script file "`sysid_ls.m`." The main portion of this code is displayed below.

```
>> % Construct observation matrix UU
>> UU = zeros(N,p);
>> for i=1:p,
>>   if i==1
>>     UU(:,1)=u;
>>   else
>>     UU(:,i)= [zeros(i-1,1)' u(1:N-i+1)']';
>>   end
>> end;
>> UU = T*UU;
>>
>> % Construct autocorrelation matrix
>> AA = UU'*UU;
>> b = UU'*y;
>>
>> % Compute estimate of h
>> h = AA\b;
>>
>> % Compute estimated standard errors of h, hse
>> e = y - U*h;
>> sigma = std(e);
>> AAinv = inv(AA);
>> hse = zeros(size(h));
>> for i=1:p,
>>   hse(i) = sqrt(AAinv(i,i))*sigma ;
>> end;
```

The above code assumes that the input and output data are contained in the $N$-element column vectors u and y, respectively. As an illustration of how one can apply the estimation algorithm, we use sample input and output "data" generated by the linear lung mechanics model discussed in Chapters 4 and 5. The following parameter values are assumed: $L = 0.01\ \mathrm{cm\,H_2O\,s^2\,L^{-1}}$, $\mathrm{R} = 1\ \mathrm{cm\,H_2O\,s\,L^{-1}}$ and $C = 0.1\ \mathrm{L\,cm\,H_2O^{-1}}$. The input in this case is a unit step in $P_{ao}$, beginning at time $t = 0$. The output is the model response in $P_A$. Gaussian white noise is added to the output to simulate the effects of measurement noise. The simulated input and output measurements are displayed in Figure 7.2a; these time series are also contained in the file labeled "data_llm.mat." The estimated impulse response is shown in Figure 7.2b, along with upper and lower bounds that reflect the estimates plus and minus one standard error. Superimposed on the estimates is the "true" impulse response, which appears as the smooth curve. The fluctuations in the estimated impulse response illustrates how sensitive it is to measurement noise, since in this case, the $p/N$ ratio of $\sim\frac{1}{2}$ was large.

Although we have confined the application of this identification method to linear systems here, it is important to point out that Equation (7.3) can be readily extended to take the form of a *Volterra* series, which also contains *nonlinear* dependences of $y(nT)$ on $u(nT)$. This formulation allows us to estimate the parameters that characterize the dynamics of a certain class of open-loop nonlinear systems. However, this topic falls beyond the scope of our present discussion. The interested reader can find further information about this type of

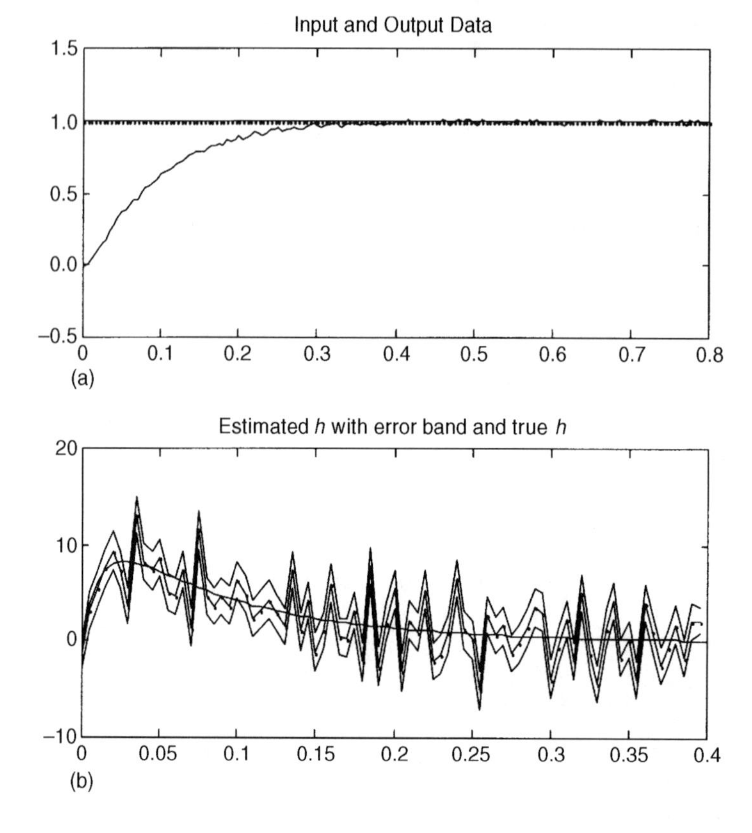


**Figure 7.2** (a) Step change in $P_{ao}$ and resulting response in $P_A$ (with noise added). (b) Estimated impulse response with error bounds superimposed on "true" impulse response (smooth curve). Horizontal axes represent time in seconds.

nonlinear system identification in Marmarelis and Marmarelis (1978), Schetzen (1980), and Korenberg and Hunter (1990).

### 7.2.3 Estimation Using Correlation Functions

Starting with Equation (7.3), for any $m \geq 0$, if we multiply both sides of the equation by $u(nT - mT)$, sum up all $N - m$ nonzero terms, and then divide through by $N - m$, we will obtain

$$\hat{R}_{uy}(mT) = \sum_{k=0}^{p-1} h(kT) \cdot \hat{R}_{uu}(mT - kT) \cdot T + \hat{R}_{ue}(mT) \tag{7.15a}$$

where

$$\hat{R}_{uy}(mT) = \frac{1}{N-m} \sum_{n=m}^{N-1} u(nT - mT)y(nT) \tag{7.16}$$

$$\hat{R}_{uu}(mT - kT) = \frac{1}{N-m} \sum_{n=m}^{N-1} u(nT - mT)u(nT - kT) \tag{7.17}$$

and

$$\hat{R}_{ue}(mT) = \frac{1}{N-m}\sum_{n=m}^{N-1} u(nT - mT)e(nT) \tag{7.18}$$

It should be noted that $\hat{R}_{uy}$, $\hat{R}_{uu}$, and $\hat{R}_{ue}$ in Equations (7.16) through (7.18) represent estimates of the cross-correlation between $u$ and $y$, the autocorrelation in $u$, and the cross-correlation between $u$ and $e$, respectively (see Section 5.3.2). We select that solution of $h$ such that $\hat{R}_{ue}(mT)$ becomes zero for all values of $m$:

$$\hat{R}_{uy}(mT) = \sum_{k=0}^{p-1} h(kT) \cdot \hat{R}_{uu}(mT - kT)T, \qquad m = 0, 1, \ldots, p-1 \tag{7.15b}$$

Equation (7.15b) may be considered the discrete-time version of Equation (5.27b), and may be solved by applying a little matrix algebra, as in Section 7.2.1.2. We define the following vector and matrix quantities:

$$\underline{\hat{R}}_{uy} = [\hat{R}_{uy}(0)\ \hat{R}_{uy}(T) \ldots \hat{R}_{uy}((p-1)T)]' \tag{7.19}$$

and

$$\hat{\mathbf{R}}_{uu} = \begin{bmatrix} \hat{R}_{uu}(0) & \hat{R}_{uu}(T) & \ldots & \hat{R}_{uu}((p-1)T) \\ \hat{R}_{uu}(T) & \hat{R}_{uu}(0) & \ldots & \hat{R}_{uu}((p-2)T) \\ \vdots & \vdots & \vdots & \vdots \\ \hat{R}_{uu}((p-1)T & \hat{R}_{uu}((p-2)T) & \ldots & \hat{R}_{uu}(0) \end{bmatrix} \tag{7.20}$$

Then, Equation (7.15b) becomes

$$\underline{\hat{R}}_{uy} = T\hat{\mathbf{R}}_{uu}\underline{h} \tag{7.21}$$

where $\underline{h}$ is defined by Equation (7.5). Since all elements of $\hat{\mathbf{R}}_{uy}$ and $\hat{\mathbf{R}}_{uu}$ can be computed from the input and output data by applying Equations (7.16) and (7.17), the unknown impulse response function is determined through the solution of Equation (7.21) by inverting $\hat{\mathbf{R}}_{uu}$:

$$\underline{h} = \frac{1}{T}\hat{\mathbf{R}}_{uu}^{-1}\underline{\hat{R}}_{uy} \tag{7.22}$$

It can be shown that, aside from possible differences in the details of computing the autocorrelation and cross-correlation functions, Equation (7.22) is essentially equivalent to Equation (7.12). As before, the feasibility of applying this approach depends on the invertibility of the autocorrelation matrix $\hat{\mathbf{R}}_{uu}$.

### 7.2.4 Estimation in the Frequency Domain

Since the Laplace transform of the impulse response is the system transfer function, carrying out the system identification process in the frequency domain should, in principle, yield the same results as any of the time-domain methods discussed earlier. The problem of transfer function identification is actually the same as that of estimating the frequency response. The underlying idea is very simple and is illustrated in Figure 7.3. At each frequency in the range of interest, apply a sinusoidal input of known amplitude and phase to the system under study; then measure the resulting output. If the system is linear, the measured output also will be a sinusoid of the same frequency. The ratio between the magnitude of the output and the magnitude of the input ($= A_{out}/A_{in}$ in Figure 7.3) would yield the system *gain* at that frequency. The phase difference between the output and input waveforms ($= 2\pi T_p/T_c$) would be the system *phase* at that frequency. By repeating this measurement over all frequencies of interest, one would be able to arrive at the frequency response of the system and therefore obtain an estimate of the transfer function.

Although the above method can provide very good estimates of the system transfer function at the frequencies investigated, one major drawback is that the entire identification procedure can be extremely time-consuming and therefore impractical for application in human or animal studies. An alternative would be to employ the spectral analysis technique presented in Section 5.3.2; the basic idea here is that the frequency response is estimated from

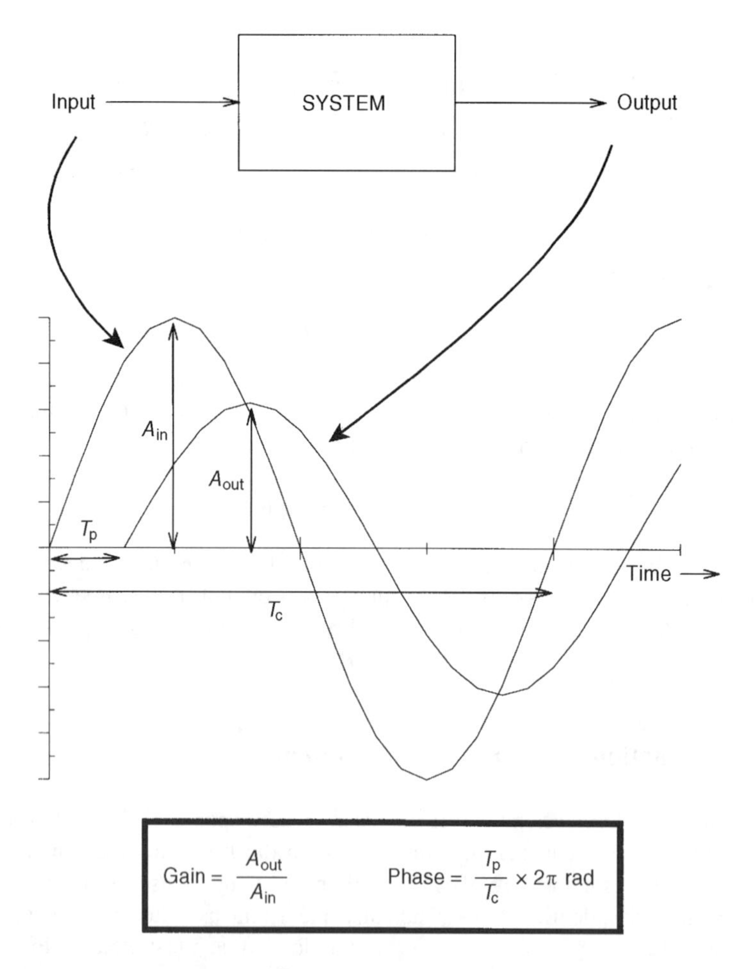


**Figure 7.3** Illustration of transfer function identification using sinusoidal inputs. System gain and phase at given frequency $\omega = 2\pi/T_c$ are as defined above.

the ratio between the input–output cross-spectrum ($S_{uy}$) and the input power spectrum ($S_{uu}$), i.e.,

$$H(\omega_k) = \frac{S_{uy}(\omega_k)}{S_{uu}(\omega_k)} \tag{7.23}$$

where $\omega_k = 2\pi k/pT$ ($k = 0, 1, \ldots, p-1$), and the spectral quantities are defined in the following way:

$$S_{uy}(\omega_k) = \sum_{m=0}^{p-1} \hat{R}_{uy}(mT)e^{-j\omega_k mT}, \qquad k = 0, 1, \ldots, p-1 \tag{7.24}$$

$$S_{uu}(\omega_k) = \sum_{m=0}^{p-1} \hat{R}_{uu}(mT)e^{-j\omega_k mT} \qquad k = 0, 1, \ldots, p-1 \tag{7.25}$$

As shown in Equations (7.24) and (7.25), $S_{uy}(\omega_k)$ and $S_{uu}(\omega_k)$ are computed by applying the discrete Fourier transform to $\hat{R}_{uy}$ and $\hat{R}_{uu}$, respectively. Equations (7.23) through (7.25) are the discrete-frequency equivalents of Equations (5.29) and (5.30), which were applied to correlation quantities based on continuous time. As mentioned in Section 5.3.2, all values of $S_{uu}$ must be positive in order for meaningful estimates of the transfer function to be obtained. Whether this condition is attained depends on the form of the input sequence, as we will see in Section 7.3.

Once $H(\omega)$ has been estimated, it is possible to interpret the results in the context of physiologically meaningful entities if a parametric model is available. As an example of how this can be done, consider the linear lung mechanics model that was discussed in Chapters 4 and 5. From Equation (5.6), the frequency response predicted from this model takes the form

$$H_{\text{model}}(\omega) = \frac{1}{(1 - LC\omega^2) + jRC\omega} \tag{7.26}$$

The values of the lung mechanical parameters ($R$, $L$, and $C$) that most closely correspond to the measured frequency response $H_{\text{meas}}(\omega)$ can be estimated by first defining a "criterion function," $J$, which represents the "distance" between $H_{\text{model}}$ and $H_{\text{meas}}$, and, secondly, by searching for the parameter values that minimize this distance. Since $H_{\text{model}}$ and $H_{\text{meas}}$ are complex-valued functions of $\omega$, a suitable criterion function might be

$$J = \sum_{k=0}^{p-1} (\text{Re}[H_{\text{meas}}(\omega_k)] - \text{Re}[H_{\text{model}}(\omega_k)])^2 + (\text{Im}[H_{\text{meas}}(\omega_k)] - \text{Im}[H_{\text{model}}(\omega_k)])^2 \tag{7.27}$$

The above expression assumes that frequency response measurements are available at the frequencies $\omega_k$, where $k = 0, 1, \ldots, p-1$. The methodology for minimizing $J$ is described in the next section.

### 7.2.5 Optimization Techniques

As we mentioned at the beginning of this chapter, the identification of "gray-box" or "parametric" models consists of two stages. First, the model structure has to be developed, consistent with prior knowledge about the physiological system in question. Frequently, this takes the form of a set of differential equations. Once the model has been formulated, the next task is to estimate the unknown model parameters by minimizing (or maximizing) some criterion that reflects the goodness of fit between the model predictions and the observed output measurements. When dealing with models represented by differential equations of high order, we mentioned in Section 2.8 that it is generally better, from the viewpoint of numerical stability, to employ a *state-space* framework. Another advantage of employing a state-space model is that the analysis can readily be extended to *nonlinear* systems. We will now illustrate how this system identification technique works by considering our favorite example of the linear lung mechanics model. A more advanced example, involving the analysis of a nonlinear model, is given in Section 7.5.1.

***7.2.5.1 State-space Model Formulation.*** The differential equation characterizing the lung mechanics model was derived in Section 4.1 and is given by

$$LC\,\frac{d^2P_A}{dt^2} + RC\,\frac{dP_A}{dt} + P_A = P_{ao} \tag{7.28}$$

Since $P_{ao}$ is the input and $P_A$ is the output of this system, we make the new variable assignments:

$$y_1 = P_A \qquad \text{and} \qquad u = P_{ao} \tag{7.29a,b}$$

Also, assume:

$$y_2 = \frac{dP_A}{dt} = \frac{dy_1}{dt} \tag{7.30}$$

Then, we can rewrite Equation (7.28) as:

$$LC\,\frac{dy_2}{dt} + RCy_2 + y_1 = u \tag{7.31}$$

Using Equations (7.30) and (7.31), rearranging terms, and writing the two equations in matrix form, we obtain:

$$\frac{d}{dt}\begin{bmatrix} y_1 \\ y_2 \end{bmatrix} = \begin{bmatrix} 0 & 1 \\ -\frac{1}{LC} & -\frac{R}{L} \end{bmatrix}\begin{bmatrix} y_1 \\ y_2 \end{bmatrix} + \begin{bmatrix} 0 \\ \frac{1}{LC} \end{bmatrix} u \tag{7.32a}$$

If we define:

$$\underline{\mathbf{y}} = \begin{bmatrix} y_1 \\ y_2 \end{bmatrix} \tag{7.33}$$

$$\mathbf{A} = \begin{bmatrix} 0 & 1 \\ -\frac{1}{LC} & -\frac{R}{L} \end{bmatrix} \tag{7.34}$$

and

$$\mathbf{B} = \begin{bmatrix} 0 \\ \frac{1}{LC} \end{bmatrix} \tag{7.35}$$

Equation (7.32a) becomes:

$$\frac{d}{dt}\underline{\mathbf{y}} = \mathbf{A}\underline{\mathbf{y}} + \mathbf{B}u \tag{7.32b}$$

Thus, we have converted the second-order scalar differential equation (Equation (7.31)) into the equivalent first-order matrix state equation. This type of equation can be conveniently solved by numerical integration using one of the MATLAB ordinary differential equation solver functions: "ode45," "ode23," "ode113," "ode15s," and "ode23s." An even easier way is to construct a state-space representation of the model within MATLAB using the "ss" function, and then use the "lsim" function (see Section 4.6) to generate the model response to a given input waveform $u(t)$. The MATLAB script file (provided as the file "sss_llm.m") that performs these tasks is displayed below.

```
>> A = [0 1; -1/L/C -R/L];
>> B = [0 1/L/C]';
>> t = [0:0.005:0.8]';
>> u = ones(size(t));

% Construct the system using state-space formulation
>> Hs = ss(A,B,[1 0],0);

% Solve state space equation using lsim and plot results
>> y = lsim(Hs,u,t);
>> plot(t,u,t,y)
```

In the above MATLAB script file, note that we have assumed the following companion "observation equation":

$$\hat{y} = \mathbf{D}\underline{\mathbf{y}} + Eu \tag{7.36}$$

where $\hat{y}$ represents the measured output of this system, i.e., $P_A$. In this case, we are able to measure the state variable $y_1(=P_A)$ directly. Thus, here, we have:

$$\mathbf{D} = [1 \quad 0] \qquad \text{and} \qquad E = 0 \tag{7.37a,b}$$

This accounts for the last two items in the argument list of the function "ss" in the above MATLAB script. One other detail is that the matrices **A** and **B** do not need to be evaluated directly. If the transfer function of the model is available, then the MATLAB function "tf2ss" can be used to convert the system representation from transfer function format to state-space format:

```
>> [A,B,D,E] = tf2ss(num,den);
```

***7.2.5.2 Optimization algorithm.*** Having constructed the model, the next step is to select the means by which the response of the model to a given input sequence can be compared to the response of the physiological system to the same input. The comparison is made through the use of a *criterion function* that provides a measure of the *goodness of fit* between the two time series. There are many possible candidates for the criterion function, but the one most commonly employed is the sum of squares of the differences between the measured and predicted outputs:

$$J = \sum_{n=0}^{N-1} \{y(nT) - y_{\text{pred}}(nT)\}^2 = \sum_{n=0}^{N-1} e(nT)^2 \tag{7.38}$$

This is the same expression as that presented in Equation (7.9) in Section 7.2.2.

With the criterion function having been defined, the problem of parameter estimation becomes transformed into a problem of optimization, where the objective is to find the combination of parameter values that minimizes the criterion function. The entire scheme of parameter estimation is illustrated in the schematic block diagram shown in Figure 7.4. It should be noted that if the model selected provides an accurate representation of the dynamics of the real system, then the residual errors $\{e(nT),\ n = 0,\ 1, \ldots,\ N-1\}$ should closely reflect the measurement noise affecting the output. On the other hand, if the selected model is largely "wrong" and does not provide an adequate description of the output dynamics, there will be a significant contribution from *structural errors* as well. This is one of the major drawbacks of opting to employ a "structural" model: erroneous information about the dynamics of the underlying physiology can translate into large errors in the parameter estimates and/or an inability of the model predictions to "fit" the data well.

The choice of the algorithm employed to perform the minimization of the criterion function is also important. There is a large array of algorithms available, but it is not within the scope of this chapter to examine all or even most of these. The most commonly applied methods employ the *gradient descent* approach. These methods are best explained by considering a problem in which two parameters need to be estimated. We will refer to these two parameters as $\theta_1$ and $\theta_2$. Since $J$ is a function of $\theta_1$ and $\theta_2$, evaluating $J$ over selected ranges of $\theta_1$ and $\theta_2$ would yield a surface in a 3-dimensional space with the Cartesian axes formed by $\theta_1$, $\theta_2$, and $J$. Suppose the surface looks like the contour map shown in Figure 7.5a, where each contour corresponds to a uniform value of $J$. In the gradient descent approach, we start off with an initial guess of the parameters, represented as the point $A$. Then, information about the slope of the local terrain is obtained, and based on this information, we move down the slope along the direction of "steepest descent." The size of the step taken in this direction differs with the different gradient descent methods, with some methods using information about the curvature (i.e., second derivatives) of the surface as well.

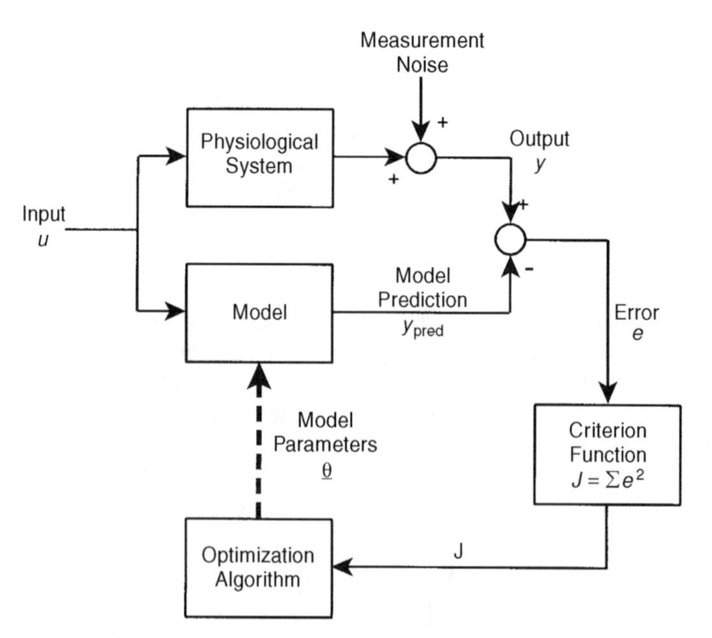


**Figure 7.4** Schematic diagram of the optimization approach to parameter estimation.

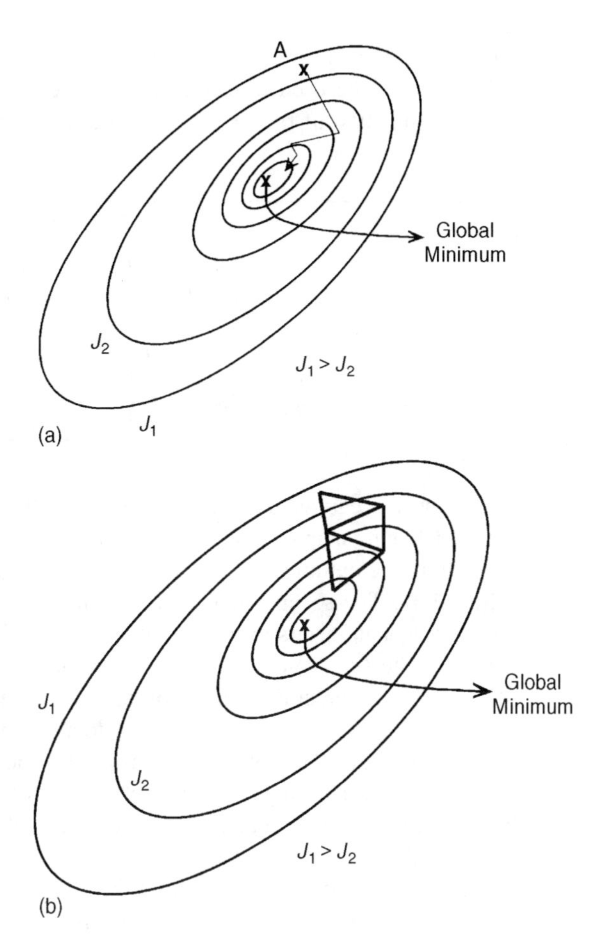


**Figure 7.5** Methods for finding the minimum of the criterion function surface: (a) steepest gradient method; (b) simplex method.

One drawback of the gradient approach is the need to evaluate the derivatives of $J$ with respect to all the parameters being estimated at every iteration step. Furthermore, these methods are generally quite susceptible to producing solutions that correspond to false local minima, if the $J$-surface is highly irregular, as would be the case when the signal-to-noise ratio is low. A popular alternative, which does not require any derivative computations at all, is the *Nelder–Mead simplex* algorithm. For a 3-parameter problem, the *simplex* takes the form of a tetrahedron, while for the 2-parameter problem, it is a triangle. Figure 7.5b shows the same criterion function surface discussed above, together with the simplex (triangle) and how the shape and position of the triangle moves over the course of a few iterations. The vertices of the triangle represent the three points on the $J$-surface that are known at any given iteration of the algorithm. Starting at the initial three points, the triangle is reflected over the two vertices with the lowest $J$-values and the height of the triangle is expanded or contracted so that the remaining vertex is located at the point of lowest $J$-value. Next, the triangle is reflected over the two of the three vertices that have the lowest $J$-values, and the new third vertex is found by stretching or shrinking the height of the triangle. This process is repeated until some tolerance for convergence towards the minimum is met.

The MATLAB function "`fmins`" employs the Nelder–Mead simplex algorithm to determine the minimum point of a given multidimensional function. The MATLAB script (contained in the m-file named "popt_llm.m") presented below shows an example of how "fmins" can be used to estimate the values of the two unknown parameters in the state-space formulation of the linear lung mechanics model (Section 7.2.5.1).

```
>> global u y
>> theta_init(1)= input(' Enter initial value of 1st parameter >>');
>> theta_init(2)= input(' Enter initial value of 2nd parameter >>');

% Perform optimization to minimize the objective function J
% defined by the function "fn_llm"
>> [theta,options] = fmins('fn_llm',theta_init);
```

Two items are required as inputs to "fmins". The first is a user-defined function that defines the model being employed and returns to "fmins" the value of the criterion function at each iteration in the optimization process. In our particular example, we have named this function "fn_llm". The second input is a column vector ("theta_init") containing an initial guess of the parameters to be estimated. "fmins" produces two sets of outputs. The first set contains the estimated parameter vector ("theta"). The second set of outputs ("options") contains information about the minimization process. For instance, the 10th element of "options" contains the total number of iterations or function calls in the optimization run. As well, in this example, the data file "data_llm.mat" has to be loaded prior to running "popt_llm" so that the input and output data is present in the workspace as vectors $u$ and $y$, respectively, for "fmins" to work on. Since the function "fn_llm" must also use these data, the "global" declaration is included in both "popt_llm" and "fn_llm" to make u and y universally accessible. The relevant portion of the MATLAB code for "fn_llm" is given below, with the complete listing given in the m-file "fn_llm.m".

```
>> function J = fn_llm(theta)
>> global u y

>> A = [0 1; -theta(1) -theta(2)];
>> B = [0 theta(1)]';
>> Hs = ss(A,B,[1 0],0);
>> ypred = lsim(Hs,u,t);
>> e = y - ypred;
>> J = sum(e.^2);
```

It should be noted from the script for "fn_llm" that the two parameters being estimated, $\theta_1$ and $\theta_2$, correspond to the lung mechanical parameters $1/LC$ and $R/L$, respectively (see Equations (7.34) and (7.35)). Application of this algorithm to the simulated data given in "data_llm.mat" leads to the parameter estimation results shown below. The estimated parameter values of 1006.3 and 100.8 agree closely with the "true" values for $1/LC$ and $R/L$ of 1000 and 100, respectively.

```
Final Parameter Values :
1.0e+003 *
 1.0063
 0.1008
Total Number of Iterations:
198
```

## 7.3 PROBLEMS IN PARAMETER ESTIMATION: IDENTIFIABILITY AND INPUT DESIGN

### 7.3.1 Structural Identifiability

The problem of structural identifiability is intimately coupled to the problem of model-building. In theory, if knowledge about the underlying physiology of the system in question is available, it should be possible for us to translate this knowledge into a parametric model by applying the basic laws of physics and chemistry. The more we know about the system, the more details we will be able to add to the model. In general, a more detailed and complex model would be expected to account for a greater range of observations under a larger variety of conditions. However, the price that one has to pay for the increased model complexity is the emergence of more model *parameters*, the values of which have to be assumed or estimated. In the several models that we have discussed in previous chapters, we assumed the parameter values to be known. For example, in the linear lung mechanics model, we assumed values for $R$, $L$, and $C$ that were considered "representative" of the population of subjects with normal lungs. This assumption ignores the fact that there is a considerable degree of variability in these lung mechanical parameters across subjects that one can consider "normal." On the other hand, we could choose to estimate the parameters in each individual subject. The problem of structural identifiability arises when the information that is required for the parameter estimation process is incomplete. This could be due to the inaccessibility of certain signals or the lack of dynamic content in the stimulus.

As an example, consider the differential equation (Equation (7.28)) that characterizes the linear lung mechanics model. Here, there are three unknown parameters: $R$, $L$, and $C$. However, the mathematical structure of this model turns out to be such that the parameters only appear as paired combinations of one another: $LC$ and $RC$. As a consequence, the dynamics of the model are determined by only two parameters ($LC$ and $RC$) and not by the original three ($R$, $L$, and $C$). This fact again becomes evident when one looks at the state-space formulation of the model in Equation (7.32a). Here, only two independent parameters determine the solution (i.e., dynamics) for the vector $\mathbf{y}$, and these are $1/LC$ and $R/L$. Thus, it is clear that, using only measurements of $P_A$ and $P_{ao}$, the linear lung mechanics model is not completely identifiable in terms of all three parameters—$R$, $L$, and $C$. We should stress that this assertion on identifiability (or rather, the lack of it) holds true regardless of whether noise is present or absent in the measurements. On the other hand, this model could become fully identifiable if an additional channel of measurement, such as airflow, were to become available. For instance, one could estimate $C$ separately from static changes, e.g., from the ratio of the change in lung volume resulting from an applied change in pressure. Then, by combining this additional piece of information with the two parameters that can be estimated from the step-response in $P_A$, we would be able to identify all three original model parameters.

### 7.3.2 Sensitivity Analysis

A model that has been found to be structurally identifiable may still turn out to be unidentifiable in practice, if the parameter estimation process is sufficiently degraded by the presence of measurement noise. Therefore, having arrived at a structurally identifiable model, the next test that we should subject the model to is the determination of whether the parameters that need to be estimated are resolvable in the presence of noise. Since the parameter estimation process requires us to find the lowest point on the multidimensional surface of the criterion function, it follows that parameter identifiability depends heavily on the quality of the $J$-surface. Figure 7.6 illustrates this statement with the help of two hypothetical examples that assume the case involving only a single parameter ($\theta_1$). In both cases, we also assume that the presence of measurement noise limits the resolvability of changes in $J$ to a value $\Delta J$. In Figure 7.6a, there is a deep minimum. The error in the parameter estimate ($\Delta\theta_1$) made in arriving at a solution that is located at a criterion function value $\Delta J$ above the global minimum is small. On the other hand, in Figure 7.6b, the $J$-surface contains a very shallow minimum. In this case, the effect of the same amount of measurement noise is a much larger error in the parameter estimate. What distinguishes case (a) from case (b) in this example is the fact that in case (a), a given change in the parameter value leads to a large (and therefore, highly observable) change in the model output or $J$. Thus, in case (a), the model possesses *high sensitivity* to parameter variations, whereas in case (b) sensitivity is low.

The inverse relationship between sensitivity and parameter estimation error can be demonstrated analytically. We begin by recalling the definition of the criterion function (see Equation (7.38)) but rewriting it in vector form:

$$J = (\underline{\mathbf{y}} - \underline{\mathbf{y}}_{\text{pred}})'(\underline{\mathbf{y}} - \underline{\mathbf{y}}_{\text{pred}}) \tag{7.39}$$

Differentiating $J$ with respect to the parameter vector $\underline{\theta}$, we obtain

$$\frac{\partial J}{\partial \underline{\theta}} = -(\underline{\mathbf{y}} - \underline{\mathbf{y}}_{\text{pred}})' \frac{\partial \underline{\mathbf{y}}_{\text{pred}}}{\partial \underline{\theta}} \tag{7.40}$$

Note that the derivative on the right-hand side of Equation (7.40) is a $N \times p$ matrix, the elements of which represents the effect of a small change in each parameter on the model output. Thus, we can refer to this entity as the *sensitivity matrix*, $\mathbf{S}$, i.e.:

$$\mathbf{S} = \frac{\partial \underline{\mathbf{y}}_{\text{pred}}}{\partial \underline{\theta}} \tag{7.41}$$

Suppose $\underline{\theta}^*$ represents the parameter vector at the global minimum point on the $J$-surface. Then, by applying a Taylor series expansion and keeping *only first-order terms*, the model output in the vicinity of the minimum point can be expressed as

$$\underline{\mathbf{y}}_{\text{pred}}(\underline{\theta}) = \underline{\mathbf{y}}_{\text{pred}}(\underline{\theta}^*) + \left.\frac{\partial \underline{\mathbf{y}}_{\text{pred}}}{\partial \underline{\theta}}\right|_{\underline{\theta}^*} (\underline{\theta} - \underline{\theta}^*) = \underline{\mathbf{y}}_{\text{pred}}(\underline{\theta}^*) + \mathbf{S}_{\underline{\theta}^*}(\underline{\theta} - \underline{\theta}^*) \tag{7.42}$$

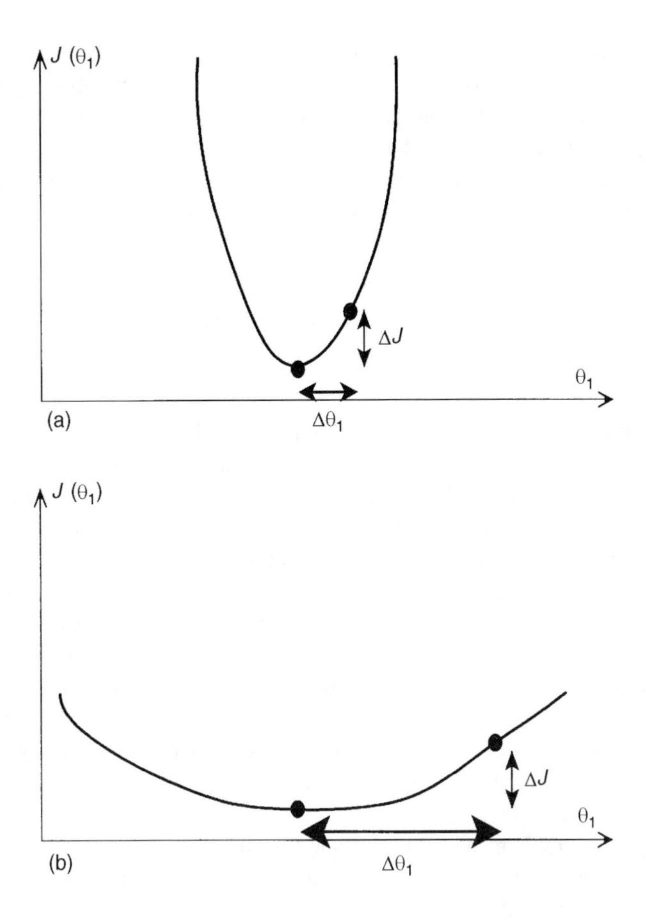


**Figure 7.6** Relationship between sensitivity to parameter variations and parameter estimation error: (a) high sensitivity; (b) low sensitivity.

where $\mathbf{S}_{\underline{\theta}^*}$ denotes the matrix $\mathbf{S}$ evaluated at the minimum point. At the minimum point, $\partial J/\partial\underline{\theta}$ in Equation (7.40) becomes a null vector. Thus, substituting Equation (7.42) into Equation (7.40), we obtain

$$-\frac{\partial J}{\partial \underline{\theta}} = 0 = \{\underline{\mathbf{y}} - \underline{\mathbf{y}}_{\text{pred}}(\underline{\theta}^*)\}'\mathbf{S}_{\underline{\theta}^*} - (\underline{\theta} - \underline{\theta}^*)'\mathbf{S}'_{\underline{\theta}^*}\mathbf{S}_{\underline{\theta}^*} \tag{7.43}$$

which can be rearranged to yield the following expression for the parameter estimate error:

$$(\underline{\theta} - \underline{\theta}^*)' = \underline{\mathbf{e}}^{*\prime}\mathbf{S}_{\underline{\theta}^*}(\mathbf{S}'_{\underline{\theta}^*}\mathbf{S}_{\underline{\theta}^*})^{-1} \tag{7.44}$$

where $\underline{\mathbf{e}}^*$ represents the vector that contains the residual errors between the measurements $\mathbf{y}$ and the predicted output values $\mathbf{y}_{\text{pred}}(\theta^*)$. Left-multiplying both sides of Equation (7.44) with $(\underline{\theta} - \underline{\theta}^*)$ and applying the expectation operator (see Equation (5.28)) to both sides of the resulting equation, we get

$$\mathbf{P}_{\underline{\theta}} \equiv E[(\underline{\theta} - \underline{\theta}^*)(\underline{\theta} - \underline{\theta}^*)'] = (\mathbf{S}'_{\underline{\theta}^*}\mathbf{S}_{\underline{\theta}^*})^{-1}\mathbf{S}'_{\underline{\theta}^*}E[\underline{\mathbf{e}}^*\underline{\mathbf{e}}^{*\prime}]\mathbf{S}_{\underline{\theta}^*}(\mathbf{S}'_{\underline{\theta}^*}\mathbf{S}_{\underline{\theta}^*})^{-1} \tag{7.45}$$

where $\mathbf{P}_{\underline{\theta}}$ is also known as the *parameter error covariance matrix*. The diagonal elements of $\mathbf{P}_{\underline{\theta}}$ contain the variances of all $p$ parameters in $\underline{\theta}$, whereas the off-diagonal elements represent the cross-covariances between the different paired combinations of the parameters. If we assume the sequence of residual errors to be white, i.e., the present error is uncorrelated with past or future errors, then the matrix $\mathbf{E}[\underline{\mathbf{e}}^*\underline{\mathbf{e}}^{*\prime}]$ reduces to the identity matrix scaled by a factor equal to the variance, $\sigma^2$, of the residual errors. Thus, Equation (7.45) simplifies to

$$\mathbf{P}_{\underline{\theta}} = \sigma^2(\mathbf{S}'_{\underline{\theta}^*}\mathbf{S}_{\underline{\theta}^*})^{-1} \tag{7.46}$$

From Equation (7.46), it is important to note that each element of the symmetric matrix $\mathbf{S}'_{\underline{\theta}^*}\mathbf{S}_{\underline{\theta}^*}$ reflects the change in model output resulting from small changes in all possible pairings of the parameters. If changes in one or more of the parameters have no effect on the model output (zero sensitivity), then one or more columns and rows of $\mathbf{S}'_{\underline{\theta}^*}\mathbf{S}_{\underline{\theta}^*}$ will be zero; as a result, $\mathbf{S}'_{\underline{\theta}^*}\mathbf{S}_{\underline{\theta}^*}$ will be singular and the parameter errors will be infinite. This occurs when the model is *not structurally identifiable*. In structurally identifiable models, $\mathbf{S}'_{\underline{\theta}^*}\mathbf{S}_{\underline{\theta}^*}$ can still become close to singular if there are strong interdependences between some of the parameters; in this case, there will be strong correlations between columns or rows of matrix $\mathbf{S}'_{\underline{\theta}^*}\mathbf{S}_{\underline{\theta}^*}$. Inversion of this close-to-singular matrix will yield variance and covariance values in $\mathbf{P}_{\underline{\theta}}$ that are unacceptably large. However, it is important to bear in mind from Equation (7.46) that, even under circumstances where model sensitivity is high, it is still possible for the parameters to be poorly estimated if the variance of the measurement noise ($\sigma^2$) is very large.

Equation (7.46) provides lower-bound estimates of the variances and cross-covariances associated with the model parameters when these are estimated from noisy measurements. However, computation of the $\mathbf{P}_{\underline{\theta}}$ matrix is based on local changes in the vicinity of the optimal set of parameter values. A common alternative method of assessing model sensitivity is to base the calculations over a larger range of parameter value changes. In this approach, the criterion function $J$ is evaluated over a selected span of values (say, $\pm 50\%$) for each parameter in turn, while holding the rest at their nominal values. Ideally, the "nominal" or "reference" values selected should correspond to the optimal set $\underline{\theta}^*$. The form of the criterion function $J$ is the same as that given in Equation (7.39), except that, in this case, the vector of observations $\underline{\mathbf{y}}$ is replaced by $\underline{\mathbf{y}}_{ref}$, where the latter represents the model predictions when the parameters are at their nominal values. Here, $\underline{\mathbf{y}}_{pred}$ corresponds to the vector of model predictions at any of the parameter combinations ($\neq \underline{\theta}^*$) being evaluated. An example of this type of sensitivity analysis is shown in Figure 7.7 for the linear lung mechanics model. The nominal parameter set in this case is: $R = 1$ cm $H_2O$ s $L^{-1}$, $L = 0.01$ cm $H_2O$ s $L^{-2}$ and $C = 0.1$ L cm $H_2O^{-1}$. The model is assumed to be perturbed by a unit step in $P_{ao}$. The plot for $R$, for instance, shows changes in $J$ that would result if $R$ were to be varied over the range 0.5–1.5 cm $H_2O$ s $L^{-1}$, while $L$ and $C$ are kept at their nominal values. The model output is reasonably sensitive to changes in $R$ and $C$, but virtually insensitive to changes in $L$. This kind of "flatness" in the sensitivity curve provides a good indication that at least one of the parameters will be not identifiable. This conclusion is consistent with our analysis of structural identifiability of this model in Section 7.3.1. The sensitivity results in Figure 7.7 were generated by the MATLAB script file "`sensanl.m`" (which also calls the function "`fn_rlc.m`").

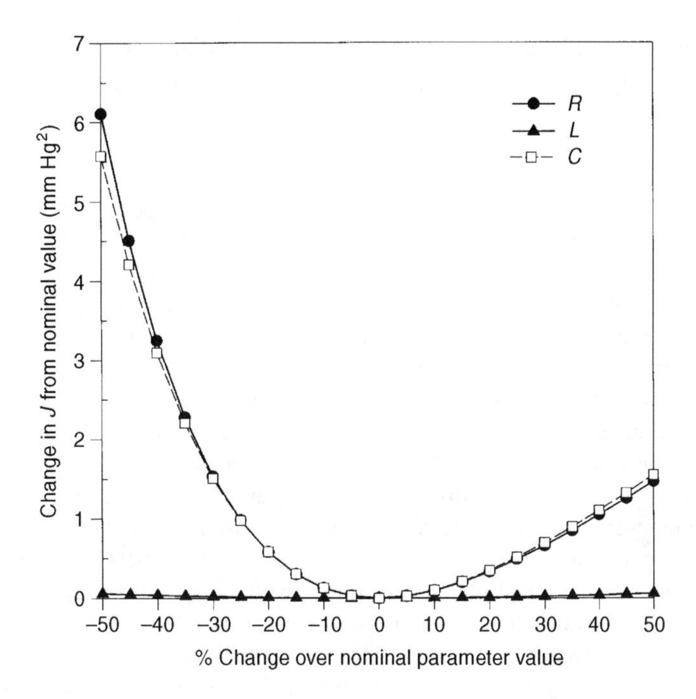


**Figure 7.7** Sensitivity of the linear lung mechanics model to variations in the model parameters about their nominal values ($R = 1, L = 0.01, C = 0.1$). Flat curve for $L$ (i.e. very low sensitivity) suggests identifiability problems for this parameter.

### 7.3.3 Input Design

The result represented by Equation (7.46) is valid for any general parametric model. Comparison of this result with Equation (7.13) shows a striking similarity between the two equations. This similarity is by no means coincidental. In fact, Equation (7.13) represents a special case of Equation (7.46) when the assumed "model" is simply the impulse response of the system under study. To demonstrate this, note that the vector $\underline{\mathbf{h}}$ containing the sampled impulse response is the unknown parameter vector $\underline{\theta}$ that we would like to estimate. Thus, the model predictions are given by

$$y(nT)_{\text{pred}} = \sum_{k=0}^{p-1} \theta_k \cdot u(nT - kT) \cdot T, \quad n = 0, 1, \ldots, N-1 \tag{7.47}$$

where

$$\theta_k = h(kT) \tag{7.48}$$

The sensitivity of the $n$th output value to changes in the $k$th parameter is

$$S_{nk} \equiv \frac{\partial y(nT)_{\text{pred}}}{\partial \theta_k} = u(nT - kT) \cdot T, \qquad n = 0, 1, \ldots, N-1; \quad k = 0, 1, \ldots, p-1 \tag{7.49}$$

Looking back at Equation (7.7), it can be easily seen that the matrix **U** is simply a special case of the sensitivity matrix **S**. Consequently, we have the following equality:

$$\mathbf{S}'_{\underline{\theta}^*}\mathbf{S}_{\underline{\theta}^*} = \mathbf{U}'\mathbf{U} \tag{7.50}$$

While it was not specifically mentioned in the previous section, it is clear from Equation (7.50) that the $p \times p$ matrix $\mathbf{S}'_{\underline{\theta}^*}\mathbf{S}_{\underline{\theta}^*}$ is a function of the input time-course. This implies that if the researcher has control over the type of stimulus that can be administered to the system in question, it should be possible to design the input waveform in such a way as to best "condition" the matrix $\mathbf{S}'_{\underline{\theta}^*}\mathbf{S}_{\underline{\theta}^*}$ so that the elements of its inverse can be minimized. From linear algebra, we know that a matrix is singular if there is linear dependence between any two or more of its columns (or rows). The best-case scenario for matrix inversion occurs when the matrix to be inverted is diagonal and all the diagonal elements are nonzero. For the matrix $\mathbf{U}'\mathbf{U}$ to become diagonal, it would be necessary to choose an input time course in which any sample in the waveform is uncorrelated with all other samples. Another way of saying this is that the input waveform should have zero autocorrelation over all lags, except at the zeroth lag (which simply measures how correlated the signal is with itself). One type of input waveform that has this kind of autocorrelation function is white noise. This is one of the reasons for the popularity of white noise as a test input. Another reason relates to the fact that the white noise time series also has a power spectrum that is essentially flat over a broad range of frequencies. This *persistently exciting* kind of stimulation allows the system to be probed over a larger range of dynamic modes.

Although white noise has been employed as a test input in many studies investigating various neural systems, it has not been used as much for identifying other physiological systems. A major reason for this is the practical difficulty of implementing this kind of input forcing. The *pseudo-random binary sequence* (PRBS) offers an attractive alternative that is very easy to implement and can lead to good estimation results in many applications. The PRBS is so named because the time series produced is actually periodic with a cycle duration of $N + 1$ samples, if $N$ is the total number of points in the sequence. However, within one period of this series, each sample is virtually uncorrelated with other samples. One of the most commonly used methods for generating the PRBS employs binary shift-registers with feedback. Figure 7.8a displays a 4-stage shift-register. The process begins with all stages assigned a value of 1. Then, at the end of each time-step ($T$), the value contained in each stage is moved to the right by one stage. The value in the rightmost stage of the shift-register (="1") becomes the first value of the PRBS. At the same time, this value is fed back toward the first stage and is added (or more precisely, "XOR-ed") to the value originally in the first stage. In this case, applying Boolean arithmetic, we get: $1 + 1 = 0$. Thus, at the end of the first time-step, the values in the shift-register are "0111". During the next time-step, the "1" value in the rightmost stage is moved to the right and becomes the second value in the PRBS. At the same time, this value is fed back to the first stage and added to its original value: $1 + 0 = 1$. The new result is assigned to the first stage. Thus, at the end of the second time-step, the values in the shift-register are "1011". This process continues on until the values in the shift-register revert to "1111", which was what it had started with. It can be easily shown that the 4-stage shift register assumes the "1111" value at the end of the 16th time-step and the whole sequence repeats itself. The output of this process is a 15-point sequence with randomlike properties, as depicted in Figure 7.8b. The autocorrelation function of this kind of sequence approximates that of white noise up to a maximum lag number of 14, as is shown in Figure 7.8c. However, beyond this range, it is clear that the sequence is periodic. The maximum autocorrelation value for a PRBS signal of amplitude $A$ is $A^2$, and the minimum

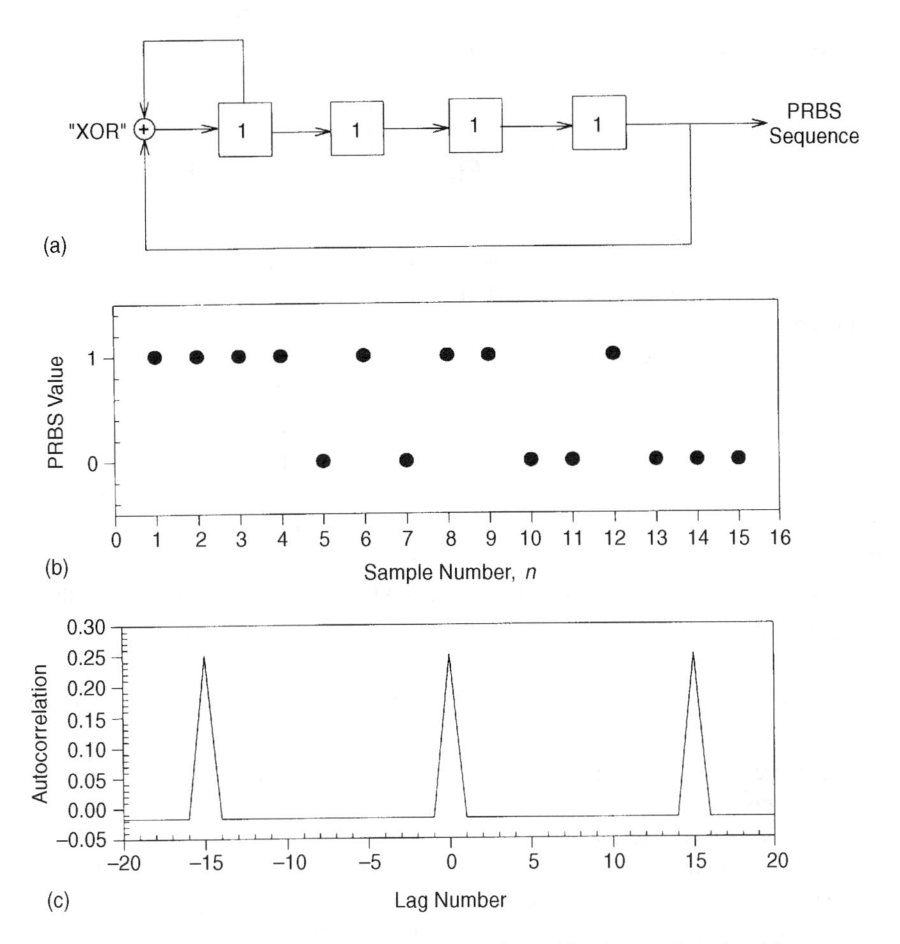


**Figure 7.8** (a) Shift register method for generating a 15-point pseudorandom binary sequence. (b) The 15-point PRBS signal generated from (a). (c) Theoretical autocorrelation function of the 15-point PRBS signal.

value is $-A^2/N$. Thus, in our example, the maximum and minimum values turn out to be 0.25 and $-0.01667$, respectively. Although the PRBS example shown here is based on a 4-stage shift-register, the latter may be extended to more stages. For an $m$-stage shift-register, the total output sequence will consist of $2^m - 1$ "random" values. The PRBS signal displayed in Figure 7.8b was generated by executing the MATLAB script file "`prbs.m`".

Using the PRBS as an input can lead to a dramatic simplification of the correlation method of system identification (see Section 7.2.3). If an $N$-sample PRBS input of amplitude $A$ is employed, the autocorrelation matrix $\hat{\mathbf{R}}_{uu}$ (of size $N \times N$) becomes

$$\hat{\mathbf{R}}_{uu} = \frac{A^2}{N}\begin{bmatrix} N & -1 & \dots & -1 \\ -1 & N & \dots & -1 \\ \vdots & \vdots & \vdots & \vdots \\ -1 & -1 & \dots & N \end{bmatrix} \tag{7.51}$$

This matrix can easily be inverted, taking the following form:

$$\hat{\mathbf{R}}_{uu}^{-1} = \frac{N}{A^2(N+1)} \begin{bmatrix} 2 & 1 & \dots & 1 \\ 1 & 2 & \dots & 1 \\ \vdots & \vdots & \vdots & \vdots \\ 1 & 1 & \dots & 2 \end{bmatrix} \tag{7.52}$$

One can verify that the right-hand side of Equation (7.52) is the inverse of $\hat{\mathbf{R}}_{uu}$ by multiplying this by the right-hand side of Equation (7.51) and showing that the result yields the identity matrix.

Then, applying Equation (7.22), we can obtain the impulse response vector:

$$\mathbf{h} = \frac{N}{A^2(N+1)T} \begin{bmatrix} 2 & 1 & \dots & 1 \\ 1 & 2 & \dots & 1 \\ \vdots & \vdots & \vdots & \vdots \\ 1 & 1 & \dots & 2 \end{bmatrix} \begin{bmatrix} \hat{R}_{uy}(0) \\ \hat{R}_{uy}(T) \\ \vdots \\ \hat{R}_{uy}((N-1)T) \end{bmatrix} \tag{7.53}$$

By evaluating the right-hand side of Equation (7.53), we can decompose the above matrix equation into the following set of equations:

$$h(kT) = \frac{N}{A^2(N+1)T} \left( \hat{R}_{uy}(kT) + \sum_{i=0}^{N-1} \hat{R}_{uy}(iT) \right), \qquad k = 0,\ 1, \dots,\ N-1 \tag{7.54}$$

The expression for $h(kT)$ in Equation (7.54) makes it necessary only to compute the cross-correlation between the input and output sequences. Explicit matrix inversion is thereby averted. However, a serious practical limitation of Equation (7.54) is that the errors associated with the estimates of the impulse response can be unacceptably large if the input and output measurements are very noisy, since $N$ values of $h(kT)$ have to be estimated from $N$ pairs of input–output data (see Section 7.2.2). A good example of the application of this technique to the identification of a physiological system is given in Sohrab and Yamashiro (1980).

## 7.4 IDENTIFICATION OF CLOSED-LOOP SYSTEMS: "OPENING THE LOOP"

The system identification methods discussed in Section 7.3 were based implicitly on the assumption of an open-loop system: the stimulus (input) to the system was assumed to be unaffected by the response (output). However, since most physiological control processes operate under closed-loop conditions, researchers have applied a variety of techniques to "open the loop" by isolating the components of interest from other components that comprise the entire system. In some cases, "opening the loop" has meant literally that: the subsystem of interest was surgically separated from the rest of the system. Denervation, ablation of certain focal areas, and the redirection of blood flow have become standard techniques in physiological investigations. Another group of methods have been less invasive, involving the use of pharmacological agents to minimize or eliminate potentially confounding influences while the component of interest is studied. A third class of techniques apply clever, noninvasive experimental manipulations to the intact physiological system in order to open the loop functionally rather than physically or pharmacologically. In this chapter, we will review several classic examples that represent the wide spectrum of these methods.

### 7.4.1 The Starling Heart–Lung Preparation

In Section 3.5, we discussed a simple closed-loop model of cardiac output regulation, consisting of essentially two major subsystems, one comprising of the heart and pulmonary circulation, and the other representing the systemic circulation. The now legendary experiments by Patterson, Piper, and Starling (1914) provided the first systematic characterization of the former subsystem, thereby enabling the measurement of the intrinsic response of the heart to changes in venous return and arterial blood pressure. As illustrated schematically in Figure 7.9, the heart and lungs were surgically isolated from the rest of the systemic circulation. By connecting the right atrium to a reservoir of blood placed above it and controlling the flow of blood from the reservoir to the heart, the researchers were able to artificially vary the right atrial pressure. Blood ejected from the left ventricle was led to an arterial capacitance and then through an adjustable resistance ("Starling resistor") back to the venous reservoir after being heated to body temperature. Adjustment of the Starling resistor or the vertical position of the arterial capacitance allowed the researchers to control arterial (or aortic) pressure. In this way,

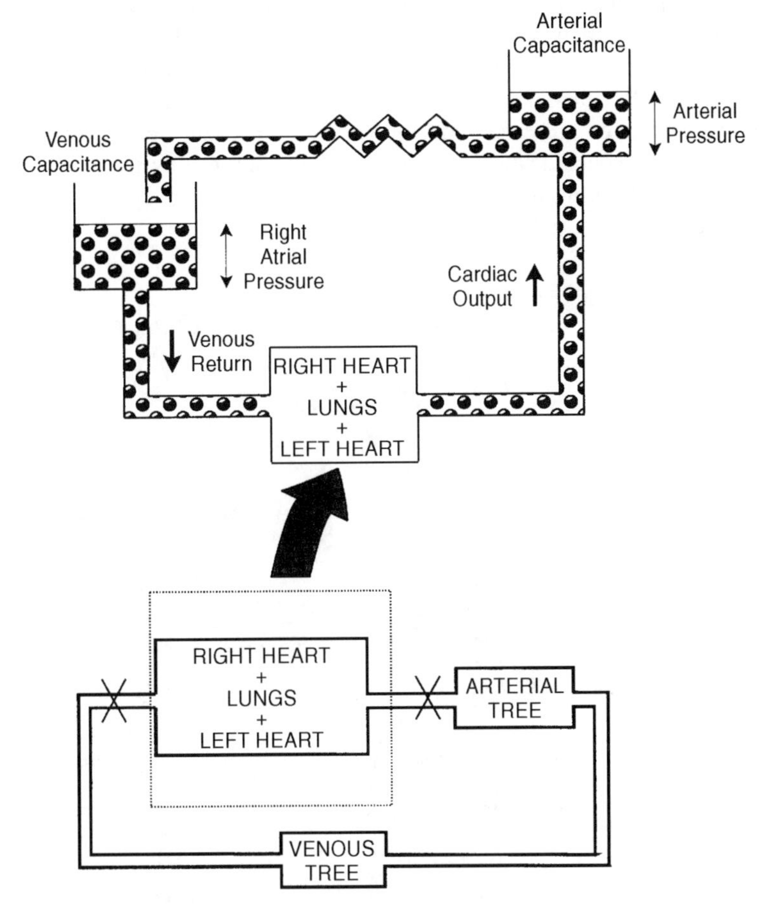


**Figure 7.9** Schematic illustration of the way in which Starling and coworkers "opened the loop" to study the control of cardiac output.

systematic changes in right atrial pressure and arterial pressure were related to the corresponding cardiac output. These data formed the basis of Guyton's cardiac function curves (see Section 3.5.1).

### 7.4.2 Kao's Cross-Circulation Experiments

Kao and Ray (1954) performed experiments on anesthetized dogs to determine whether the increase in cardiac output observed during exercise was due to neural or humoral (blood-borne) factors. In order to separate the neural from humoral effects, their experiments were designed in the following way. In each experiment, two anesthetized dogs were used. The hind-limbs of the "neural dog" were stimulated electrically so that muscular work was induced. However, arterial blood perfusing the hind-limbs of this dog came from the second dog, and venous blood leaving the limbs was directed back to the "humoral dog." The basic experimental design is displayed in Figure 7.10. The authors hypothesized that: (1) if the exercise-induced cardiac output increase was due solely to neural feedback from the exercising limbs, the "neural dog" would continue to show this increase while the "humoral dog" should not respond at all; and (2) if the exercise-induced cardiac output increase was due solely to humoral factors, the "humoral dog" should show this increase, while there should be no response in the "neural dog." Based on the results of nine pairs of these animals, it was found that cardiac output increased significantly in both "neural" and "humoral" dogs. This led the authors to conclude that both neural and humoral factors are involved in the regulation of cardiac output during muscular activity.

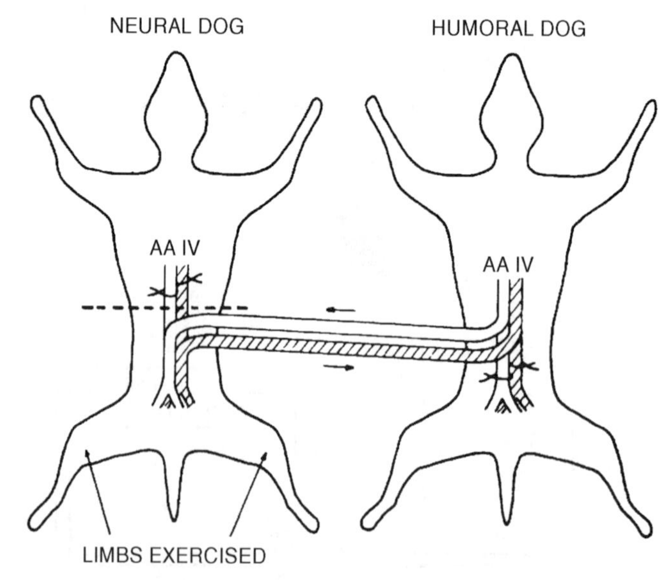


**Figure 7.10** Kao's experimental design for separating neuromuscular feedback from humoral effects on exercise-induced hyperpnea. Reproduced from Kao and Ray (1954).

### 7.4.3 Artificial Brain Perfusion for Partitioning Central and Peripheral Chemoreflexes

In Section 6.7, we examined the stability properties of a model of the chemoreflex regulation of ventilation. The analysis showed that the gains and time constants associated with the central and peripheral chemoreflexes are important determinants of respiratory stability. The question of being able to measure the dynamics of these two chemoreflexes in isolation from one another was addressed by Berkenbosch and colleagues (1979) in a series of experiments that employed the clever technique of artificial brain perfusion. This method is illustrated schematically in Figure 7.11. In anesthetized cats, the researchers directed blood from one of the femoral arteries through an extracorporeal circuit in which the blood was equilibrated in a foamer with a gas mixture of known composition, defoamed, and then returned to the cat through a cannulated vertebral artery. The other vertebral artery was clamped, so that the brain was perfused only by the blood leaving the extracorporeal circuit. This allowed the $P_{CO_2}$, $P_{O_2}$, and pH of the blood perfusing the medullary chemosensitive regions to be maintained at constant levels set by the researchers. This effectively "opened" the central chemoreflex loop. Consequently, the effects of dynamic changes in arterial $P_{CO_2}$ or $P_{O_2}$ (produced by inhalation of hypercapnic or hypoxic gas mixtures) on the peripheral chemoreflex contribution to ventilation could be measured in isolation from the central contribution.

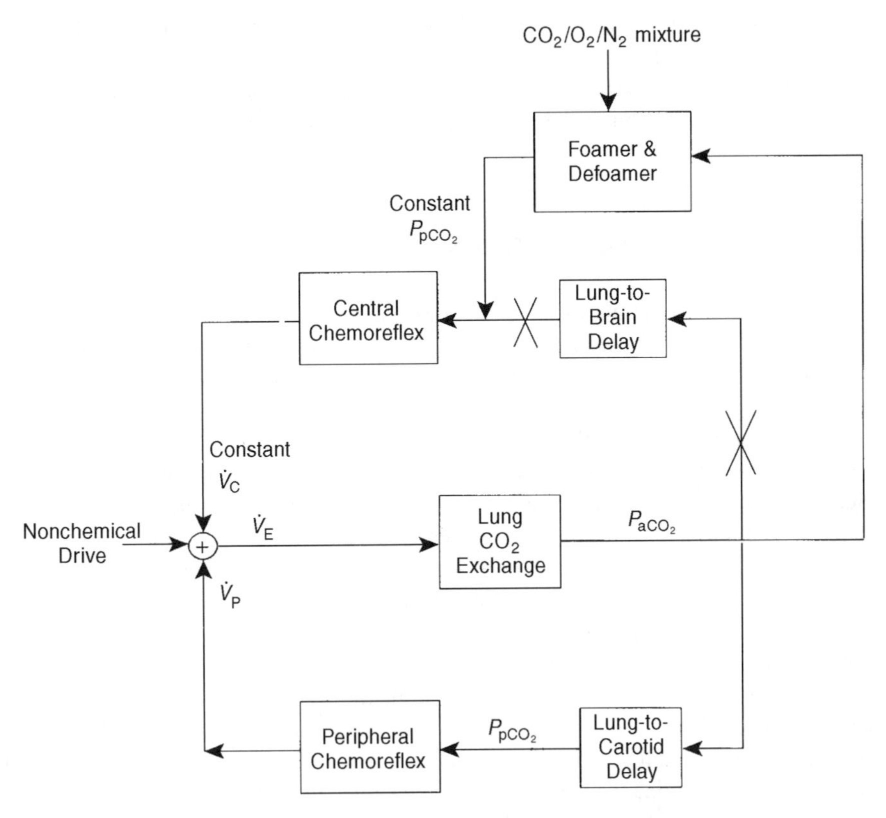


**Figure 7.11** Schematic representation of the artificial brain perfusion setup for separating central and peripheral chemoreflex drives.

### 7.4.4 The Voltage Clamp

The basic mechanism for the neuronal action potential is a classic example of a physiological process in which both negative and positive feedback occur. Consider the Hodgkin–Huxley model shown in circuit form in Figure 7.12a and block diagram form in Figure 7.12b; the model has been simplified here to exclude the leakage channel due to the chloride ions. Normally, potassium ions ($K^+$) tend to leak out of the nerve cell because of the much larger $K^+$ concentration in the axoplasm relative to the extracellular fluid. The opposite occurs with the sodium ions ($Na^+$). When the membrane is depolarized by a presynaptic stimulus, the variable $Na^+$ conductance increases rapidly and considerably, allowing a large influx of $Na^+$ ions from the extracellular fluid, which depolarizes the cell membrane even further. Thus, the positive feedback dominates this initial phase of the action potential. Fortunately, the increase in $Na^+$ conductance is short-lived and the influx of $Na^+$ ions slows after a fraction of a millisecond. At the same time, the $K^+$ conductance starts to increase, following a short delay. This allows $K^+$ ions to flow out of the axoplasm, acting to reverse the depolarization of the membrane. The repolarization speeds up the decline in $Na^+$ conductance which, in turn, promotes the repolarization process (Figure 7.12b). Through the insertion of an electrode into the axoplasm, it is possible to control precisely the voltage inside the nerve cell. By applying a step depolarization through this electrode and keeping the applied voltage constant, one is effectively "opening" both the positive and negative feedback loops. Then, by measuring the current flowing across the membrane and by altering the composition of the extracellular fluid to isolate the $Na^+$ from $K^+$ effects, it is possible to deduce the time-courses of the $Na^+$ and $K^+$ conductances to the step depolarization. This was the basic methodology employed by Hodgkin, Huxley, and Katz (1952), as well as researchers after them, to study the mechanisms underlying the generation of the action potential.

### 7.4.5 Opening the Pupillary Reflex Loop

The model of the pupillary light reflex that we employed in Section 6.6 to demonstrate stability analysis was based largely on Stark's ingenious experiments in which he developed techniques to functionally open the reflex loop. The two basic means by which this was done are illustrated in Figure 7.13. In the normal closed-loop state, an increase in total light flux impinging on the retina results in a reduction in pupil area which, assuming the light intensity remains constant, would decrease total light flux to offset the initial increase. In Figure 7.13a, Stark used a pupillometer to measure the size of the pupil and an adjustable light source that delivered light at intensities that were inversely proportional to the pupil area. By introducing these devices into the feedback loop, he was able to offset the effect of changing pupil area on total light flux by raising the light intensity. In this way, the total light flux could be controlled quite precisely, enabling him to deduce the loop transfer function characteristics of the reflex. Another technique that he used to effectively "open the loop" is illustrated in Figure 7.13b. Here, he applied a very narrow beam of light through the pupil. By restricting the cross-sectional area of the beam to a size that was smaller than the residual area of the pupil, total light flux was rendered completely independent of pupil area, since the area of the light beam impinging on the retina was not affected by changes in pupil size. Under these conditions, it was possible to completely control the time-course of the input (total light flux) and measure the corresponding response (pupil area) of the "opened" reflex loop.

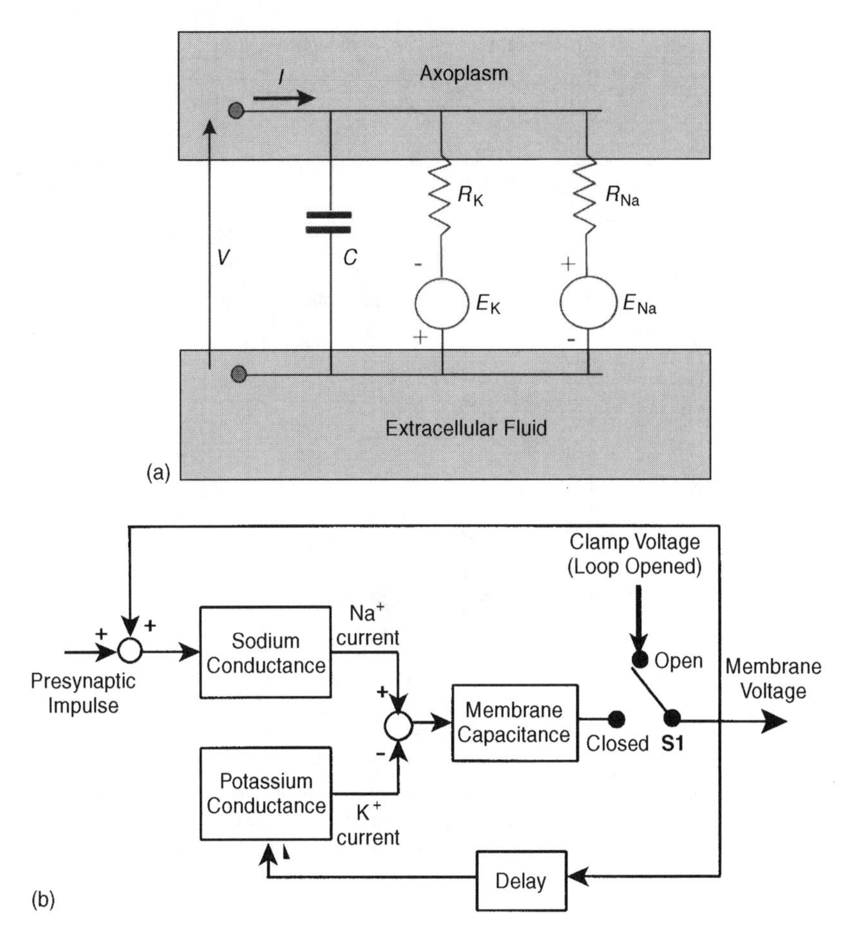


**Figure 7.12** (a) Simplified schematic of the Hodgkin–Huxley model. (b) "Opening the loop" via application of the voltage clamp technique.

### 7.4.6 Read Rebreathing Technique

Under normal operating circumstances, ventilation ($\dot{V}_E$) and arterial $P_{CO_2}$ ($P_{aCO_2}$) are tightly coupled through the powerful negative feedback loops of the chemoreflexes: any increases in $P_{aCO_2}$ lead rapidly to increases in $\dot{V}_E$ that act to offset the initial rise in $P_{aCO_2}$. However, Read (1967) found a simple experimental technique of functionally breaking this closed-loop relationship. The subject breathes into and out of a small (4–6 liters) rebreathing bag which is filled with an initial gas mixture containing 7% $CO_2$ in oxygen. After an initial transient phase, an equilibrium is established between arterial blood, oxygenated mixed venous blood, and gas in the lungs and rebreathing bag. Thereafter, the $P_{CO_2}$ in both blood and gas phases increases linearly with time, and $\dot{V}_E$ also increases proportionally, without reversing the rise in $P_{CO_2}$ as one would expect in the closed-loop situation. The way in which

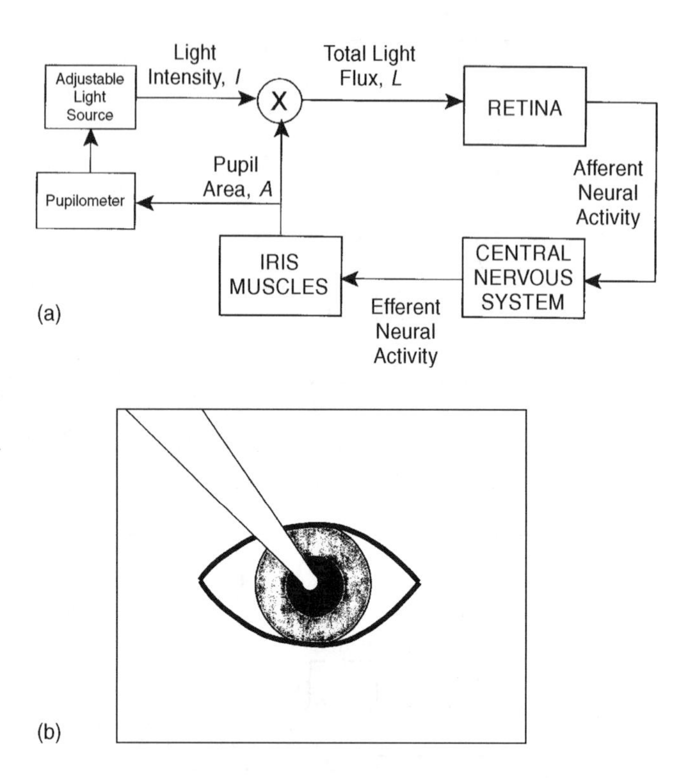


**Figure 7.13** Two methods of functionally opening the pupillary reflex loop: (a) modulation of applied light intensity using measurements of pupil area; (b) application of narrow light beam to residual area of pupil.

the technique works is best appreciated from a modeling perspective. The following differential equation provides the simplest dynamic characterization of $CO_2$ exchange at the level of the body tissues:

$$\frac{V_t}{863}\frac{dP_{vCO_2}}{dt} = QK_{CO_2}(P_{aCO_2} - P_{vCO_2}) + \dot{V}_{CO_2} \tag{7.55}$$

Equation (7.55) assumes the capacitance effect of the body tissues to be lumped into the volume $V_t$. $\dot{V}_{CO_2}$ is the metabolic production rate of $CO_2$. Following the establishment of the equilibrium between the arterial and mixed venous blood and gas in the lungs and bag, it can be seen that in Equation (7.55) the arteriovenous gradient disappears and the derivative becomes a constant proportional to the $CO_2$ metabolic production rate. Integrating Equation (7.55) results in $P_{vCO_2}$ assuming a linear dependence on time. Since $P_{aCO_2}$, alveolar $P_{CO_2}$ ($P_{ACO_2}$) and the bag $P_{CO_2}$ are equilibrated with $P_{vCO_2}$, these variables also increase linearly with time during the rest of the rebreathing process. With the linearly rising arterial and tissue $P_{CO_2}$, brain tissue $P_{CO_2}$ will also increase in linear fashion, driving $\dot{V}_E$ along a similar time-course. The increasing $\dot{V}_E$ is mediated almost completely by the central chemoreceptors, since the high oxygenation levels suppress peripheral chemoreception. However, because of the equilibration between the bag (inspired) $P_{CO_2}$ and $P_{ACO_2}$ (see Equation (6.42) in Section 6.7), the increasing $\dot{V}_E$ is prevented from influencing $P_{ACO_2}$, hence breaking the negative

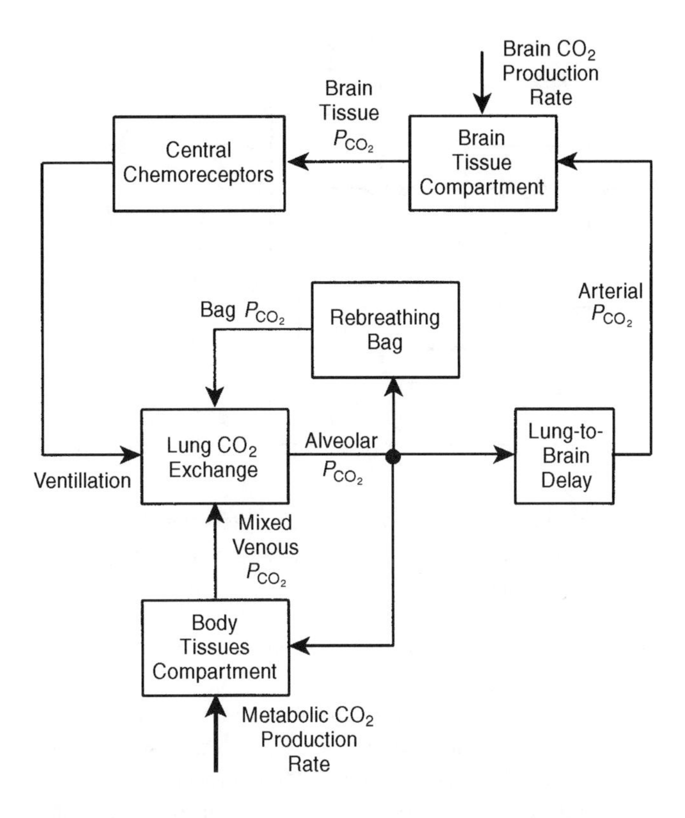


**Figure 7.14** Schematic block diagram of respiratory control during rebreathing.

feedback in this closed-loop system. A schematic block diagram of the rebreathing model is shown in Figure 7.14.

## 7.5 IDENTIFICATION UNDER CLOSED-LOOP CONDITIONS: CASE STUDIES

Although we have seen a wide range of physiological examples in which experimental interventions were employed to "open the loop," such techniques are not always applicable. Moreover, a major criticism leveled against this kind of approach is that the system under study is placed under nonphysiological conditions and subjected to nonphysiological inputs when these interventions are applied. Ideally, we would like to identify the physiological system under "normal operating conditions" when its feedback loops are functionally intact. However, consider the problem involved with identifying the impulse response $h(t)$ of the closed-loop system component shown in Figure 7.15. The unknown (and unobservable) disturbance, $u(t)$, that enters the closed loop represents both a "measurement" and "process" noise input. It is considered "measurement noise" since it corrupts the measurements $x(t)$, which otherwise would be related solely to $y(t)$, the input to the system component. This is clear from the mathematical expression relating $x(t)$ to $y(t)$ and $u(t)$:

$$x(t) = \int_0^{\infty} h(\tau)y(t-\tau)\, dt + u(t) \tag{7.56}$$

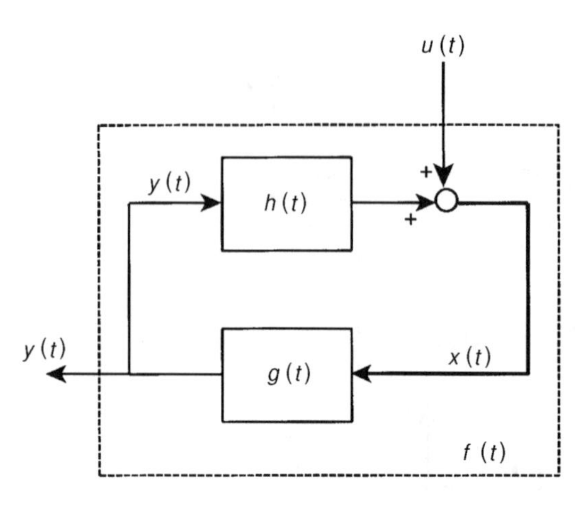


**Figure 7.15** Correlation of the process noise input, $u(t)$, with $y(t)$ complicates the identification $h(t)$ from closed-loop measurements, $y(t)$ and $x(t)$. $f(t)$ is the impulse response of the closed-loop system.

$u(t)$ also takes the form of "process noise" since it enters the closed-loop system and becomes correlated with $y(t)$. If we consider $u(t)$ the input and $y(t)$ the output of the overall system (as defined by the dashed rectangle in Figure 7.15), we obtain

$$y(t) = \int_0^\infty f(\tau)u(t-\tau)\, d\tau \tag{7.57}$$

In order to obtain an unbiased estimate of $h(t)$ from Equation (7.56) by the least squares approach using $y(t)$ as input and $x(t)$ as output, the final solution must be such that $u(t)$ becomes uncorrelated with (or orthogonal to) $y(t)$. However, as Equation (7.57) clearly shows, $y(t)$ is correlated with $u(t)$. Thus, the direct application of open-loop system identification methods to this problem will not yield accurate estimates of $h(t)$. A couple of approaches for circumventing this problem are described in the following two subsections. One other approach is to impose constraints on the orthogonality condition, but this falls outside the scope of the present discussion. For further information about this last method, the interested reader is referred to a study by Khoo (1989).

### 7.5.1 Minimal Model of Blood Glucose Regulation

One effective way of partitioning the effects of the feedforward and feedback components of a closed-loop system from one another is to assume a model structure for at least one of these components. Then, if the effects of all other extraneous influences (process noise) entering the closed-loop system are small relative to the magnitude of the system responses, the parameters of the assumed model can be estimated. The "minimal model" of blood glucose regulation, developed by Bergman and colleagues (1979), represents a good example of this kind of approach. Referring to Figure 7.15, suppose that $x(t)$ and $y(t)$ correspond to the plasma glucose and insulin concentrations at time $t$, respectively. Then, the impulse response function, $h(t)$, would represent glucose regulation kinetics, while $g(t)$ would reflect the dynamics of insulin production and utilization. The closed-loop system is perturbed by an impulsive input $u(t)$, consisting of an intravenous injection ($300\,\mathrm{mg\,kg^{-1}}$ in dogs) of glucose. By using the resulting time-courses in $y(t)$ and $x(t)$ as *input* and *output*, respectively, the model of *glucose dynamics* can be identified. Subsequently, by using $x(t)$ as *input* and $y(t)$ as *output*, the parameters of the model of *insulin dynamics* can be estimated. Bergman and coworkers have referred to this methodology as *partition analysis*, since both halves of the closed-loop system are identified as if they were in the open-loop state. It should be emphasized that the key assumptions that make this kind of closed-loop estimation possible are: (a) the imposition of structure and causality on the dynamics characterizing glucose and insulin production and utilization; and (b) relatively large signal-to-noise ratios in the measurements.

In this section, we will discuss only the estimation of the minimal model of glucose regulation, i.e., how insulin affects glucose. The estimation of the converse model in which glucose affects insulin will not be considered. Thus, the input here is the measured plasma insulin concentration $y(t)$ following the intravenous glucose injection, while the output is the corresponding measured blood glucose concentration $x(t)$. The model employed by Bergman contains the features incorporated in the glucose kinetics model proposed by Stolwijk and Hardy (see Sections 3.6 and 5.5), but is more realistic in that it allows for the delayed effect of insulin on glucose disappearance, a feature that has been observed. The insulin concentration, $y(t)$, does not affect glucose dynamics directly. Instead, it acts through a "remote compartment," so that the *effective* insulin concentration, $y_{\mathrm{eff}}(t)$ is given by

$$\frac{dy_{\mathrm{eff}}}{dt} = k_2 y(t) - k_3 y_{\mathrm{eff}}(t) \tag{7.58a}$$

where $k_2$ and $k_3$ represent the fractional rate parameters for insulin transport into and elimination from the remote compartment. This compartment is "remote" in that $y_{\mathrm{eff}}$ is not directly measurable. It should also be noted that the volume of the remote compartment has been factored into the rate constants $k_2$ and $k_3$. The rate of change of glucose in the blood plasma is given by

$$\frac{dx}{dt} = \begin{matrix}\text{Net rate of glucose}\\ \text{production by the liver}\end{matrix} - \begin{matrix}\text{Rate of glucose utilization}\\ \text{by other tissues}\end{matrix} \tag{7.59a}$$

where

$$\text{Net rate of glucose production by the liver} = B_0 - k_5 x(t) - k_6 y_{\mathrm{eff}}(t) x(t) \tag{7.60}$$

and

$$\text{Rate of glucose utilization by other tissues} = R_{d0} + k_1 x(t) + k_4 y_{\mathrm{eff}}(t) x(t) \tag{7.61}$$

In Equation (7.60), $B_0$ represents the rate of glucose production by the liver. The rate of glucose uptake by the liver is assumed to be proportional to an insulin-independent component (through rate constant $k_5$) and an insulin-dependent component (through rate constant $k_6$). Similarly, in Equation (7.61), the rate of glucose utilization by nonhepatic tissues is assumed to have a constant component, a component proportional to glucose concentration and a component sensitive to both glucose and effective insulin concentration. Substituting Equations (7.60) and (7.61) into Equation (7.59a) and rearranging terms, we obtain the result:

$$\frac{dx}{dt} = [B_0 - R_{\mathrm{d0}}] - [k_5 + k_1]x(t) - [k_6 + k_4] y_{\mathrm{eff}}(t) x(t) \tag{7.59b}$$

As in Equation (7.58a), the effective plasma glucose capacitance is factored into the parameters on the right-hand side of Equation (7.59b).

Equations (7.58a) and (7.59b) provide a complete characterization of glucose kinetics. However, it is obvious that there are too many redundant parameters. For instance, in Equation (7.59b), it would not be possible to estimate $B_0$ and $R_{\mathrm{d0}}$ separately; only the combined term $[B_0 - R_{\mathrm{d0}}]$ can be identified. The same is true for $[k_5 + k_1]$ and $[k_6 + k_4]$. In addition, since $y_{\text{eff}}(t)$ is not measurable, a further reduction in parametrization can be achieved by defining the new variable $z(t)$ that is proportional to $y_{\text{eff}}(t)$:

$$z(t) \equiv [k_6 + k_4]y_{\text{eff}}(t) \tag{7.62}$$

Substituting Equation (7.62) into Equations (7.58a) and (7.59b) we obtain

$$\frac{dz}{dt} = -p_2 z(t) + p_3 y(t) \tag{7.58b}$$

and

$$\frac{dx}{dt} = p_4 - p_1 x(t) - z(t)x(t) \tag{7.59c}$$

where $p_1 = k_1 + k_5$, $p_2 = k_3$, $p_3 = k_2(k_4 + k_6)$, and $p_4 = B_0 - R_{d0}$. Equations (7.58b) and (7.59c) provide the same dynamic characterization of glucose regulation for the minimum number of unknown parameters that have to be estimated from the input–output data. For this reason, it is referred to as a *minimal model.*

The way in which the unknown parameters $p_1$, $p_2$, $p_3$, and $p_4$ are estimated is as follows. First, we begin with initial guesses for the unknown parameters. Using the measured input time-course, $y(t)$, and the initial parameter values, Equation (7.58b) is first solved to obtain the value of $z$ at the current time-step. Using this value of $z$ in Equation (7.59c) and integrating this equation, the glucose concentration at the next time step can be computed. This process is repeated until predictions for $x(t)$ have been made for the entire duration of the experiment. The predictions are compared to the actual blood glucose measurements, and the value of the criterion function (sum of squares of the differences between measured and predicted glucose values) is computed. An optimization algorithm is used to search for another combination of the four unknown parameters that would produce a lower value of the criterion function. Using the new combination of parameter values, $z(t)$ and $x(t)$ are again solved using Equations (7.58b) and (7.59c), and the whole process is repeated until the incremental reduction in criterion function is considered insignificant.

An example of the results achieved with minimal model estimation is displayed in Figure 7.16. "Data" required for the estimation were generated using a SIMULINK implementation (named "`gmm_sim.mdl`") of Bergman's models of both glucose and insulin subsystems (Bergman et al., 1979; Bergman et al., 1985; Toffolo et al., 1980). These were combined and made to operate in closed-loop mode. Random perturbations were added to the glucose concentration, $x(t)$, predicted by the model to simulate "measurement noise" in the glucose observations. The SIMULINK model, shown in Figure 7.17, produced samples of $x(t)$ and the plasma insulin concentration, $y(t)$, at intervals of 1 minute to mimic the blood sampling conducted in the real experiments. These "measurements" are shown as the solid circles in both upper and lower panels of Figure 7.16. This particular set of "measurements" has also been saved in the MATLAB data file: "`data_gmm.mat`". Parameter estimation is performed using the Nelder–Mead simplex algorithm, which is implemented in MATLAB with the function "`fmins`". The primary command lines of the MATLAB script file, labeled "`gmm_est.m`", are:

```
>> options(1)=1;
>> [p,options] = fmins('fn_gmm',p_init,options,[]);
```

where the function "`fn_gmm`" (in MATLAB file "`fn_gmm.m`") is used by "`gmm_est.m`" to produce values of the criterion function $J$ with each iteration of the algorithm. For each new set of parameter values, "`fn_gmm`" solves the model equations given in Equations (7.58b) and (7.59c) using the Euler method of integration (with time-steps of 0.01 minute) and computes the sum of squares of the differences between the "observed" glucose concentration samples and the values predicted by model solution. The parameters to be estimated are the four unknown coefficients in Equations (7.58b) and (7.59c): $p_1$, $p_2$, $p_3$, and $p_4$; in addition, the "true" glucose concentration at time zero, $x(0)$, is treated as the fifth unknown parameter. In the above MATLAB command lines, note that the array "`options`" is used as both an input and output argument. This is done (specifically, the first element of "`options`" is set equal to "1") to allow the algorithm to display the value of $J$, along with the parameter values associated with the simplex vertices, at each stage of the computations. The final estimated set of parameter values are $p_1 = 0.068$, $p_2 = 0.091$, $p_3 = 6.72 \times 10^{-5}$, $p_4 = 6.03$ and $x(0) = 284.9$ mg/100 ml. These may be compared with the "true" parameter values used in the SIMULINK program: $p_1 = 0.049$, $p_2 = 0.091$, $p_3 = 8.96 \times 10^{-5}$, and $p_4 = 4.42$. These latter values were selected from the results obtained by Bergman and coworkers from their experiments on dogs. The "best-fit" model prediction is shown as the solid curve in Figure 7.16.

The minimal model has been employed successfully in many clinical studies to quantitate insulin sensitivity and glucose effectiveness in various populations at risk for diabetes. More details on the relationship between these indices and the model parameters may be found in the original papers by Bergman and coworkers listed in the Bibliography section of this chapter.

### 7.5.2 Closed-Loop Identification of the Respiratory Control System

In this section, we illustrate a somewhat different approach to closed-loop identification. In the previous example, an optimization technique was employed for parameter estimation. As we mentioned earlier, one disadvantage of this kind of iterative method is the possibility of convergence to a local minimum instead of the global solution. Here, we take the alternative approach of least squares estimation, where the optimal solution is arrived at in one computational step. Another difference that we will highlight here is the use of a *persistently* exciting input to stimulate the closed-loop system, instead of the brief but potent impulsive disturbance employed in the minimal model of glucose regulation. Practical considerations dictate the use of the former type of input in the case of the respiratory control system. A potent impulsive disturbance in this case would take the form of an inhaled breath of gas with very high $CO_2$ content. Such a potent stimulus would be certain to evoke a behavioral response in addition to the chemoreflex-mediated changes, thereby allowing the measurement process to affect the system under observation. In this case, the stimulus takes the form of a pseudorandom binary sequence (PRBS) in the inhaled $P_{CO_2}$. This allows the system to be excited with relatively low $CO_2$ concentrations over a broad range of frequencies within the limited experimental duration. An example of the practical implementation of this kind of PRBS time-course in inhaled $P_{CO_2}$ ($P_{ICO_2}$) and the resulting effects on alveolar $P_{CO_2}$ ($P_{ACO_2}$) and ventilation ($\dot{V}_E$) in a normal human subject is displayed in Figure 7.18. As in the previous section, partition analysis is employed in the identification procedure. The first stage of the analysis involves the estimation of the parameters of the plant (i.e., gas exchange in the

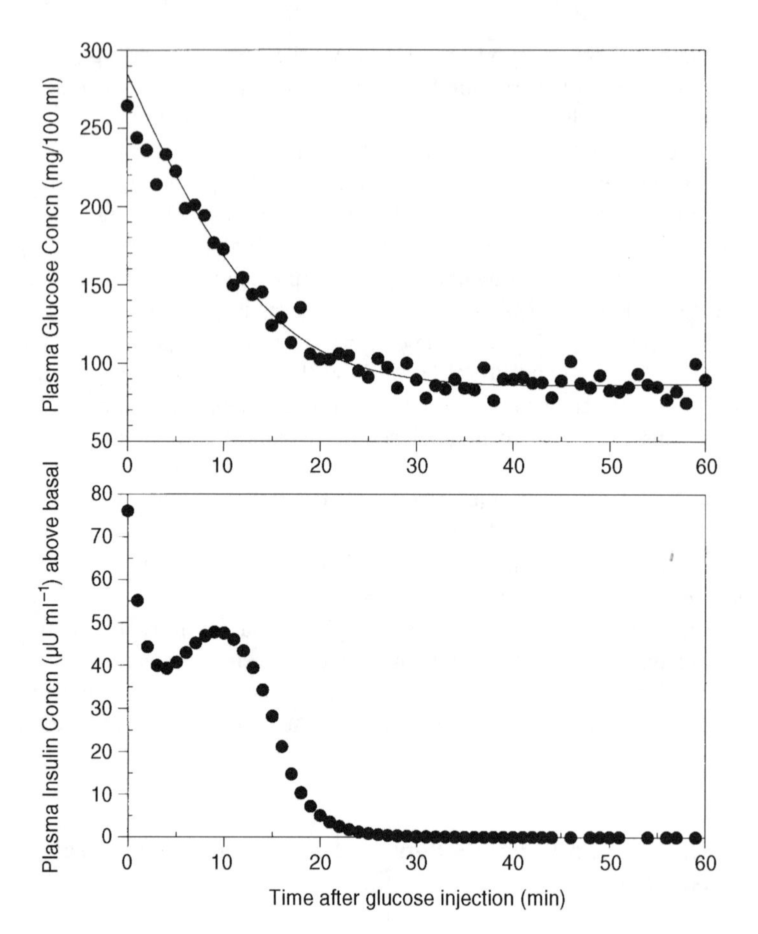


**Figure 7.16** "Measurements" of blood glucose and insulin levels (solid circles) following intravenous bolus infusion of 300 mg kg$^{-1}$ of glucose. The best-fit prediction for glucose is shown as the solid curve.

lungs), using measurements of $\dot{V}_E$ and $P_{ICO_2}$ as inputs, and $P_{ACO_2}$ as output. The second stage consists of the estimation of the controller and lung-to-chemoreceptor delay using $P_{ACO_2}$ as the input and $\dot{V}_E$ as the output (Figure 7.19).

*7.5.2.1 **Identification of the plant.*** The model employed to represent the $CO_2$ exchange in the lungs is the small-signal expression derived in Equation (6.43b) in Section 6.7.1. However, in the expression shown below, we have also allowed for perturbations in $P_{ICO_2}$ (which, in Equation (6.43b), was kept constant)

$$\tau_{\text{lung}} \frac{d(\Delta P_{ACO_2})}{dt} + \Delta P_{ACO_2} = G_1\, \Delta P_{ICO_2} - G_2 \Delta \dot{V}_E \tag{7.63}$$

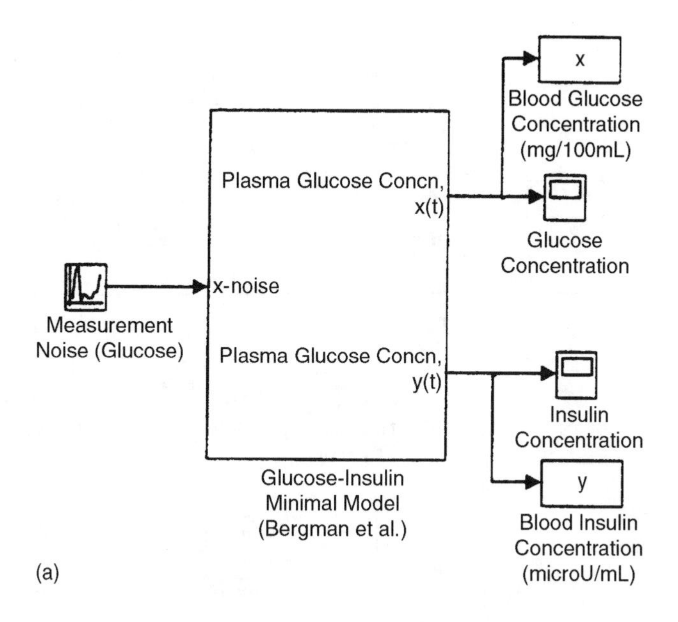


**Figure 7.17** SIMULINK model "`gmm_sim.mdl`" of combined glucose–insulin kinetics based on Bergman's minimal models: (a) model input and outputs.

where $\Delta P_{ACO_2}$, $\Delta \dot{V}_E$ and $\Delta P_{ICO_2}$ represent small changes in $P_{ACO_2}$ (assumed equal to $P_{aCO_2}$), $\dot{V}_E$, and $P_{ICO_2}$ about their equilibrium values, and

$$G_1 = \frac{\dot{V}_E - \dot{V}_D}{\dot{V}_E - \dot{V}_D + 863QK_{CO_2}} \tag{7.64}$$

$$G_2 = \frac{P_{ACO_2} - P_{ICO_2}}{\dot{V}_E - \dot{V}_D + 863QK_{CO_2}} \tag{7.65}$$

Since, in this analysis, we are limiting our attention to the characterization of how small *changes* in $P_{ICO_2}$ elicit *changes* in $P_{ACO_2}$ and $\dot{V}_E$, we assume, to a first approximation, that the operating values of $P_{ACO_2}$, $\dot{V}_E$, $\dot{V}_D$, and $P_{ICO_2}$ are constant. Hence, we regard the two factors $G_1$ and $G_2$ to be constant-valued parameters, which have to be estimated from the measurements, as we demonstrate below.

Since the measurements of $P_{ACO_2}$, $P_{ICO_2}$ and $\dot{V}_E$ are not made continuously in time but are obtained on a breath-by-breath basis, for purposes of parameter estimation it is more useful to express the plant equation in the form of a difference equation with a discrete-time base (with "breaths" as the unit of time). Furthermore, $P_{ACO_2}$ cannot be directly sampled; instead, we assume that end-tidal $P_{CO_2}$ (the highest value of $P_{CO_2}$ measured in the exhaled stream during expiration) reliably reflects $P_{ACO_2}$. By integrating Equation (7.63) from the end of the previous breath to the end of the current breath, the differential equation can be converted into a difference of the following form:

$$\Delta P_{ACO_2}(n) + \alpha\, \Delta P_{ACO_2}(n-1) = \beta_1\, \Delta P_{ICO_2}(n) - \beta_2\, \Delta \dot{V}_E(n) + e(n) \tag{7.66}$$

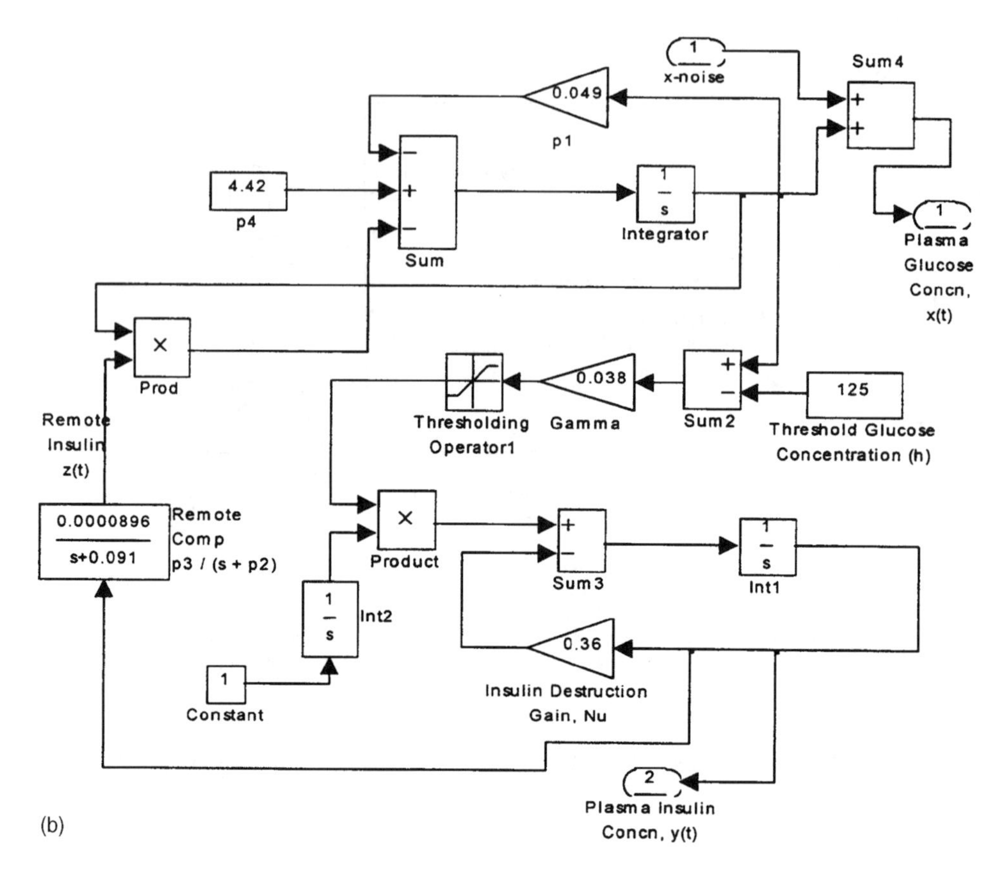


**Figure 7.17** SIMULINK model "gmm_sim.mdl" of combined glucose–insulin kinetics based on Bergman's minimal models: (b) details of the SIMULINK implementation.

where $n$ represents the current breath number, and $0 \le n \le N-1$, $N$ being the total number of breaths used for data analysis. This type of continuous-time to discrete-time conversion is also known as the *impulse invariance* method; further details on this method may be found in most texts on signal processing, e.g., Jackson (1995). It can be further shown that $\alpha$, $\beta_1$, and $\beta_2$ in Equation (7.66) are related to the parameters $G_1$, $G_2$, and $\tau_{\text{lung}}$ of Equation (7.63) through the following relations:

$$\beta_1 = \frac{G_1}{\tau_{\text{lung}}} \tag{7.67}$$

$$\beta_2 = \frac{G_2}{\tau_{\text{lung}}} \tag{7.68}$$

$$\alpha = -e^{-T/\tau_{\text{lung}}} \tag{7.69}$$

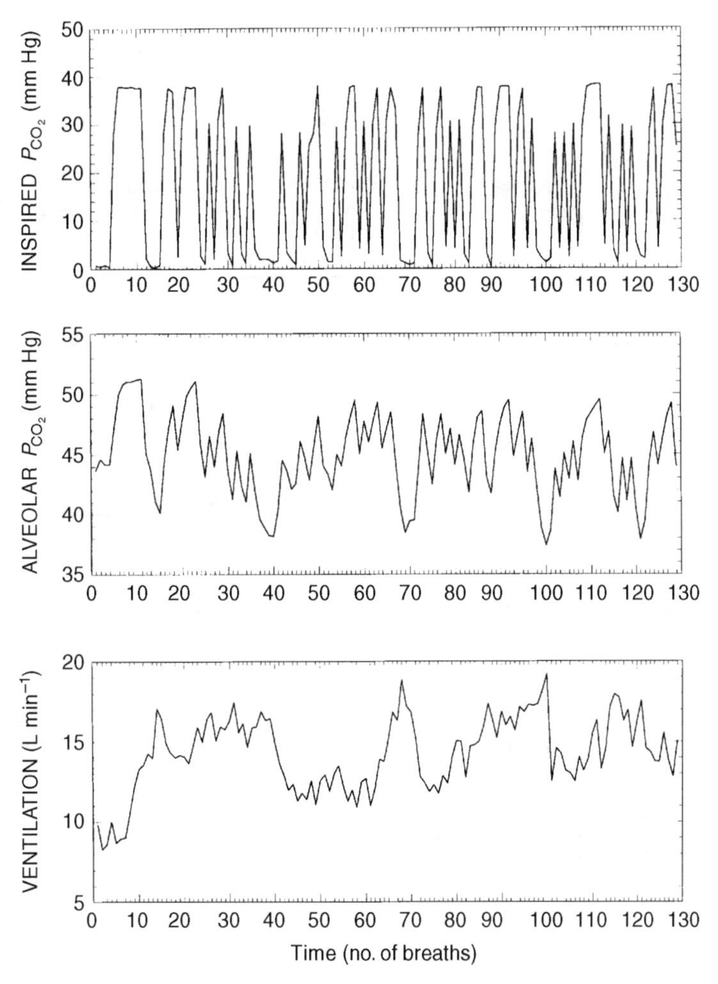


**Figure 7.18** Responses in $P_{ACO_2}$ (middle panel) and ventilation (bottom panel) produced in a normal subject during inhalation of 6% $CO_2$ in air, administered on a pseudorandom binary basis ($P_{ICO_2}$, top panel). Reproduced from Ghazanshahi and Khoo (1997).

where $T$ is the "sampling interval" which, in this case, would be the breath duration. Strictly speaking, $T$ would vary from breath to breath, since the breathing frequency is somewhat variable. However, previous studies in this field have demonstrated that assuming $T$ to be constant and equal to the *average breath duration* simplifies matters considerably without affecting the outcome of the analysis significantly in most experimental situations.

In Equation (7.66), the last term $e(n)$ is added to account for the residual error between the measured $\Delta P_{ACO_2}$ and model-predicted $\Delta P_{ACO_2}$. Equation (7.66) is a special case of the general class of models known as *ARX* (autoregressive with exogenous input) models (Ljung, 1987). In this special case, $\alpha$, $\beta_1$, and $\beta_2$ are the unknown parameters to be estimated using $\Delta P_{ACO_2}$ as the output measurement, and $\Delta P_{ICO_2}$ and $\Delta \dot{V}_E$ as the inputs. Estimation of these

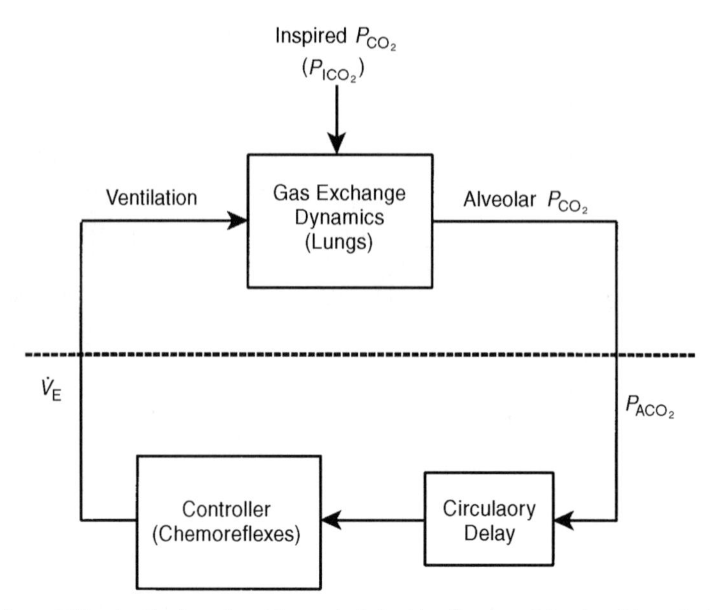


**Figure 7.19** Application of partition analysis for identification of the plant (above dashed line) and controller (below dashed line) portions of the closed loop.

parameters can be easily achieved using least squares minimization, as the following equations illustrate. Rewriting Equation (7.66) for all values of $n$ in vector form, we have

$$\begin{bmatrix} \Delta P_{ACO_2}(0) \\ \Delta P_{ACO_2}(1) \\ \vdots \\ \Delta P_{ACO_2}(N-1) \end{bmatrix} = \begin{bmatrix} 0 & \Delta P_{ICO_2}(0) & -\Delta \dot{V}_E(0) \\ -\Delta P_{ACO_2}(0) & \Delta P_{ICO_2}(1) & -\Delta \dot{V}_E(1) \\ \cdots & \cdots & \cdots \\ \cdots & \cdots & \cdots \\ -\Delta P_{ACO_2}(N-2) & \Delta P_{ICO_2}(N-1) & -\Delta \dot{V}_E(N-1) \end{bmatrix} \begin{bmatrix} \alpha \\ \beta_1 \\ \beta_2 \end{bmatrix} + \begin{bmatrix} e(0) \\ e(1) \\ \vdots \\ e(N-1) \end{bmatrix} \tag{7.70}$$

It can be seen that Equation (7.70) is of the form displayed in Equation (7.8), i.e.,

$$\underline{\mathbf{y}} = \mathbf{U}\underline{\mathbf{h}} + \underline{\mathbf{e}} \tag{7.71}$$

where $\underline{\mathbf{y}}$ represents the column vector on the left-hand side of Equation (7.70), $\mathbf{U}$ is the $N \times 3$ matrix on the other side of the equation, and $\underline{\mathbf{h}}$ is the vector containing the unknown parameters. Thus, $\underline{\mathbf{h}}$ can be estimated using Equation (7.12), which we rewrite below:

$$\underline{\mathbf{h}} = (\mathbf{U}'\mathbf{U})^{-1}\mathbf{U}'\underline{\mathbf{y}} \tag{7.72}$$

***7.5.2.2 Identification of the Controller and Circulatory Delay.*** To model the controller and circulatory delay, we assume the form proposed by Bellville et al. (1979)

(see Section 6.7.3). Using the Laplace transform version of this model, we have (from Equation (6.50a,b)):

$$\Delta\dot{V}_E(s) = \left(\frac{G_c}{\tau_c s + 1} + \frac{G_p}{\tau_p s + 1}\right) e^{-sT_d}\, \Delta P_{ACO_2}(s) \tag{7.73a}$$

Here, we have made the simplifying assumption of using only one common circulatory delay, $T_d$, in place of the separate central ($T_c$) and peripheral ($T_p$) delays assumed in the chemoreflex model of Section 6.7.3. Employing a common denominator for both terms in the summation of Equation (7.73a), we can rewrite the equation in the following form:

$$\Delta\dot{V}_E(s) = \left(\frac{(G_c\tau_p + G_p\tau_c)s + (G_c + G_p)}{\tau_c\tau_p s^2 + (\tau_c + \tau_p)s + 1}\right) e^{-sT_d} \Delta P_{ACO_2}(s) \tag{7.73b}$$

When inverse-Laplace-transformed back into the time-domain, Equation (7.73b) takes the following differential equation form:

$$\tau_c\tau_p \frac{d^2(\Delta\dot{V}_E)}{dt^2} + (\tau_c + \tau_p)\frac{d(\Delta\dot{V}_E)}{dt} + \Delta\dot{V}_E = (G_c\tau_p + G_p\tau_c)\frac{d(\Delta P_{ACO_2}(t - T_d))}{dt} + (G_c + G_p)\, \Delta P_{ACO_2}(t - t_d) \tag{7.74}$$

As in Section 7.5.2.1, since the measurements are made on a breath-by-breath basis, it is more convenient to assume a discrete-time base and recast the model in finite difference form, as in Equation (7.66) for the plant. In this case, the corresponding finite difference equation is

$$\Delta\dot{V}_E(n) + a_1\, \Delta\dot{V}_E(n-1) + a_2\, \Delta\dot{V}_E(n-2) = b_0\, \Delta P_{ACO_2}(n - N_d) + b_1\, \Delta P_{ACO_2}(n - 1 - N_d) + \epsilon(n) \tag{7.75}$$

where, as in Equation (7.66), $n$ represents the current breath number and $0 \leq n \leq N-1$. $N_d$ represents the circulatory delay in number of breaths, i.e., $N_d = T_d/T$. In this case, $\epsilon(n)$ is added to account for the discrepancy between the model-predicted $\Delta\dot{V}_E$ and the measured $\Delta\dot{V}_E$. Equation (7.75) can also be cast in the form

$$\Delta\dot{V}_E = \left[-\Delta\dot{V}_E(n-1) \;\; -\Delta\dot{V}_E(n-2) \;\; \Delta P_{ACO_2}(n - N_d) \;\; \Delta P_{ACO_2}(n - N_d - 1)\right] \begin{bmatrix} a_1 \\ a_2 \\ b_0 \\ b_1 \end{bmatrix} + \epsilon(n) \tag{7.76}$$

By applying Equation (7.76) to all $N$ sets of data points, we can again construct a matrix equation of the form displayed in Equation (7.70), and thus estimate the unknown parameters $a_1$, $a_2$, $b_0$, and $b_1$ using least squares minimization. However, in order to solve for the unknown parameters, it is necessary to know what $N_d$ is. Determination of $N_d$ is done in the following way. We first select a range of physiologically feasible values for $N_d$. The lung-to-ear delay in most normals is generally in the range of 6 to 12 seconds. Thus, a reasonable range for $N_d$ might be 1 to 4. For each of these values of $N_d$, we solve the least squares minimization problem and estimate $a_1$, $a_2$, $b_0$, and $b_1$. For each case, we compute $J$, the residual sum of squares of the differences between the measured and predicted $\Delta\dot{V}_E$ (as given in Equation (7.35)). The "best" estimate of $N_d$ is that value that yields the lowest value of $J$.

Having estimated the unknown parameters, $a_1$, $a_2$, $b_0$, and $b_1$, it is possible in principle to relate them to the gains ($G_c$ and $G_p$) and time constants ($\tau_c$ and $\tau_p$) that characterize the corresponding differential equation (Equation (7.74)) in a way similar to Equations (7.67) through (7.69). However, in this case, the relations will be nonlinear and the latter group of parameters would generally tend to be sensitive to errors in the estimates of $a_1$, $a_2$, $b_0$, and $b_1$. A more robust alternative approach is to characterize the controller dynamics in terms of its corresponding unit impulse response. This can be achieved quite easily by setting $\Delta P_{ACO_2}(0)$ to 1 and $\Delta P_{ACO_2}(n)$ to 0 for all $n > 0$, and computing $\Delta \dot{V}_E$ recursively from Equation (7.75). The error terms, $\epsilon(n)$, are set equal to zero during this computation. Note that this would produce results similar to the PRBS technique described in Section 7.3.3. However, one can expect much less noisy estimates of the impulse response function from the present method, since $h(n)$ in this case would be derived from only four parameters estimated from a large number (128 or higher) of datapoints. Further details on the application of this approach to human respiratory data and the results obtained may be found in journal papers by Khoo et al. (1995) and Ghazanshahi and Khoo (1997). In the problems given at the end of this chapter, the reader will be able to explore this technique in greater detail by applying the accompanying MATLAB script file "`rcs_est.m`" to a number of datasets ("`prbs1.mat`", "`prbs2.mat`", "`prbs3.mat`", "`prbs4.mat`") obtained from human experiments.

## BIBLIOGRAPHY


Bekey, G.A. System identification: an introduction and a survey. *Simulation* **16**: 151–166, 1970.

Bellville, J.W., B.J. Whipp, R.D. Kaufman, G.D. Swanson, K.A. Aqleh, and D.M. Wiberg. Central and peripheral chemoreflex loop gain in normal and carotid body-resected subjects. *J. Appl. Physiol.* **46**: 843–853, 1979.

Bergman, R.N., Y. Ziya Ider, C.R. Bowden, and C. Cobelli. Quantitative estimation of insulin sensitivity. *Am. J. Physiol.* **236**: E667–E677, 1979.

Bergman, R.N., D.T. Finegood, and M. Ader. Assessment of insulin sensitivity in vivo. *Endocrine Rev.* **6**: 45–86, 1985.

Bergman, R.N., and J.C. Lovejoy (Eds.). *The Minimal Model Approach and Determinants of Glucose Tolerance*. Louisiana State University Press, Baton Rouge, LA, 1997.

Berkenbosch, A., J. Heeringa, C.N. Olievier, and E.W. Kruyt. Artificial perfusion of the ponto-medullary region of cats: A method for separation of central and peripheral effects of chemical stimulation of ventilation. *Respir. Physiol.* **37**: 347–364, 1979.

Ghazanshahi, S.D., and M.C.K. Khoo. Estimation of chemoreflex loop gain using pseudorandom binary $CO_2$ stimulation. *IEEE Trans. Biomed. Eng.* **44**: 357–366, 1997.

Hill, J.D., and G.J. McMurtry. An application of digital computers to linear system identification. *IEEE Trans. Auto. Cont.* **AC-9**: 536–538, 1964.

Hodgkin, A.L., A.F. Huxley, and B. Katz. Measurement of current–voltage relations in the membrane of the giant axon of Loligo. *J. Physiol. (London)* **116**: 424–448, 1952.

Jackson, L.B. *Digital Filters and Signal Processing*, 3d ed. Kluwer Academic, Boston, 1995.

Kao, F.F., and L.H. Ray. Regulation of cardiac output in anesthetized dogs during induced muscular work. *Am. J. Physiol.* **179**: 255–260, 1954.

Khoo, M.C.K. Noninvasive tracking of peripheral ventilatory response to $CO_2$. *Int. J. Biomed. Comput.* **24**: 283–295, 1989.

Khoo, M.C.K. Estimation of chemoreflex gain from spontaneous breathing data. In: *Modeling and Parameter Estimation in Respiratory Control* (edited by M.C.K. Khoo), Plenum Press, New York, 1989; pp. 91–105.

Khoo, M.C.K., F. Yang, J.W. Shin, and P.R. Westbrook. Estimation of dynamic chemoresponsiveness in wakefulness and non-rapid-eye-movement sleep. *J. Appl. Physiol.* **78**: 1052–1064, 1995.

Korenberg, M.J., and I.W. Hunter. The identification of nonlinear biological systems: Wiener kernel approaches. *Ann. Biomed. Eng.* **18**: 629–654, 1990.

Ljung, L. *System Identification: Theory for the User.* Prentice-Hall, Englewood Cliffs, NJ, 1987.

Marmarelis, P.Z., and V.Z. Marmarelis. *Analysis of Physiological Systems.* Plenum Press, New York, 1978.

Patterson, S.W., H. Piper, and E.H. Starling. The regulation of the heart beat. *J. Physiol. (London)* **48**: 465, 1914.

Read, D.J.C. A clinical method for assessing the ventilatory response to carbon dioxide. *Australas. Ann. Med.* **16**: 20-32, 1967.

Schetzen, M. *The Volterra and Wiener Theories of Nonlinear Systems.* Wiley, New York, 1980.

Sohrab, S., and S.M. Yamashiro. Pseudorandom testing of ventilatory response to inspired carbon dioxide in man. *J. Appl. Physiol.* **49**: 1000–1009, 1980.

Stark, L. *Neurological Control Systems: Studies in Bioengineering.* Plenum Press, New York, 1968.

Swanson, G.D. Biological signal conditioning for system identification. *Proc. IEEE* **65**: 735–740, 1977.

Toffolo, G., R.N. Bergman, D.T. Finegood, C.R. Bowden, and C. Cobelli. Quantitative estimation of beta cell sensitivity to glucose in the intact organism: A minimal model of insulin kinetics in the dog. *Diabetes* **29**: 979–990, 1980.

## PROBLEMS

**P7.1.** One technique that has been used to assess lung mechanical function is known as the method of "forced oscillations." In one variant of this method, a loudspeaker system is used to generate random pressure perturbations ($P_{ao}$, considered the "input") that are directed into the subject's airways. The resulting fluctuations in airflow ($\dot{V}$, considered the "output") are measured. Using these input–output measurements, it is possible to deduce the quantities that represent airway resistance ($R$), airway inertance ($L$) and respiratory compliance ($C$) by assuming a linear model of respiratory mechanics, such as the linear model that we have considered previously (see Figure 4.1). Using the MATLAB script file "`sensanl.m`," perform a sensitivity analysis to assess parameter identifiability. Plot sensitivity curves such as those displayed in Figure 7.7, assuming nominal parameter values of $R = 1.5$ cm $H_2O$ s $L^{-1}$, $L = 0.01$ cm $H_2O$ $s^2$ $L^{-1}$, and $C = 0.1$ L cm $H_2O^{-1}$. Use the MATLAB "`randn`" function to generate the white noise sequence that represents the applied forcing in $P_{ao}$. Assume a time step of 0.01 s and a total duration of 25 s for each experimental trial. (Note: You will need to implement and solve the model differential equation in a function that will be called by "`sensanl.m`.")

**P7.2.** The dataset provided in the file "`data_fo.mat`" contains measurements of the input (labeled "`Pao`") and output (labeled "`Flow`") signals measured during an application of the method of forced oscillations, described in Problem P7.1. Assuming the respiratory mechanics model structure shown in Figure 4.1, estimate the model parameters ($R$, $L$, and $C$) from the input–output data. Use the optimization technique discussed in Section 7.2.5.2. It is expected that you will modify and apply the MATLAB script file "`popt_llm.m`." Perform the minimization using different starting parameter estimates in order to obtain several sets of final parameter estimates. The differences in values of each parameter will give you some idea of the estimation error.

**P7.3.** Using the dataset in "data_fo.mat," and assuming "Pao" to be the input and "Flow" to be the output, apply least squares estimation (see Section 7.2.2) to deduce the impulse response of the corresponding system. It is expected that you will modify the MATLAB script file "sysid_ls.m." The sampling interval is 0.01 s. Assume the number of points in the impulse response to be 50. Compute also the error band associated with the impulse response estimate.

**P7.4.** Use the SIMULINK model file "gmm_sim.mdl" to generate 10 sets of insulin–glucose "data": in all cases, set the variance of the "measurement noise" at 36 $mg^2/100\ ml^2$, but in each case set the random generator seed to a different integer. For each dataset, use "gmm_est.m" to estimate the parameters $p_1$, $p_2$, $p_3$, and $p_4$ of the minimal model. Then, from the results of all 10 datasets, compute the mean and standard error associated with each of the model parameters.

**P7.5.** The datasets provided in the files "prbs1.mat," "prbs2.mat," "prbs3.mat," and "prbs4.mat" represent measurements of $P_{ACO_2}$ and $\dot{V}_E$ obtained from four human subjects who were breathing from a gas mixture, the composition of which was alternated between air and 6% $CO_2$ in air on a pseudorandom binary basis. One set of these measurements is displayed in Figure 7.18. Using the MATLAB script file "rcs_est.m," estimate in each case: (a) the impulse response that characterizes the dynamics of gas exchange in the lungs; (b) the impulse response that characterizes the dynamics of the chemoreflexes; and (c) the lung-to-chemoreceptor delay. Assume the time-scale to be expressed in numbers of breaths. By applying the fast Fourier transform to the impulse responses in (a) and (b), deduce the corresponding frequency responses. How much intersubject variability is there in the responses?

# 8

# Optimization in Physiological Control

## 8.1 OPTIMIZATION IN SYSTEMS WITH NEGATIVE FEEDBACK

Consider the linear, negative-feedback control system shown in Figure 8.1a. Suppose our task is to select a suitable controller transfer function, $G_c(s)$, that would enable the control system to minimize the effects of different types of disturbances, $\delta$, on the steady-state system output, $y(t \to \infty)$. From simple analysis, we find that

$$\Delta Y(s) = \frac{G_p(s)}{1 + G_c(s)G_p(s)}\,\delta(s) \tag{8.1}$$

Then, as was discussed for the case of the steady-state error in Section 4.5.3, the steady-state response to the disturbance can be deduced from

$$\Delta y(t \to \infty) = \lim_{s \to 0} s\,\Delta Y(s) = \lim_{s \to 0}\left(\frac{sG_p(s)}{1 + G_c(s)G_p(s)}\,\delta(s)\right) \tag{8.2}$$

We first consider the situation where $G_c(s)$ takes the form of a proportional gain, $K$. Then, the steady-state response to a unit step disturbance can be deduced by setting $\delta(s) = 1/s$ and evaluating the right-hand side of Equation (8.2):

$$\Delta y(t \to \infty) = \lim_{s \to 0}\left(\frac{G_p(s)}{1 + KG_p(s)}\right) = \frac{G_{pss}}{1 + KG_{pss}} \tag{8.3}$$

Thus, in order for the system to be minimally perturbed by the step disturbance, it would be necessary to set the proportional controller gain $K$ to as large a value as possible. However, raising this gain would also predispose the system to instability, as we discussed in Chapter 6. Now, consider an alternative: Suppose our choice for $G_c(s)$ is an integral controller, $K/s$.

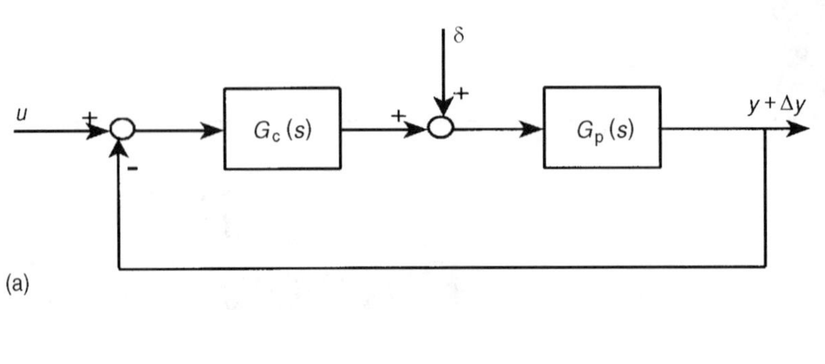


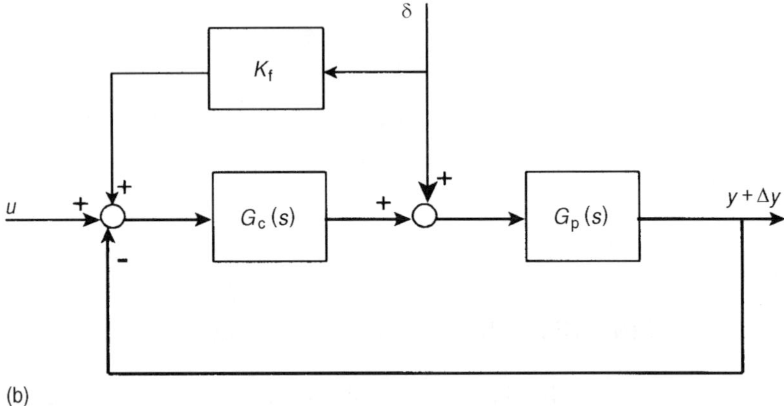


**Figure 8.1** (a) Control system utilizing only feedback. (b) Feedback control system with added feedforward element to cancel out effect of plant disturbance.

If we repeated the calculations performed above, we would obtain the following results. The closed-loop response to a unit step perturbation would be

$$\Delta y(t \to \infty) = \lim_{s \to 0} \left( \frac{G_p(s)}{1 + \frac{K}{s} G_p(s)} \right) = 0 \tag{8.4}$$

Thus, in this case, the steady-state response to the step perturbation is zero. However, the incorporation of the integral controller will slow the transition to the steady state. Therefore, in terms of minimizing the steady-state response of the system to step disturbances, we would choose an integral controller for $G_c(s)$. However, if the disturbance were to take the form of a ramp, $\delta(s) = 1/s^2$, the integral controller would result in a system response given by

$$\Delta y(t \to \infty) = \lim_{s \to 0} \left( \frac{sG_p(s)}{1 + \frac{K}{s} G_p(s)} \frac{1}{s^2} \right) = \frac{G_{pss}}{KG_{pss}} = \frac{1}{K} \tag{8.5}$$

Here, although the steady-state response of the system with integral controller to impulsive and step disturbances is zero, ramplike disturbances would still affect the system response, unless the gain of the controller were to be increased to infinity.

An alternative scheme for buffering the closed-loop response to external disturbances is shown in Figure 8.1b. Here we have added an element that senses the disturbance and produces a proportional output that is "fed forward" into the closed-loop system. This component is known as a *feed-forward* element. The system response to the disturbance is now given by

$$\Delta Y(s) = \frac{G_p(s)(1 + K_f G_c(s))}{1 + G_c(s)G_p(s)} \delta(s) \tag{8.6}$$

Again, if $G_c(s)$ is a proportional controller with gain $K$, and $\delta(s)$ takes the form of a unit step, then the steady-state change in the output becomes

$$\Delta y(t \to \infty) = \lim_{s \to 0} s\, \Delta Y(s) = \lim_{s \to 0} \left( \frac{sG_p(s)(1 + K_f K)}{1 + KG_p(s)} \frac{1}{s} \right) = \frac{G_{pss}(1 + K_f K)}{1 + KG_{pss}} \tag{8.7}$$

The steady-state response to the disturbance can be made to become zero if the feedforward gain $K_f$ is set equal to the negative inverse of the controller gain, i.e.,

$$K_f = \frac{-1}{K} \tag{8.8}$$

The discussion above illustrates the kind of considerations that a designer of an engineering control system would make if the sole purpose of the design were to eliminate the steady-state influences of external disturbances to the system. However, it is also clear that these considerations represent an oversimplification of what would have be taken into account in the design of a "real" control system. In the examples discussed, we used as the measure of "performance" the steady-state system response to an external disturbance: the "better" system would be the one in which the disturbance produced the smaller steady-state change in output. Instead of this criterion, we could have based our performance measure on the complete (i.e., transient+steady state) response to the disturbance. A different performance measure would be likely to have led us to a different conclusion as to which controller design is the most meritorious. Thus, the *optimal* solution is not necessarily a unique one and depends heavily on the *criterion* used to measure performance.

These considerations inevitably lead to the question of whether "optimality" may be found in physiological control systems. It would seem reasonable to hypothesize that evolutionary changes and natural selection have endowed the organisms of today with the more robust of designs. However, it is unclear whether we would expect to find optimal solutions at all levels of organization—subcellular, cellular, tissue, organ, whole-body—and whether such optimal solutions are adhered to at all times. Furthermore, what would be the performance criteria involved in the optimization? The more complex systems tend to show greater redundancy in structure, which provides for greater versatility in function. As a consequence, there might be a "price" that has to be paid in the optimality of certain functions relative to others. Although the concept of biological optimization has been explored for several decades, there remains no clear consensus whether it is a general principle that is embedded in the design of all physiological systems. The problem has been that it is extremely difficult to "prove" the existence of an optimal design, given the considerable variability that exists across individual subjects. Nevertheless, the idea of optimization and its application in various physiological systems are interesting of themselves and warrant some attention. Optimization principles become particularly useful, however, in the design of "smart" devices that are to be used for artificial control or replacement of physiological organ systems or some of their components. In the rest of this chapter, we will explore some of the models that have proposed the minimization of power as the basis on which respiratory and cardiac flow patterns are optimally controlled. This survey is also intended to provide the reader with a sense of the methodology employed to deduce the optimal solutions. As well, we will introduce the notion of adaptive control as a means of optimization and illustrate, in a given example, the application of this principle to the problem of on-line closed-loop physiological control.

## 8.2 SINGLE-PARAMETER OPTIMIZATION: CONTROL OF RESPIRATORY FREQUENCY

The notion that the respiratory frequency may be chosen by the controller on the basis that it will minimize the work rate of breathing was first proposed by Rohrer (1925). Subsequently, Otis, Fenn, and Rahn (1950) conducted a quantitative analysis to determine the optimal frequency that would be predicted through such a hypothesis. The Otis model assumes that the expiratory phase is completely passive and, therefore, that the work done in each breath occurs only in inspiration. The pattern of airflow is assumed sinusoidal, so that

$$\dot{V} = \dot{V}_{\max} \sin 2\pi f t \tag{8.9a}$$

Since the integral of airflow over the inspiratory duration equals the tidal volume, $V_{\mathrm{T}}$, we have

$$V_T = \int_0^{1/2f} \dot{V}_{\max} \sin 2\pi f t \, dt = \frac{\dot{V}_{\max}}{\pi f} \tag{8.10}$$

The work expended per breath is given by

$$W = \int_0^{V_{\mathrm{T}}} p \, dV \tag{8.11}$$

where $p$, the pressure developed by the inspiratory muscles at any given instant during inspiration, is given by

$$p = KV + K'\dot{V} + K''\dot{V}^2 \tag{8.12}$$

In Equation (8.12), the first term on the right-hand side represents that portion of the inspiratory muscle pressure that goes to overcoming the elastic forces that resist the expansion in lung volume, $V$, that occurs above the relaxation lung volume (or functional residual capacity). The constant $K$, therefore, represents lung stiffness, and its inverse would be equal to lung compliance. The second and third terms represent the components of pressure required to overcome the resistive forces as the air is drawn through the large and small airways. The coefficient of the second term, $K'$, is simply the linear viscous resistance to the flow through the airways as well as the nonelastic deformation of lung tissue. The third coefficient, $K''$, multiplied by the square of the airflow represents the nonlinear part of the resistance due to turbulence in airflow through the larger airways as well as the nonlinearity in lung deformation. Substituting Equation (8.12) into Equation (8.11) yields

$$W = \int_0^{V_{\mathrm{T}}} (KV + K'\dot{V} + K''\dot{V}^2) \, dV \tag{8.13a}$$

However, note that since $\dot{V} \equiv dV/dt$, Equation (8.9a) can be rewritten as

$$dV = \dot{V}_{\max} \sin 2\pi f t \, dt \tag{8.9b}$$

Then, using Equation (8.10) to replace $\dot{V}_{\max}$, we obtain:

$$dV = \pi f V_{\mathrm{T}} \sin 2\pi f t \, dt \tag{8.9c}$$

Using Equation (8.9c), we change the integration variable from volume ($V$) to time ($t$) in the second and third terms of Equation (8.13a). This leads to

$$W = \int_0^{V_T} KV\,dV + \pi f V_T \int_0^{1/2f} K'\dot{V} \sin 2\pi f t\,dt + \pi f V_T \int_0^{1/2f} K''\dot{V}^2 \sin 2\pi f t\,dt \quad (8.13b)$$

Then, substituting Equation (8.9a) into Equation (8.13b) and evaluating the resulting definite integrals, we obtain

$$W = \tfrac{1}{2}KV_T^2 + \tfrac{1}{4}K'\pi^2 f V_T^2 + \tfrac{2}{3}K''\pi^2 f^2 V_T^3 \quad (8.14)$$

Since this is the work expended per breath, the total inspiratory work per unit time can be derived by multiplying $W$ by the respiratory frequency $f$. Thus, the work rate is given by

$$\dot{W} = \tfrac{1}{2}KfV_T^2 + \tfrac{1}{4}K'\pi^2 f^2 V_T^2 + \tfrac{2}{3}K''\pi^2 f^3 V_T^3 \quad (8.15)$$

Equation (8.15) highlights the fact that the work rate is a function of both respiratory frequency and tidal volume. However, the relationship between $f$ and $V_T$ is constrained by their having to satisfy the requirement of maintaining alveolar ventilation ($\dot{V}_A$), and thus gas exchange, at a constant level. Since

$$\dot{V}_A = f(V_T - V_D) \quad (8.16)$$

where $V_D$ represents the volume of the non-gas-exchanging dead space, we can substitute Equation (8.16) into Equation (8.15) to eliminate $V_T$ and to obtain

$$\dot{W} = \frac{1}{2}Kf\left(\frac{\dot{V}_A}{f} + V_D\right)^2 + \frac{1}{4}K'\pi^2(\dot{V}_A + fV_D)^2 + \frac{2}{3}K''\pi^2(\dot{V}_A + fV_D)^3 \quad (8.17)$$

It may be noted from Equation (8.17) that, for given $\dot{V}_A$ and $V_D$, the resistive contributions (second and third terms on the right-hand side) to work rate increase with frequency, whereas, as frequency increases, the elastic contribution (first term) decreases. To find the "optimal" frequency that minimizes work rate, we differentiate Equation (8.17) with respect to frequency and set the result equal to zero. This yields the following fourth-order polynomial expression in $f$:

$$2K''\pi^2 V_D^2 f^4 + \pi^2\left(\frac{1}{2}K'V_D^2 + 4K''\dot{V}_A V_D\right)f^3 + \left(\frac{KV_D^2}{2} + \frac{1}{2}K'\pi^2\dot{V}_A V_D + 2K''\pi^2\dot{V}_A^2\right)f^2 - K\dot{V}_A^2 = 0 \quad (8.18)$$

The optimal respiratory frequency, $f_{opt}$, for a given set of parameter values ($\dot{V}_A$, $V_D$, $K$, $K'$ and $K''$) can be deduced by solving Equation (8.18).

The values of the lung mechanics parameters measured in three normal subjects by Otis and colleagues were $K = 8.5$ cm $H_2O$ $L^{-1}$, $K' = 0.0583$ cm $H_2O$ min $L^{-1}$, and $K'' = 4.167 \times 10^{-4}$ cm $H_2O$ min$^2$ $L^{-2}$. If we assume $\dot{V}_A$ and $V_D$ to take on typical values for a resting subject, i.e., 6 L min$^{-1}$ and 0.2 L, respectively, the dependence of work rate and its three components on respiratory frequency may be found from Equation (8.17) to take the form shown in Figure 8.2. Note that the work-rates displayed have been expressed in units of calories per minute. The model predicts that work rate is minimized at the respiratory frequency of 14 breaths min$^{-1}$. This optimal frequency is determined primarily by the elastic and linear resistive components of work-rate; the effect of the nonlinear resistance term is very small over the range of frequencies shown. A subsequent model by Mead (1960), using average muscle force as the criterion for optimization, found the optimal frequency to be at

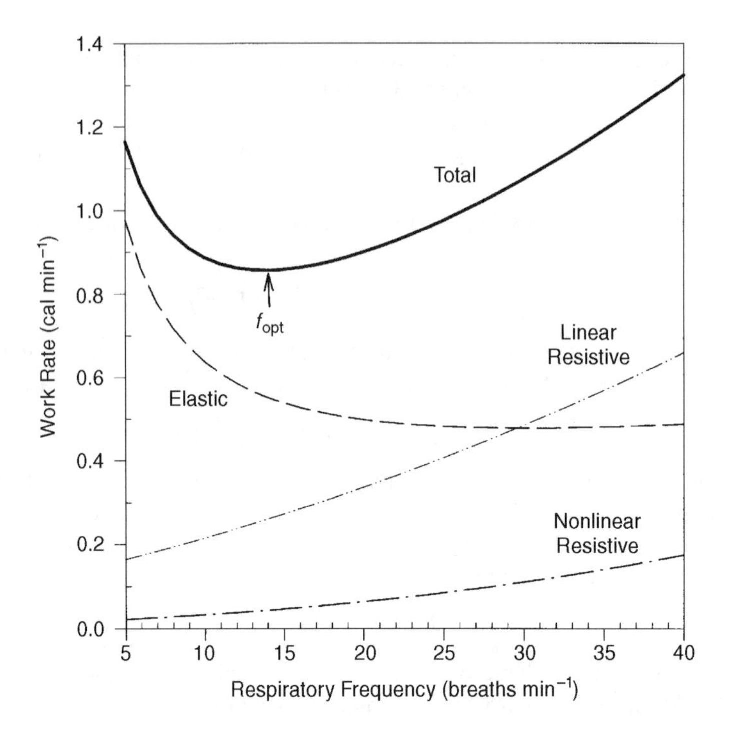


**Figure 8.2** The work-rate of breathing and its components as predicted by the Otis model. Minimization occurs at 14 breaths min$^{-1}$ ($= f_{opt}$).

the somewhat higher value of 23 breaths min$^{-1}$. Both predictions fall within the normal range of breathing frequencies observed in resting humans.

## 8.3 CONSTRAINED OPTIMIZATION: AIRFLOW PATTERN REGULATION

### 8.3.1 Lagrange Multiplier Method

In the Otis model, the somewhat restrictive assumption of a sinusoidal airflow pattern was made in order to simplify the mathematics of the problem. Since breathing patterns are generally not sinusoidal, one can take a more realistic approach by relaxing this requirement and posing the question: Is there an optimal combination of frequency and respiratory airflow pattern that minimizes the work rate of breathing? The inclusion of more unknown parameters (i.e., those characterizing airflow pattern) clearly adds greater complexity to the original problem. The approach adopted by Yamashiro and Grodins (1971) to solve this more complicated problem involved the introduction of an additional unknown parameter known as the *Lagrange multiplier*. This method of solution involves the following steps. Suppose the criterion to be optimized is $J(t, \theta_1, \theta_2, \ldots, \theta_p)$, where $\theta_i$ $(i = 1, \ldots, p)$ represent the parameters that have to be adjusted to produce the optimal value of $J$. At the same time, however, the following equality constraint must also be satisfied:

$$F(\theta_1, \theta_2, \ldots, \theta_p) = \gamma \tag{8.19}$$

where $F(\cdot)$ represents some function of the parameters and $\gamma$ is a constant. Then, a necessary condition for the required extremum is

$$\frac{\partial H}{\partial \theta_1} = \frac{\partial H}{\partial \theta_2} = \cdots = \frac{\partial H}{\partial \theta_p} = 0 \tag{8.20}$$

where the Hamiltonian, $H$, is defined as

$$H = J + \lambda F \tag{8.21}$$

and $\lambda$ is the unknown Lagrange multiplier, which needs to be determined along with the parameters $\theta_i$ $(i = 1, \ldots, p)$ by simultaneous solution of the system of equations represented by Equations (8.19) and (8.20).

### 8.3.2 Optimal Control of Airflow Pattern

In the problem at hand, the criterion to be minimized, the work-rate of breathing, is given by

$$\dot{W} = f \int_0^{V_\mathrm{T}} P\, dV = f \int_0^{V_\mathrm{T}} (KV + K'\dot{V})\, dV \tag{8.22}$$

As in the Otis model, Equation (8.22) assumes that expiration is passive; however, since the contribution from nonlinear resistive contribution is relatively small, this is omitted here. In addition, the airflow pattern is periodic and the inspiratory duration is the same as the expiratory duration. Since the airflow pattern is periodic, $\dot{V}$ can be expressed in the form of a Fourier series, i.e., an infinite number of sinusoidal components:

$$\dot{V} = \sum_{i=1}^{\infty} a_i \sin 2\pi i f t \tag{8.23}$$

Using the fact that $\dot{V} = dV/dt$ and substituting Equation (8.23) into Equation (8.22), we obtain

$$\begin{aligned} \dot{W} &= \tfrac{1}{2} K f V_\mathrm{T}^2 + K' f \int_0^{1/2f} \dot{V} \sum_{i=1}^{\infty} a_i \sin 2\pi i f t\, dt \\ &= \tfrac{1}{2} K f V_\mathrm{T}^2 + K' f \int_0^{1/2f} \sum_{i=1}^{\infty} \sum_{k=1}^{\infty} a_i a_k \sin 2\pi i f t \sin 2\pi k f t\, dt \end{aligned} \tag{8.24a}$$

The orthogonality property of sinusoids can be used to simplify the last term in Equation (8.24a):

$$\int_0^{1/2f} \sin 2\pi i f t \sin 2\pi k f t\, dt = \begin{cases} 0 & i \neq k \\ \dfrac{1}{4f} & i = k \end{cases} \tag{8.25}$$

Thus, using Equation (8.25) in Equation (8.24a) yields the result:

$$\dot{W} = \frac{K f V_\mathrm{T}^2}{2} + \frac{K'}{4} \sum_{i=1}^{\infty} a_i^2 \tag{8.24b}$$

$V_T$ can also be expressed as a function of the dead space, $V_D$, and alveolar ventilation, $\dot{V}_A$:

$$V_T = V_D + \frac{\dot{V}_A}{f} \tag{8.26}$$

Substituting this into Equation (8.24b), we have

$$\dot{W} = \frac{Kf}{2}\left(V_D + \frac{\dot{V}_A}{f}\right)^2 + \frac{K'}{4}\sum_{i=1}^{\infty} a_i^2 \tag{8.24c}$$

Since $V_T$ is a function of the airflow pattern,

$$V_T = \int_0^{1/2f} \sum_{i=1}^{\infty} a_i \sin 2\pi i f t \, dt = \sum_{i=1}^{\infty} \frac{a_{2i-1}}{(2i-1)\pi f} \tag{8.27}$$

The constraint that alveolar ventilation must remain constant can also be written

$$\dot{V}_A - \frac{1}{\pi}\sum_{i=1}^{\infty} \frac{a_{2i-1}}{2i-1} + f V_D = 0 \tag{8.28}$$

Thus, we can now restate the problem in a way in which the Lagrange multiplier method can be applied. We need to minimize the work rate in Equation (8.24c), subject to the constraint in Equation (8.28). To apply the Lagrange multiplier technique, we first form the Hamiltonian, which in this case is given by

$$H(f, a_1, a_2, \ldots) = \frac{Kf}{2}\left(V_D + \frac{\dot{V}_A}{f}\right)^2 + \frac{K'}{4}\sum_{i=1}^{\infty} a_i^2 + \lambda\left[\dot{V}_A - \frac{1}{\pi}\sum_{i=1}^{\infty} \frac{a_{2i-1}}{2i-1} + f V_D\right] \tag{8.29}$$

Next, we differentiate $H$ with respect to each of these parameters and set the result in each case equal to zero:

$$\frac{\partial H}{\partial f} = \frac{K}{2}\left(V_D^2 - \frac{\dot{V}_A^2}{f^2}\right) + \lambda V_D = 0 \tag{8.30}$$

and

$$\frac{\partial H}{\partial a_i} = \frac{K' a_i}{2} = 0 \qquad \text{for even values of } i \tag{8.31a}$$

$$= \frac{K' a_i}{2} - \frac{\lambda}{\pi i} = 0 \qquad \text{for odd values of } i \tag{8.31b}$$

The optimal parameter values are deduced by simultaneous solution of Equations (8.28), (8.30), and (8.31). Note from Equation (8.31a) that all the Fourier coefficients that represent even harmonics of the fundamental frequency $f$ (i.e., $a_{2m}$) equal zero. Thus, from Equation (8.31b), we have

$$a_{2m-1} = \frac{2\lambda}{K'\pi(2m-1)} \tag{8.32}$$

Substituting this result into Equation (8.28) yields

$$\frac{2\lambda}{K'\pi^2}\sum_{i=1}^{\infty} \frac{1}{(2i-1)^2} = \dot{V}_A + f V_D \tag{8.33}$$

However, a useful expansion result that helps to greatly simplify Equation (8.33) is

$$\sum_{i=1}^{\infty} \frac{1}{(2i-1)^2} = \frac{\pi^2}{8} \tag{8.34}$$

Substituting Equation (8.34) into Equation (8.33) gives the final result for $\lambda$:

$$\lambda = 4K'(\dot{V}_A + fV_D) \tag{8.35}$$

The optimal frequency, $f_{opt}$, is deduced by substituting Equation (8.35) into Equation (8.30):

$$f_{opt} = \frac{K}{16K'}\left(\sqrt{1 + \frac{32K'\dot{V}_A}{KV_D}} - 1\right) \tag{8.36}$$

To obtain the optimal airflow pattern coefficients, we employ Equation (8.35) in Equation (8.32):

$$a_{2i-1} = \frac{8}{\pi(2i-1)}(\dot{V}_A + fV_D) \tag{8.37}$$

The optimal airflow pattern is thus characterized by the Fourier series expansion:

$$\dot{V} = 2(\dot{V}_A + fV_D)\left(\frac{4}{\pi}\sin 2\pi ft + \frac{4}{3\pi}\sin 6\pi ft + \frac{4}{5\pi}\sin 10\pi ft + \cdots\right) \tag{8.38}$$

which corresponds to a square wave of amplitude $2(\dot{V}_A + fV_D)$. One should recall, however, that we had assumed expiratory flow to be completely passive. Therefore, the above pattern applies only during inspiration. During expiration, the inspiratory muscle pressure is zero, and thus

$$P = KV + K'\frac{dV}{dt} = 0 \tag{8.39}$$

Integrating Equation (8.39) with respect to time ($1/2f_{opt} < t < 1/f_{opt}$), with the initial condition (at $t = 1/2f_{opt}$ or end inspiration) of $V = V_T$, we get

$$V(t) = V_T e^{-K(t-1/2f)/K'} \qquad \frac{1}{2f} \le t < \frac{1}{f} \tag{8.40}$$

Finally, differentiating $V(t)$ with respect to time, we obtain the airflow time-course during expiration:

$$\dot{V} = -\frac{KV_T}{K'}e^{-K(t-1/2f)/K'} \qquad \frac{1}{2f} \le t < \frac{1}{f} \tag{8.41}$$

The complete airflow pattern of the optimal breath is shown in Figure 8.3a. Using Equation (8.24b), the work-rates of breathing corresponding to this airflow pattern at different frequencies can be computed and are displayed in Figure 8.3b. For comparison, the corresponding prediction from the Otis model is also shown. For the same lung mechanical parameter values, and for $\dot{V}_A = 6$ L min$^{-1}$ and $V_D = 0.2$ L, an optimal respiratory frequency of 16 breaths min$^{-1}$ is predicted. This is slightly higher than the optimal frequency (14 breaths min$^{-1}$) predicted by the Otis model, but falls within the range of physiological variability. However, observed inspiratory airflow patterns resemble the model prediction only during exercise or when subjects breathe through external resistances. Under such conditions, the assumption of a completely passive expiration may be violated. If expiration is assumed

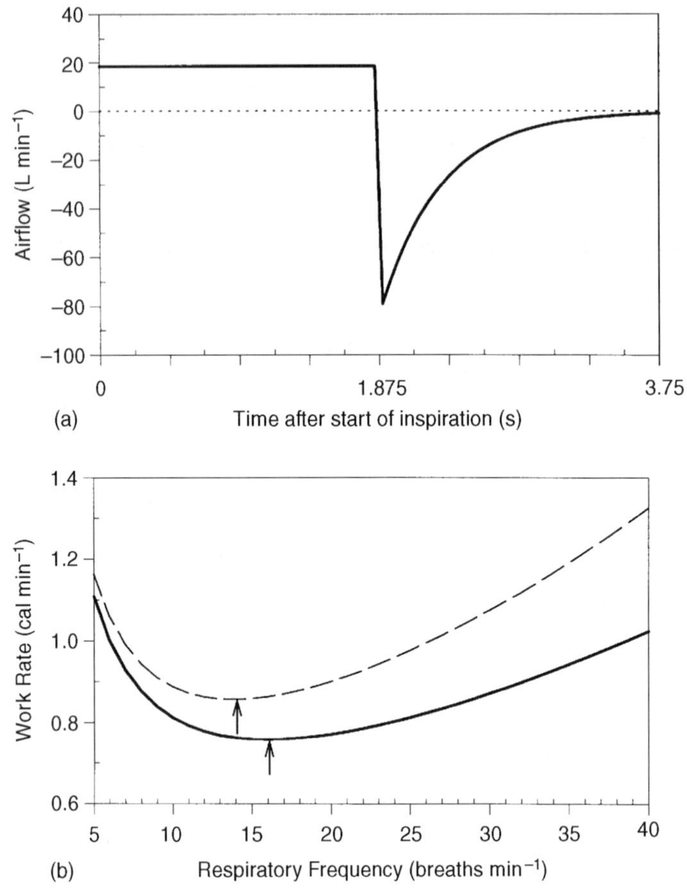


**Figure 8.3** (a) Optimal airflow pattern predicted by the Yamashiro and Grodins model for $\dot{V}_A = 6\,\mathrm{L\,min^{-1}}$ and $V_D = 0.2\,\mathrm{L}$. (b) Comparison of predicted dependence of work-rate on frequency between the Yamashiro and Grodins model (solid curve) and the Otis model (broken curve). Arrows indicate the location of the optimal frequencies.

active, the model predicts an optimal square-wave pattern of breathing (in expiration as well as inspiration); however, there is no optimal frequency. At higher rates of breathing and flows, the effects of airflow turbulence and fluid inertia may not be negligible, as was assumed here. On the other hand, Ruttimann and Yamamoto (1972) have demonstrated that inclusion of these factors does not change the optimal square-wave strategy for the regulation of airflow.

## 8.4 CONSTRAINED OPTIMIZATION: CONTROL OF AORTIC FLOW PULSE

### 8.4.1 Calculus of Variations

In situations where the criterion function to be minimized and the constraints are in the form of integral equations, the method of *calculus of variations* constitutes a useful approach.

In fact, this approach represents a generalization of the Lagrange multiplier method discussed in Section 8.3. Suppose the criterion to be optimized is of the form

$$I = \int_{t_0}^{t_1} J(t, \theta_1, \dot{\theta}_1, \theta_2, \dot{\theta}_2, \ldots, \theta_p, \dot{\theta}_p)\, dt \tag{8.42}$$

where $\theta_i(t)$; $(i = 1, 2, \ldots, p)$ are the functions that produce the optimal value of $I$. At the same time, we assume for generality that two types of equality constraints have to be satisfied. The first kind takes the form of $M$ definite integrals, each of which equals some known constant value, $\gamma_k$ $(1 \leq k \leq M)$:

$$\int_{t_0}^{t_1} F_k(t, \theta_1, \dot{\theta}_1, \theta_2, \dot{\theta}_2, \ldots, \theta_p, \dot{\theta}_p)\, dt = \gamma_k, \qquad \text{where } k = 1, \ldots, M \tag{8.43}$$

The second kind of equality constraints take the form of nonintegral type equations:

$$G_i(\theta_1, \dot{\theta}_1, \theta_2, \dot{\theta}_2, \ldots, \theta_p, \dot{\theta}_p) = 0, \qquad \text{where } i = 1, \ldots, N \tag{8.44}$$

Then, the necessary conditions for the required extremum are given by the set of Euler equations:

$$\begin{gathered} \frac{\partial H}{\partial \theta_1} - \frac{d}{dt}\left(\frac{\partial H}{\partial \dot{\theta}_1}\right) = 0 \\ \vdots \\ \frac{\partial H}{\partial \theta_p} - \frac{d}{dt}\left(\frac{\partial H}{\partial \dot{\theta}_p}\right) = 0 \end{gathered} \tag{8.45}$$

where $H$, the *adjoined criterion function*, is defined as

$$H = J + \sum_{k=1}^{M} \lambda_k F_k + \sum_{i=1}^{M} \mu_i(t) G_i \tag{8.46}$$

In Equation (8.46), $\lambda_k$ $(1 \leq k \leq M)$ are unknown Lagrange multiplier constants that have to be estimated along with the other parameters. In addition to these, there are also $N$ time-dependent Lagrange multipliers, $\mu_i$ $(1 \leq i \leq N)$, that have to be determined. Thus, the unknown functions $\theta_i(t)$ $(i = 1, 2, \ldots, p)$ can be deduced along with these two sets of Lagrange multipliers through simultaneous solution of Equations (8.43), (8.44), and (8.45).

### 8.4.2 Optimal Left Ventricular Ejection Pattern

The method of calculus of variations has been applied to the problem of predicting the optimal left ventricular ejection pattern that would minimize the power expended during the cardiac cycle for a given level of cardiac output (Yamashiro et al., 1979). The electrical analog of the simple model assumed in this problem is shown in Figure 8.4. Blood from the left ventricle is ejected through the aortic valve, which has resistance $R_C$, into the systemic circulation, which has a lumped resistance and compliance of $R_P$ and $C_A$, respectively. $Q(t)$ and $q(t)$ represent the time-courses of blood flow leaving the left ventricle and entering the systemic circulation, respectively. We assume that the duration of systole is $T_S$ and the total

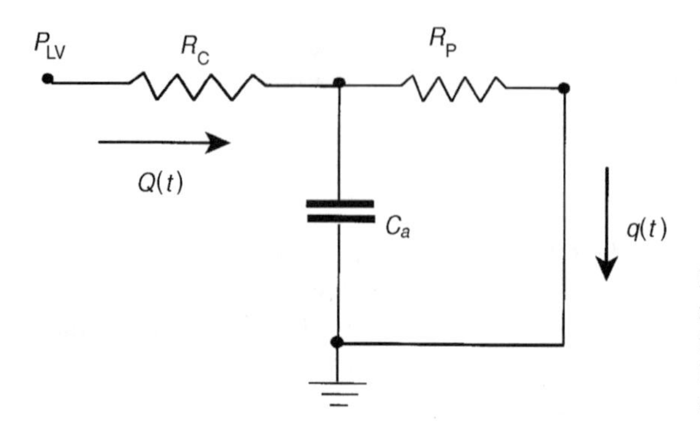


**Figure 8.4** Electrical analog of cardiovascular model used to determine the optimal aortic root flow pattern for given cardiac output. $P_{LV}$ = left ventricular pressure. Other symbols are defined in the text.

duration of the cardiac cycle is $T$. For constant cardiac output, the stroke volume, $V_S$, must also be constant, i.e.,

$$V_S = \int_0^{T_S} Q(t)\, dt = \text{constant} \tag{8.47}$$

The above equation constitutes one constraint that has to be met. A second constraint arises from the conservation of mass principle (or equivalently, Kirchhoff's law for currents entering and leaving the middle node in the electric analog in Figure 8.4):

$$Q(t) = R_P C_a \frac{dq}{dt} + q(t) \tag{8.48}$$

It is also assumed that during diastole, the aortic valve is closed, so that

$$Q(t) = 0, \qquad T_S \leq t < T \tag{8.49}$$

During this interval, the part of the original ventricular output that accumulated in the arterial capacitance in systole is discharged into the rest of the circulation. Thus, the peripheral flow, $q(t)$, during the interval $T_S \leq t < T$ is determined by the solution to Equation (8.48) with $Q(t)$ set equal to zero, i.e.:

$$q(t) = q_{T_S} e^{-(t-T_S)/R_P C_a} \tag{8.50}$$

But, at $t = T$, the value of $q(t)$ must be equal to its value at the start of the cycle. Thus,

$$q(T) = q_0 = q_{T_S} e^{-(T-T_S)/R_P C_a} \tag{8.51a}$$

or

$$q_{T_S} = q(T_S) = q_0 e^{(T-T_S)/R_P C_a} \tag{8.51b}$$

Thus, for given $T$, $T_S$, and initial condition $q_0$, $q_{T_S}$ in Equation (8.51b) provides the boundary condition that must be applied for the determination of the unique optimal solution of $q(t)$.

Similar considerations also apply to the energy balance. During systole, part of the energy expended is stored as elastic energy in the arterial capacitance. However, during diastole, all this stored energy is dissipated as continuing flow through the systemic circulation. Therefore, over the entire cardiac cycle, the total energy dissipated is given by

$$W = R_C \int_0^{T_S} Q^2(t)\, dt + R_P \int_0^{T_S} q^2(t)\, dt + R_P \int_{T_S}^{T} q^2(t)\, dt \tag{8.52}$$

However, the energy dissipated during diastole (last term of Equation (8.52)) is independent of the time-course of $Q(t)$, since Equations (8.50) and (8.51b) show that $q(t)$ during this interval is a function only of $q_0$ and the cardiovascular mechanical parameters. Thus, it is possible to simplify the problem by defining the criterion function that we wish to minimize to be

$$W_S = R_C \int_0^{T_S} Q^2(t)\,dt + R_P \int_0^{T_S} q^2(t)\,dt \tag{8.53}$$

The problem can now be restated as follows: determine the time-courses of $Q(t)$ and $q(t)$ that would minimize $W_S$ in Equation (8.53), while satisfying the constraints given by Equations (8.47) and (8.48). This can clearly be solved by applying the method of calculus of variations (Section 8.4.1). Note that Equation (8.47) takes the form of a definite integral and is therefore a constraint of the type represented by Equation (8.43). Equation (8.48), on the other hand, does not explicitly contain any integrals; thus, in this case, it is an equality constraint of the type represented by Equation (8.44). These considerations lead to the following adjoined criterion function ($H$) that provides us with the capability of taking both types of equality constraints into account:

$$\begin{aligned} H &= W_S + \lambda Q(t) + \mu(t)[Q(t) - R_P C_a \dot{q}(t) - q(t)] \\ &= R_C Q^2(t) + R_P q^2(t) + \lambda Q(t) + \mu(t)[Q(t) - R_P C_a \dot{q}(t) - q(t)] \end{aligned} \tag{8.54}$$

The corresponding Euler equations are

$$\frac{\partial H}{\partial Q} - \frac{d}{dt}\left(\frac{\partial H}{\partial \dot{Q}}\right) = 2R_C Q(t) + \lambda + \mu(t) = 0 \tag{8.55}$$

and

$$\frac{\partial H}{\partial q} - \frac{d}{dt}\left(\frac{\partial H}{\partial \dot{q}}\right) = 2R_P q(t) - \mu(t) + R_p C_a \dot{\mu}(t) = 0 \tag{8.56}$$

We use Equations (8.55) and (8.56) to eliminate $\mu(t)$ and $\dot{\mu}(t)$, yielding the following result:

$$R_P q(t) + R_C Q(t) + \frac{\lambda}{2} - R_C \tau \dot{Q}(t) = 0 \tag{8.57}$$

where we now define

$$\tau = R_P C_a \tag{8.58}$$

Equation (8.48) is used to eliminate $Q(t)$ and $\dot{Q}(t)$ from Equation (8.57), leading to the following result:

$$\ddot{q}(t) - \alpha^2 q(t) = \frac{\lambda}{2\tau^2 R_C} \tag{8.59}$$

where

$$\alpha = \frac{1}{\tau}\sqrt{\frac{R_P + R_C}{R_C}} \tag{8.60}$$

Solution of Equation (8.59) yields

$$q(t) = \left(q_0 + \frac{\lambda}{2(R_P + R_C)}\right) \cosh \alpha t + \frac{\dot{q}(0)}{\alpha} \sinh \alpha t - \frac{\lambda}{2(R_P + R_C)} \tag{8.61}$$

In the above result, note that $\dot{q}(0)$ and $\lambda$ are unknown quantities that need to be determined. The former can be found by evaluating Equation (8.61) at $t = T_S$, and using Equation (8.51b) to relate $q(T_S)$ to $q_0$ and the other known quantities. This leads to the following result for the optimal $q(t)$:

$$q(t) = A_1 \cosh(\alpha t) + A_2 \sinh(\alpha t) - A_0 \tag{8.62}$$

where

$$A_0 = \frac{\lambda}{2(R_P + R_C)} \tag{8.63}$$

$$A_1 = q_0 + A_0 \tag{8.64}$$

and

$$A_2 = \frac{1}{\sinh(\alpha T_S)} \left[q_0 e^{(T-T_S)/\tau} + A_0 - A_1 \cosh(\alpha T_S)\right] \tag{8.65}$$

Using Equation (8.62), we can evaluate Equation (8.48) to obtain an expression for the optimal $Q(t)$:

$$Q(t) = (\tau\alpha A_1 + A_2) \sinh \alpha t + (\tau\alpha A_2 + A_1) \cosh \alpha t - A_0 \tag{8.66}$$

Note, however, that the expressions for optimal $q(t)$ and $Q(t)$ still contain an unknown parameter: $\lambda$. The latter can be eliminated by using the remaining constraint equation that has not been employed thus far, i.e., Equation (8.47). Therefore, in accordance with Equation (8.47), we integrate $Q(t)$ in Equation (8.66) with respect to time over the interval $t = 0$ to $t = T_S$ and equate the result to the stroke volume $V_S$. This allows us to solve for $\lambda$:

$$\lambda = \frac{2(R_C + R_P)\left(V_S - q_0 B_1 - \dfrac{B_2 B_3}{\sinh \alpha T_S}\right)}{B_1 + B_3\left(\dfrac{1 - \cosh \alpha T_S}{\sinh \alpha T_S}\right) - T_S} \tag{8.67}$$

where

$$B_1 = \tau(\cosh \alpha T_S - 1) + \frac{1}{\alpha} \sinh \alpha T_S \tag{8.68}$$

$$B_2 = q_0(e^{(T-T_S)/\tau} - \cosh \alpha T_S) \tag{8.69}$$

and

$$B_3 = \frac{1}{\alpha}(\cosh \alpha T_S - 1) + \tau \sinh \alpha T_S \tag{8.70}$$

Thus, for given $T$, $V_S$, $R_C$, $R_P$, $C_a$, and $q_0$, Equations (8.66) and (8.67) can be used to deduce the time-course of $Q(t)$ as a function of the duration of systole, $T_S$. One implicit constraint remains to be satisfied. To obtain the optimal $T_S$ and $Q(t)$, the appropriate value of $T_S$ must be selected by some search procedure so that $Q(T_S)$ becomes zero. This is illustrated in Figure 8.5. Here, $Q(T_S)$ is plotted against $T_S$ for three different levels of $R_P$. As $T_S$

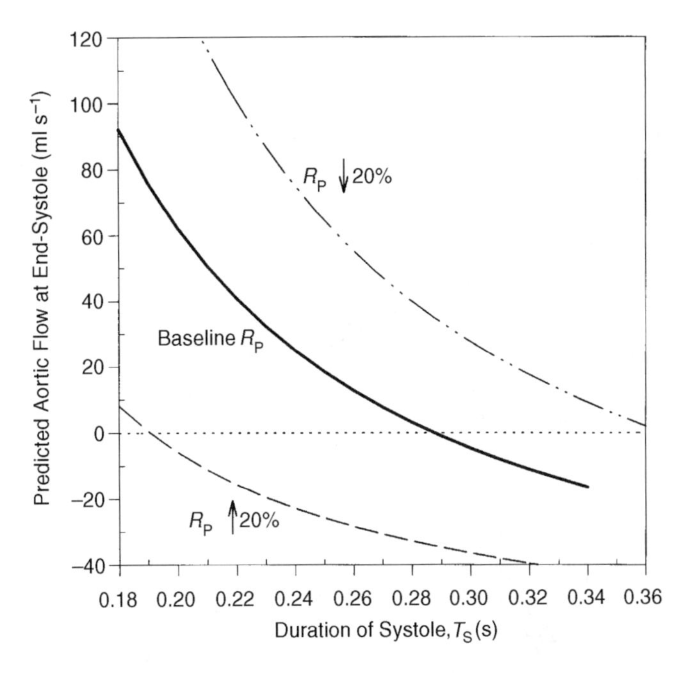


**Figure 8.5** Variation of predicted end-systolic aortic flow with duration of systole at different levels of peripheral resistance. The optimal systolic duration is the value that produces zero end-systolic aortic flow.

increases, $Q(T_S)$ decreases from positive values towards negative values. Thus, the value of $T_S$ at which $Q(T_S)$ becomes zero is the optimal value of $T_S$ that allows the systolic portion of the solution to be continuous with the diastolic portion. For baseline levels of $R_P$, optimal $T_S$ is 0.29 s. However, as $R_P$ increases, optimal $T_S$ decreases (i.e., the duration of systole becomes shorter), and vice versa.

The optimal time-courses for $Q(t)$ and $q(t)$ predicted by the model are displayed in Figures 8.6a and 8.6b, respectively. The solid curves represent the predictions made with the assumption of "nominal" parameter values for a 10 kg dog at rest. These parameter values are: $T = 0.5$ s, $q_0 = 19.16$ ml s$^{-1}$, $V_S = 12$ ml, $R_C = 0.033$ mm Hg s ml$^{-1}$, $C_a = 0.178$ ml mm Hg$^{-1}$, and $R_P = 4.17$ mm Hg s mL$^{-1}$. The computation of the optimal flow patterns, as well as the initial determination of optimal $T_S$, was performed with the use of MATLAB script file "`OptLVEF.m`." As shown in Figure 8.5, the optimal duration of systole in this case is 0.29 s or 58% of the cardiac period. When $R_P$ is increased, energy expended per cardiac cycle is minimized by shortening the duration of systole and allowing the systolic aortic flow pattern to be more triangular with substantially higher flow in the early part of the cycle. With lower $R_P$, the initial high level of flow is not required and systolic duration is lengthened. These patterns resemble those observed in measurements of aortic flow pulses, except that, in the latter, aortic flow at the start of systole does not instantaneously increase to its maximum value. This prediction is due largely to the omission of fluid inertia in the model. An improved optimization model by Livnat and Yamashiro (1981), which includes blood inertia, nonlinear aortic valve resistance, and ventricular compliance, produces more realistic left ventricular pressure and flow patterns. However, the inclusion of more complexity makes it impossible to obtain a closed-form analytic solution to the problem, and a numerical method of solution is necessary.

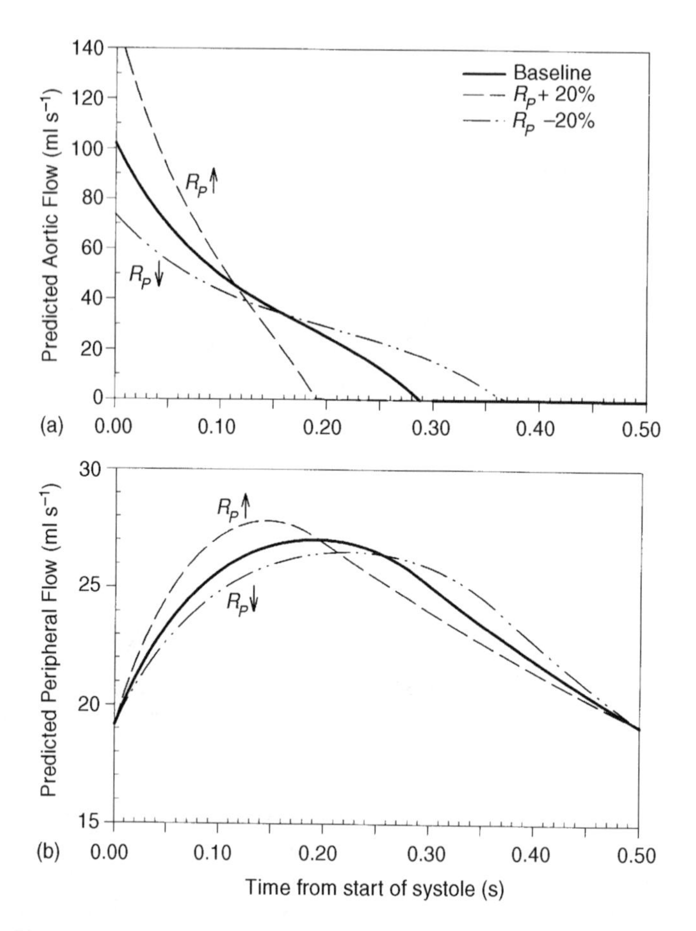


**Figure 8.6** Predicted optimal patterns of (a) aortic root flow (left ventricular output), and (b) peripheral or systemic flow at different levels of peripheral resistance.

## 8.5 ADAPTIVE CONTROL OF PHYSIOLOGICAL VARIABLES

### 8.5.1 General Considerations

Although the optimization hypothesis is teleologically attractive and has been a useful means of "explaining" specific control strategies in a number of physiological processes, whether it can be regarded as a general principle that is embedded in all biological systems at all levels of organization remains unclear at this time. On the other hand, what is clear is that all physiological systems are time-varying, i.e., their structure and governing parameters change over time. So, even if optimization has been demonstrated to exist, the next question that arises is whether this optimization is adaptive. Conversely, is the adaptive behavior of a given physiological system aimed at optimizing some measure of performance? These questions are complex and remain at the forefront of contemporary physiological (in particular, neurophysiological) research. At the same time, however, the development of medical prosthetic and assistive devices has led to substantial progress in applying the concepts of adaptive control to the problem of designing schemes for automatically

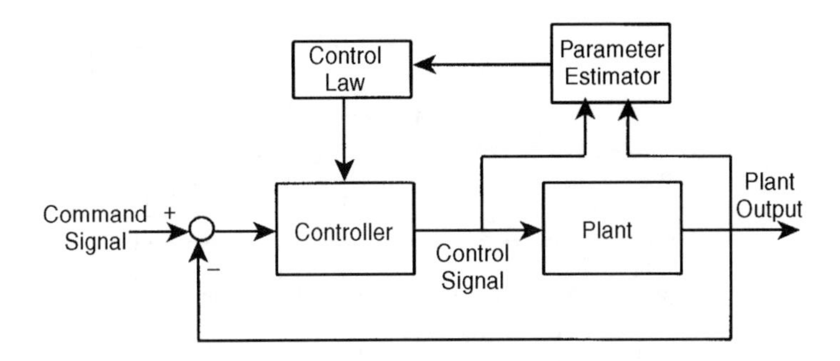


**Figure 8.7** Schematic block diagram of the general adaptive control system.

controlling various physiological variables. It is toward these kinds of adaptive control systems that we will now turn our attention, since the theory and engineering applications of such systems are well developed.

Figure 8.7 shows a block diagram for the general adaptive control system. The basic features that distinguish adaptive control from simple feedback control are the addition of a parameter estimator to determine the changes in dynamics of the unknown plant and a control law that uses an optimization algorithm to select the control signal that is optimally adjusted for the altered plant dynamics. Most adaptive control schemes require a model of the plant. Therefore, the accuracy and reliability with which this model characterizes plant dynamics are key factors that govern how well the adaptive control system will work in practice.

The two major types of adaptive controllers employed in online physiological control and closed-loop drug delivery schemes are illustrated in Figure 8.8. In the clinical setting, there is always considerable variability in plant dynamics across subjects as well as within an individual subject at different times. The *multiple model adaptive control* (MMAC) system, shown in Figure 8.8a, allows for a finite range of representations of plant states by containing a model bank. Constraints are placed on the parameters of each model employed, so that the controller responses remain reasonable and bounded. One disadvantage of this approach is that it requires significant knowledge of the plant dynamics. Another is that the controller may not be able to handle plant behavior that lies beyond the range specified in the model bank. The *model reference adaptive control* (MRAC) system, on the other hand, uses a single general model of the plant (Figure 8.8b). Thus, it can be more versatile. However, there is no guarantee of stability for the parameter estimates and for the physiological variable being controlled. Both these types of adaptive control schemes have been used in a variety of closed-loop drug delivery applications, including blood pressure control, neuromuscular blockade, and control of blood glucose level (Katona, 1982; Martin et al., 1987; Olkkola and Schwilden, 1991; Fischer et al., 1987).

### 8.5.2 Adaptive Buffering of Fluctuations in Arterial $P_{CO_2}$

A detailed consideration of adaptive control theory and its applications lies beyond the scope of this text. For this, the reader is referred to a number of excellent volumes, such as Astrom and Wittenmark (1989) and Harris and Billings (1981). However, in this section, we will illustrate an example of how this theory can be implemented in practice. The problem at hand concerns the considerable degree of breath-to-breath variability that has been observed in spontaneous ventilation. Accompanying this ventilatory variability are the corresponding

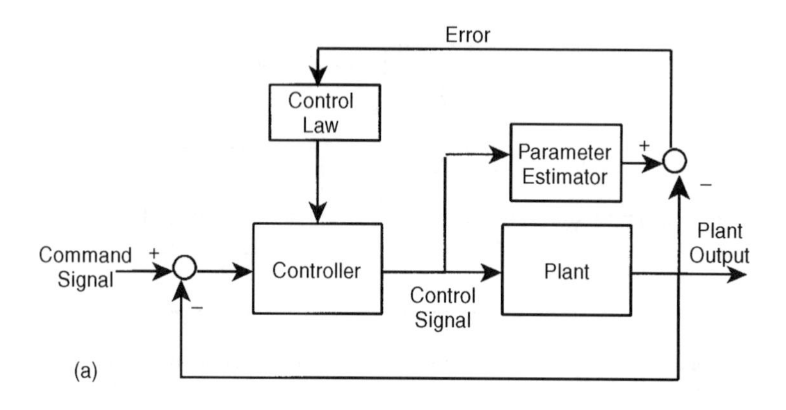


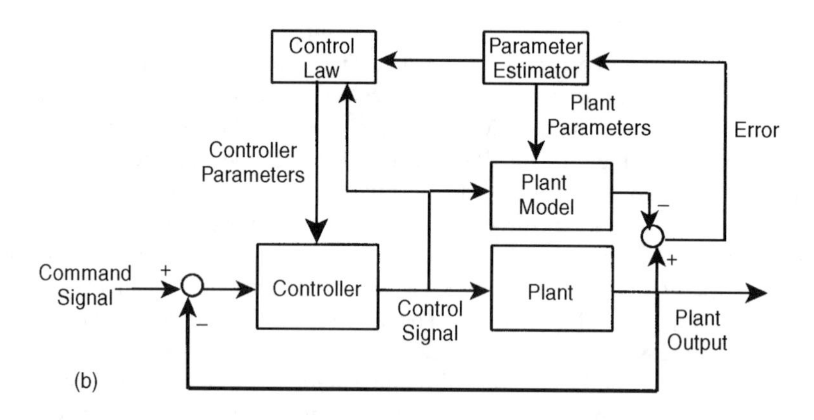


**Figure 8.8** Schematic block diagrams of (a) a multiple-model adaptive control (MMAC) system, and (b) a model reference adaptive control (MRAC) system.

fluctuations in alveolar and, therefore, arterial $P_{CO_2}$. Modarreszadeh et al. (1993) addressed the issue of buffering these fluctuations in arterial $P_{CO_2}$ in an optimal manner by changing the $CO_2$ composition of the inhaled gas ($F_{ICO_2}$) on a breath-by-breath basis. Figure 8.9 shows a block diagram of the scheme employed for achieving this goal. The bottom portion of the block diagram represents the respiratory control system. Fluctuations in arterial $P_{CO_2}$, which we assume are measured in the form of fluctuations in the end-tidal $CO_2$ fraction ($F_{ETCO_2}$), result from "ventilatory noise" entering the closed-loop system as well as changes in gas exchange dynamics in the lungs. A simple linearized model of the gas exchange process is assumed, and based on measurements of ventilation ($\dot{V}_E$) and $F_{ETCO_2}$, the plant model parameters are identified. However, since gas exchange dynamics can change with the sleep–wake state or other conditions of the subject, the estimation of plant model parameters has to be performed adaptively. At any given breath, the estimated plant model parameters are used along with existing measurements of $\dot{V}_E$ and $F_{ETCO_2}$ to predict what $F_{ICO_2}$ should be applied next to minimize the fluctuation of $F_{ETCO_2}$ in the next breath.

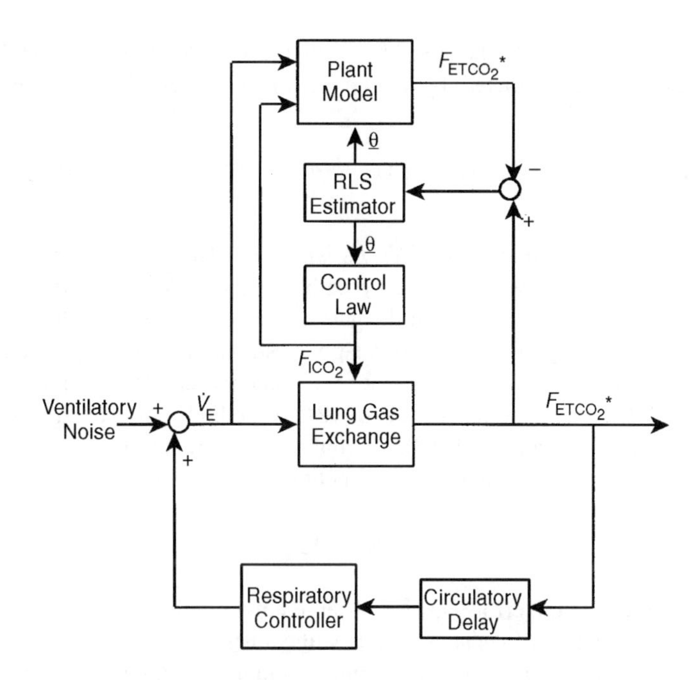


**Figure 8.9** Control scheme for adaptive buffering of spontaneous fluctuations in end-tidal $CO_2$ in humans (Modarreszadeh et al., 1993).

***8.5.2.1. Plant Model.*** Since the purpose of the scheme is to minimize the fluctuations in $F_{ETCO_2}$ about its mean level, we define the new variables $\Delta F_{ETCO_2}$, $\Delta\dot{V}_E$, and $\Delta F_{ICO_2}$ as the deviations in $F_{ETCO_2}$, $\dot{V}_E$, and $F_{ICO_2}$, respectively, about their corresponding means. Since negative $F_{ICO_2}$ cannot be realized in practice, it is assumed that the "resting" or mean level of $F_{ICO_2}$, prior to the application of adaptive control, is 2.5% and not zero (i.e., the subject is breathing a gas mixture that resembles air but contains a small amount of $CO_2$). The linearized plant model takes a form very similar to that given in Equation (7.64a) in Section 7.5.2:

$$\Delta F_{ETCO_2}(n) = a(n)\,\Delta F_{ETCO_2}(n-1) + b(n)\,\Delta\dot{V}_E(n-1) + c(n)\,\Delta F_{ICO_2}(n-1) + e(n) \tag{8.71}$$

Apart from the fact that $\Delta F_{ETCO_2}$ and $\Delta F_{ICO_2}$ are now used in place of $\Delta P_{ACO_2}$ and $\Delta P_{ICO_2}$, respectively, a large difference between Equation (8.71) and Equation (7.64a) is that the unknown parameters here ($a$, $b$, and $c$) are assumed to be time-varying; $n$ refers to the current breath. Thus, $\Delta F_{ETCO_2}(n-1)$ and $\Delta\dot{V}_E(n-1)$ represent the changes in $F_{ETCO_2}$ and $\dot{V}_E$ at the *previous* breath. However, the exception is with $\Delta F_{ICO_2}(n-1)$, which represents the change in $F_{ICO_2}$ of the *current* breath. This peculiar assignment of breath number is made because, in the real-time implementation of this scheme, the measurement of $F_{ICO_2}$ in the preceding inspiration is already available for use in the algorithm during the expiratory phase of the $n$th breath. By the same token, $\Delta\dot{V}_E(n-1)$ has to be used in place of $\Delta\dot{V}_E(n)$ in Equation (8.71) (note the use of $\Delta\dot{V}_E(n)$ in Equation (7.64a)), because the computation of $\Delta\dot{V}_E(n)$ requires knowledge of the tidal volume and total period of the $n$th breath, and the latter measurement is not available until the end of the expiratory phase in the current breath ($n$). Finally, $e(n)$ in

Equation (8.71) represents the error between the measured $\Delta F_{ETCO_2}$ at breath $n$ and the corresponding value predicted by using the plant model. Thus, the predicted $\Delta F_{ETCO_2}$ ($\Delta F^*_{ETCO_2}$) is given by

$$\Delta F^*_{ETCO_2}(n) = a(n)\,\Delta F_{ETCO_2}(n-1) + b(n)\,\Delta \dot{V}_E(n-1) + c(n)\,\Delta F_{ICO_2}(n-1) \quad (8.72a)$$

***8.5.2.2. Plant Model Parameter Estimation.*** The plant model parameters are estimated adaptively, so that with each new breath, the new set of measurements obtained can be used to update our knowledge about the gas exchange process. To achieve this task, one can choose among a variety of methods for adaptive estimation. For this problem, Modarreszadeh et al. (1993) selected one that has been commonly employed in physiological system identification: the *recursive least squares* (RLS) method. For other methods, the reader is referred to texts by Orfanidis (1988), Astrom and Wittenmark (1989), and Akay (1994).

In the RLS method, the criterion function to be minimized assumes the form

$$J(n) = \sum_{i=0}^{n} \lambda^{n-i} e(i)^2 \quad (8.73)$$

where $e(i)$ represents the error between observed and predicted values of $F_{ETCO_2}$, as given in Equation (8.71), and $\lambda$ is a constant factor (with values between 0 and 1). The latter, in effect, provides a means of assigning relative weights to more recent versus more remote events. For instance, if $\lambda$ is given a value close to zero, then $J(n)$ will be determined predominantly by $e(n)$ and virtually nothing else. At the other extreme, when $\lambda$ is close to unity, $J(n)$ will be essentially equal to the sum of squares of the residuals for the current and past $n$ breaths. For this reason, $\lambda$ is often referred to as the *forgetting factor.*

Since the estimation involves three parameters, we can recast Equation (8.72a) in the form of a vector equation:

$$\Delta F^*_{ETCO_2}(n) = \underline{\theta}(n)'\underline{\mathbf{y}}(n-1) \quad (8.72b)$$

where

$$\underline{\theta}(n) = [a(n) \quad b(n) \quad c(n)]' \quad (8.74)$$

and

$$\underline{y}(n-1) = [\Delta F_{ETCO_2}(n-1)\ \Delta \dot{V}_E(n-1)\ \Delta F_{ICO_2}(n-1)]' \quad (8.75)$$

Thus, Equation (8.73) can be rewritten as

$$J(n) = \sum_{i=0}^{n} \lambda^{n-i}\Big(\Delta F_{ETCO_2}(n) - \underline{\theta}(n)'\underline{\mathbf{y}}(n-1)\Big)^2 \quad (8.76)$$

We deduce the optimal value of $\underline{\theta}(n)$ that minimizes $J(n)$ by differentiating Equation (8.76) with respect to $\underline{\theta}(n)$ and setting the result equal to the null vector:

$$\frac{\partial J(n)}{\partial \underline{\theta}(n)} = -2\sum_{i=0}^{n} \lambda^{n-i}\Big(\Delta F_{ETCO_2}(i) - \underline{\theta}(n)'\underline{\mathbf{y}}(i-1)\Big)\underline{\mathbf{y}}(i-1)' = 0 \quad (8.77)$$

Rearranging terms in Equation (8.77), we obtain the following expression for the parameter vector:

$$\underline{\theta}(n) = \mathbf{R}(n)^{-1}\underline{r}(n) \quad (8.78)$$

where

$$\mathbf{R}(n) = \sum_{i=0}^{n} \lambda^{n-i} \underline{\mathbf{y}}(i-1)\underline{\mathbf{y}}(i-1)' \tag{8.79a}$$

and

$$\underline{r}(n) = \sum_{i=0}^{n} \lambda^{n-1} \Delta F_{ETCO_2}(i)\underline{\mathbf{y}}(i-1) \tag{8.80a}$$

Note that, by separating the last term from the rest of the summation on the right-hand side, we can rewrite Equation (8.79a) in recursive form:

$$\mathbf{R}(n) = \underline{\mathbf{y}}(n-1)\underline{\mathbf{y}}(n-1)' + \lambda \mathbf{R}(n-1) \tag{8.79b}$$

Similarly, $r(n)$ in Equation (8.80a) can also be expressed in recursive form:

$$\underline{r}(n) = \Delta F_{ETCO_2}(n)\underline{\mathbf{y}}(n-1) + \lambda \underline{r}(n-1) \tag{8.80b}$$

If we define the matrix $\mathbf{P}(n)$ to be the inverse of $\mathbf{R}(n)$:

$$\mathbf{P}(n) \equiv \mathbf{R}(n)^{-1} \tag{8.81}$$

then, using the matrix inversion lemma, it can be shown (through a somewhat complicated proof) that $\mathbf{P}(n)$ can be expressed in the recursive form:

$$\mathbf{P}(n) = \frac{1}{\lambda}\left(\mathbf{P}(n-1) - \frac{\mathbf{P}(n-1)\underline{y}(n-1)\underline{y}(n-1)'\mathbf{P}(n-1)}{\lambda + \underline{y}(n-1)'\mathbf{P}(n-1)\underline{y}(n-1)}\right) \tag{8.82}$$

Using Equation (8.80b) and Equation (8.82) in Equation (8.78), we obtain, after some algebraic manipulation, the following parameter update equation:

$$\underline{\theta}(n) = \underline{\theta}(n-1) + \underline{\mathbf{K}}(n)e(n) \tag{8.83}$$

where

$$\underline{K}(n) = \frac{P(n-1)\underline{y}(n-1)}{\lambda + \underline{y}(n-1)'P(n-1)\underline{y}(n-1)} \tag{8.84}$$

$\underline{K}(n)$ is a gain vector that determines the relative contribution of the prediction error, $e(n)$, to the estimate of the parameter vector $\underline{\theta}(n)$. $\underline{K}(n)$ depends on the $\mathbf{P}(n-1)$, which turns out to be the *parameter error covariance matrix* (see Section 7.3.2).

***8.5.2.3. Adaptive Control Law.*** Having estimated the most current values of the plant model parameters, the next task in each time step (breath) would be to determine the optimal value of $\Delta F_{ICO_2}$ in the current breath (i.e., $\Delta F_{ICO_2}(n-1)$) that would minimize the predicted $\Delta F_{ETCO_2}$ at the end of the expiration phase of the current breath (i.e., $\Delta F_{ETCO_2}(n)$). However, the criterion function to be minimized, in this case, was selected by Modarreszadeh et al. (1993) to be

$$I(n) = \left[\alpha\, \Delta F^{*}_{ETCO_2}(n)\right]^2 + \left[\beta\, \Delta F_{ICO_2}(n-1)\right]^2 \tag{8.85}$$

By including the term with $\Delta F_{ICO_2}(n-1)$ in Equation (8.85), the fluctuations in inhaled $CO_2$ concentration are minimized along with those in alveolar $CO_2$. This allows for a solution that does not lead to minimal $\Delta F_{ETCO_2}$ at the "expense" of employing large $\Delta F_{ICO_2}$. This is important from a practical point of view, since high values of $F_{ICO_2}$ can be a source of unpleasant sensation to the subject. The relative contributions of $\Delta F_{ETCO_2}$ and $\Delta F_{ICO_2}$ to the criterion function are determined by the weights $\alpha$ and $\beta$. Based on their experience, the autnors chose values of 1 for $\alpha$ and 0.5 for $\beta$.

To determine the "optimal" $\Delta F_{ICO_2}(n-1)$, Equation (8.72a) is substituted into Equation (8.85) and this is differentiated with respect to $\Delta F_{ICO_2}(n-1)$. The result of the differentiation is set equal to zero. After rearranging terms, we obtain

$$\Delta F_{ICO_2}(n-1) = \frac{-\alpha^2 c(n-1)}{\alpha^2 c(n-1)^2 + \beta^2}\left[a(n-1)\,\Delta F_{ETCO_2}(n-1) + b(n-1)\,\Delta \dot{V}_E(n-1)\right] \tag{8.86}$$

In Equation (8.86), note that the plant parameter estimates from the previous breath (i.e., $a(n-1)$, $b(n-1)$, and $c(n-1)$) are used since, from Equations (8.83) and (8.84), the

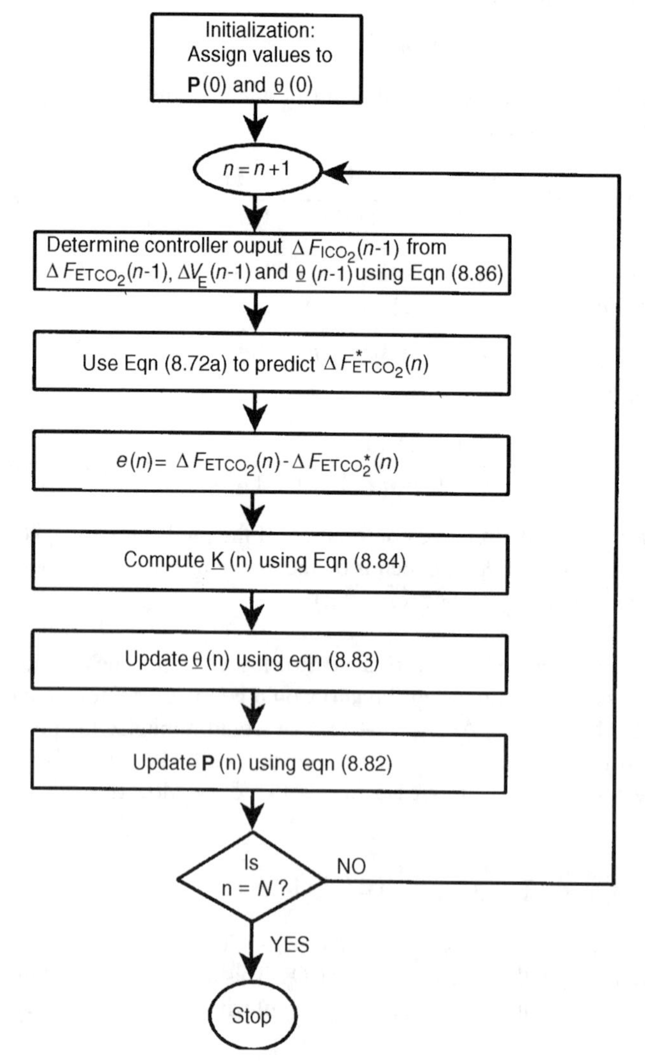


**Figure 8.10** Flowchart of adaptive control algorithm.

parameter estimates from the current breath (i.e., $a(n)$, $b(n)$, and $c(n)$) are determined in part by $\Delta F_{ICO_2}(n-1)$.

The overall algorithm for the adaptive control scheme is displayed in the form of a flowchart in Figure 8.10. When the algorithm is initiated from starting conditions ($n = 0$), it is useful to employ initial values for the plant parameters that are not too far from their "true" values. For this reason, Modarreszadeh et al. estimated these values beforehand by measuring the subjects' responses to a dynamic $CO_2$ stimulus that was altered in pseudorandom binary fashion (see Section 7.5.2.1). The average values of $a$, $b$, and $c$ in eight subjects were found to be 0.66, −0.02, and 0.69, respectively. The value for $b$ assumes that ventilation is measured in L $\text{min}^{-1}$ and $F_{ETCO_2}$ is expressed as a percentage. A starting value for the parameter error covariance matrix, **P**, is also required. A common procedure is to set **P**(0) equal to the identity matrix scaled by a factor of 100.

***8.5.2.4. Performance of the Adaptive Controller.*** Breath-by-breath measurements of $F_{ETCO_2}$ in a normal subject during spontaneous breathing are shown in Figure 8.11a. These can be compared to the corresponding measurements of $F_{ETCO_2}$ in the same subject during adaptive buffering of the end-tidal $CO_2$ fluctuations, shown in Figure 8.11b. It is clear that the adaptive controller produced a significant reduction of the fluctuations in $F_{ETCO_2}$, particularly in the lower-frequency region. It would have been possible reduce $\Delta F_{ETCO_2}$ further by setting $\beta$ in the criterion function I(n) to zero. But this would be achieved at the expense of incurring larger fluctuations in $F_{ICO_2}$.

The properties of this adaptive control scheme can be studied further by executing the MATLAB script file "`acs_CO2.m`" included with this book. However, instead of obtaining

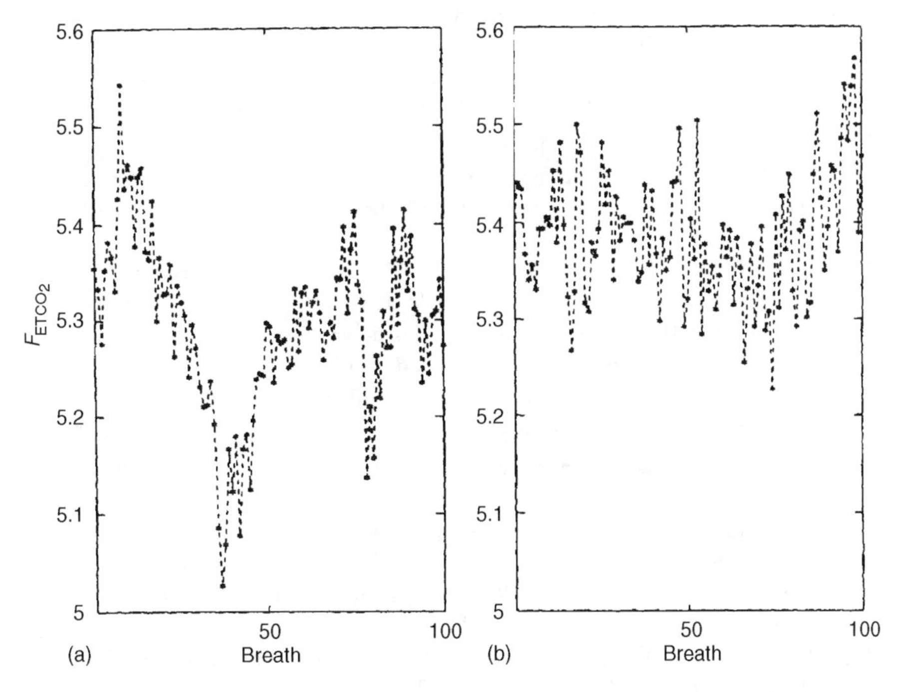


**Figure 8.11** (a) Breath-by-breath measurements of $F_{ETCO_2}$ in an adult human, showing considerable spontaneous variability. (b) $F_{ETCO_2}$ measurements in the same subject during application of the adaptive buffering scheme (Reproduced from Modarreszadeh et al., 1993)

measurements for a real human subject, a simple dynamic simulation of the chemoreflex control system is used to generate "data" and to interact with the adaptive controller. The fluctuations in $\dot{V}_E$, $F_{ICO_2}$, and $F_{ETCO_2}$ are assumed to occur around their corresponding mean levels.

## BIBLIOGRAPHY

Akay, M. *Biomedical Signal Processing*. Academic Press, New York, 1994.

Astrom, K.J., and B. Wittenmark. *Adaptive Control*. Addison-Wesley, Reading, MA, 1989.

Clarke, D.W., and P.J. Gawthrop. Self-tuning controller. *Proc. IEEE* **122**: 922–934, 1975.

Elsgolc, L.E. *Calculus of Variations*. Addison-Wesley, Reading, MA, 1962.

Fischer, U., W. Schenk, E. Salzsieder, G. Albrecht, P. Abel, and E.J. Freyse. Does physiological blood glucose control require an adaptive control strategy. *IEEE Trans. Biomed. Eng.* **BME-34**: 575–582, 1987.

Harris, C.J., and S.A. Billings (eds.). *Self-tuning and Adaptive Control: Theory and Applications*. Institute of Electrical Engineers (U.K.), London, 1981.

Katona, P.G. On-line control of physiological variables and clinical therapy. *CRC Rev. Biomed. Eng.* **8**: 281–310, 1982.

Livnat, A., and S.M. Yamashiro. Optimal control evaluation of left ventricular systolic dynamics. *Am. J. Physiol.* **240**: R370–R383, 1981.

Martin, J.F., A.M. Schneider, and N.T. Smith. Multiple-model adaptive control of blood pressure using sodium nitroprusside. *IEEE Trans. Biomed. Eng.* **BME-34**: 603–611, 1987.

Mead, J. Control of respiratory frequency. *J. Appl. Physiol.* **15**: 325–336, 1960.

Milsum, J.H. Optimization aspects in biological control theory. In: *Advances in Biomedical Engineering and Medical Physics*, Vol. 1 (edited by S.N. Levine). Wiley Interscience, New York, 1968; pp. 243–278.

Modarreszadeh, M., K.S. Kump, H.J. Chizeck, D.W. Hudgel, and E.N. Bruce. Adaptive buffering of breath-by-breath variations of end-tidal $CO_2$ in humans. *J. Appl. Physiol.* **75**: 2003–2012, 1993.

Olkkola, K.T., and H. Schwilden. Adaptive closed-loop feedback control of vecuronium-induced neuromuscular relaxation. *Eur. J. Anaesth.* **8**: 7, 1991.

Orfanidis, S. *Optimum Signal Processing*. McGraw-Hill, New York, 1988.

Otis, A.B., W.O. Fenn, and H. Rahn. Mechanics of breathing in man. *J. Appl. Physiol.* **2**: 592–607, 1950.

Poon, C.-S. Ventilatory control in hypercapnia and exercise: optimization hypothesis. *J. Appl. Physiol.* **62**: 2447–2459, 1987.

Rohrer, F. Physiologie der Atembewegung. In: A. Bethe et al. (Eds.), *Handbuch der normalen und pathologischen Physiologie*. Springer, Berlin, 1925. Vol. 2, p. 70–127.

Ruttimann, U.E., and W.S. Yamamoto. Respiratory airflow patterns that satisfy power and force criteria of optimality. *Ann. Biomed. Eng.* **1**: 146–159, 1972.

Yamashiro, S.M., J.A. Daubenspeck, and F.M. Bennett. Optimal regulation of left ventricular ejection pattern. *Appl. Math. Comp.* **5**: 41–54, 1979.

Yamashiro, S.M., and F.S. Grodins. Optimal regulation of respiratory airflow. *J. Appl. Physiol.* **30**: 597–602, 1971.

## PROBLEMS

**P8.1.** The rate at which energy is expended in walking, as measured by oxygen consumption, has been found to take the form:

$$P = 0.267V^2 + 2160$$

where the velocity of walking, $V$, is given in meters per minute, and $P$ is in calories per minute. Determine the value of the walking speed that would minimize the *energy expended per unit distance.* Also, determine the power consumption rate at this optimal speed.

**P8.2.** In the Otis model discussed in Section 8.2, derive an expression for the optimal respiratory frequency, $f_{opt}$, that would minimize the work-rate of breathing if the nonlinear resistance term in Equation (8.12) were to be omitted. $f_{opt}$ should be expressed in terms of the lung mechanical parameters $K$ and $K'$, as well as the alveolar ventilation $\dot{V}_A$ and dead space volume $V_D$. Determine the optimal breathing frequency (in breaths $min^{-1}$) when $K = 8.5$ (cm $H_2O$) $L^{-1}$, $K' = 0.0583$ (cm $H_2O$) min $L^{-1}$, $\dot{V}_A = 6$ L $min^{-1}$, and $V_D = 0.2$ L.

**P8.3.** On the basis of minimization of work-rate, the Yamashiro and Grodins (1971) model predicted a rectangular inspiratory airflow pattern that is not generally observed in resting humans. This suggested to them that, during resting conditions, the respiratory system may be minimizing a different criterion function. They proposed this other criterion function to be the mean squared acceleration, *MSA*, defined as

$$MSA \equiv \frac{1}{T}\int_0^T \left(\frac{d\dot{V}}{dt}\right)^2 dt$$

where $T = 1/f$ and $f$ = respiratory frequency. Using the Lagrange multiplier method, show that the Fourier coefficients of the optimal airflow waveform that minimizes *MSA* (given the constraint of constant $\dot{V}_A$ and $V_D$) are given by

$$a_{2i-1} = \frac{384(fV_D + \dot{V}_A)}{4\pi^3(2i-1)^3}$$

To arrive at the above result, you will need to use the following equality:

$$\sum_{i=1}^{\infty}\frac{1}{(2i-1)^4} = \frac{\pi^4}{96}$$

Plot one cycle of the optimal airflow waveform, using only the first four harmonics of the Fourier series. Assume $\dot{V}_A = 6$ L $min^{-1}$ and $V_D = 0.2$ L.

**P8.4.** Use the method of calculus of variations to derive an expression for the optimal respiratory frequency in Otis's model, assuming the nonlinear resistance term can be neglected. The constraint to be imposed is that of constant $\dot{V}_A$ and $V_D$. You should arrive at the same result as in Problem P8.2.

**P8.5.** Using the MATLAB script file "`OptLVEF.m`," determine the optimal value of systolic duration and the optimal left ventricular output pattern that would minimize the energy dissipated per cardiac cycle in a typical human. Determine also the power consumption (in calories per second) of this optimal flow pattern. As well, determine the optimal time-course of left ventricular pressure over a cardiac cycle. Assume the following representative parameter values for an adult human: stroke volume = 80 ml, heart rate = 72 beats $min^{-1}$, $R_P = 1.2$ mm Hg s $ml^{-1}$, $R_C = 0.01$ mm Hg s $ml^{-1}$, $C_a = 2.3$ ml mm $Hg^{-1}$, $q_0 = 80$ ml $s^{-1}$.

**P8.6.** In mild to moderate exercise, there is an increase in ventilation ($\dot{V}_E$) that is approximately proportional to the increase in metabolic rate. Yet, in humans, arterial $P_{CO_2}$ ($P_{aCO_2}$) remains relatively constant at its resting level of ~40 mm Hg. This observation has sparked many theories about the underlying mechanism for this "exercise hyperpnea," since the increase in ventilation apparently is not due to chemoreflex mediation (for otherwise, $P_{aCO_2}$ should increase). Poon (1987) has proposed that the isocapnic exercise response can be accounted for if one postulates that $\dot{V}_E$ is governed by an optimization policy of the respiratory controller. He suggested that $\dot{V}_E$ is selected so as to minimize a criterion function that includes both the chemical and mechanical "cost" of breathing:

$$J = (\alpha P_{aCO_2} - \beta)^2 + 2\ln(\dot{V}_E)$$

while satisfying the constraint imposed by gas exchange considerations, i.e.,

$$P_{aCO_2} = \frac{863\dot{V}_{CO_2}}{\dot{V}_E\left(1 - \frac{V_D}{V_T}\right)}$$

where $V_D/V_T$, the ratio of dead space to tidal volume, is assumed constant, and $\dot{V}_{CO_2}$ is the metabolic production rate for $CO_2$ (assumed to be proportional to exercise intensity). By performing a constrained minimization, show that this model predicts a level of $P_{aCO_2}$ that is independent of $\dot{V}_{CO_2}$. Assuming model parameters of $\alpha = 0.1$ mm Hg$^{-1}$, $\beta =$ 37 mm Hg, and $V_D/V_T = 0.2$, plot $J$ versus $\dot{V}_E$ under resting ($\dot{V}_{CO_2} = 0.2$ L min$^{-1}$) and exercise ($\dot{V}_{CO_2} = 1$ L min$^{-1}$) conditions, and determine the optimal $\dot{V}_E$ and minimum $J$ in each case.

**P8.7.** Using the MATLAB script file "`acs_CO2.m`," determine how the ratio of the variance (= standard deviation$^2$) of the fluctuations in $F_{ETCO_2}$ during adaptive buffering to the corresponding variance of $F_{ETCO_2}$ during spontaneous breathing (no buffering) will change as the weighting factor $\beta$ for $F_{ICO_2}$ in the control law is changed from 0 to 1; plot this relative variance as a function of $\beta$. Determine also how these changes in $\beta$ would affect the fluctuations in $F_{ICO_2}$. Use increments of 0.1 in $\beta$.

# 9

# Nonlinear Analysis of Physiological Control Systems

## 9.1 NONLINEAR VERSUS LINEAR CLOSED-LOOP SYSTEMS

Thus far, the methods we have employed to analyze the dynamics of physiological control systems have been primarily linear methods, although it is clear that nonlinearity is the rule and not the exception in biology. As a first approximation, linear models work surprisingly well in many instances, but one can find many more instances in which the nonlinear features are critical for the functioning of the system in question. Classic examples of this include the mechanism through which the nerve action potential is generated, as modeled by the Hodgkin–Huxley equations, and various phenomena associated with nonlinear oscillators, such as frequency entrainment and phase resetting. These will be discussed in detail later.

A key disadvantage in the analysis of nonlinear systems is that the principle of superposition can no longer be applied. This has profound consequences, for it means that in contrast to linear systems, where the dynamics can be fully characterized in terms of the impulse response, the same concise means of description cannot be applied to the nonlinear system. Another consequence of the inapplicability of the superposition principle is that local solutions cannot be extrapolated to the global scale. As an example, consider the comparison of the linear lung mechanics model described previously (in Chapters 4 and 5) with a version that contains nonlinear feedback. Both models are illustrated in Figure 9.1. The responses of the linear and nonlinear models to input steps in $P_{ao}$ are displayed in Figures 9.2a and 9.2b, respectively. In each case, the responses in $P_A$ to step inputs of amplitude 1 and 2 mm Hg in $P_{ao}$ (starting at time 0.5 s) are shown as the light and bold tracings, respectively. In the linear case, the response to a step input that is twice as large leads to a proportionately scaled version of the response to the unit step. However, in the nonlinear case, the steady-state response to the larger step is clearly less than twice the steady-state response to the unit step in $P_{ao}$. Furthermore, the response to the larger step is more oscillatory than the unit step response. Thus, knowing the unit impulse response or unit step response of the nonlinear system does not enable us to predict the responses to input steps of other amplitudes.

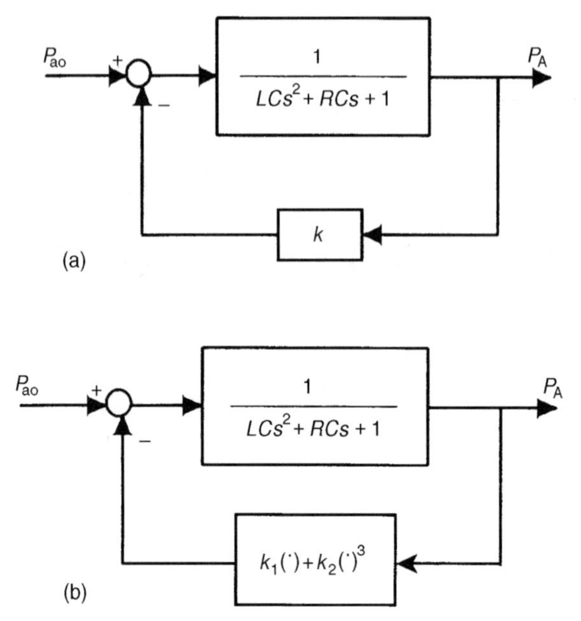


**Figure 9.1** (a) Linear lung mechanics model. (b) Lung mechanics model with nonlinear feedback.

There is another major difference between the dynamics of linear and nonlinear systems. As one might recall from Chapter 5, the dynamics of linear systems can also be characterized in terms of their frequency responses, since sinusoidal perturbation of a linear system results in a sinusoidal output of the same frequency. In nonlinear systems, however, sinusoidal perturbation can give rise to a response that contains not only the fundamental frequency of the perturbation but also higher harmonics of that frequency. Figure 9.3 shows a comparison between responses elicited from the linear and nonlinear versions of the lung mechanics model displayed in Figure 9.1. The linear response to an input sine wave of unit amplitude and frequency 0.3 Hz shows a sinusoidal output of amplitude 0.5 and the same frequency (Figure 9.3a). The same input forcing produces a nonlinear response of the same fundamental frequency and with an amplitude of approximately 0.5. However, the shape of the response is clearly nonsinusoidal but appears more "squarish," since it contains higher frequency components (Figure 9.3c). The difference becomes much more apparent when these responses are viewed in terms of their corresponding *phase-plane* plots. As described further in the next section, signals generated by systems that are governed by second-order differential equations can be completely characterized by plotting the first time-derivative versus the variable in question. In the examples given, $dP_{alv}/dt$ is plotted against $P_{alv}$. In the linear case, the phase-plane plot of the system response is an ellipse (Figure 9.3b). In the nonlinear case, the phase-plane plot also shows a closed-loop figure, but the structure of the plot is much more irregular than the ellipse (Figure 9.3d).

Another feature that illustrates the dynamic complexity of nonlinear systems is that, under certain conditions, the response to periodic stimulation at a given frequency can change dramatically if the *amplitude* of the stimulus is varied. Again, we illustrate this point with the

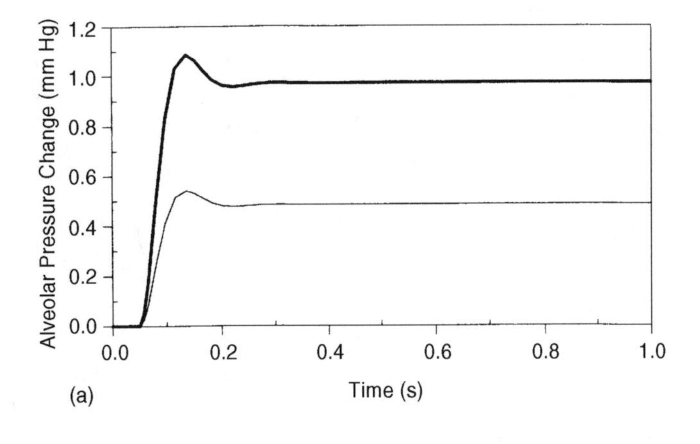


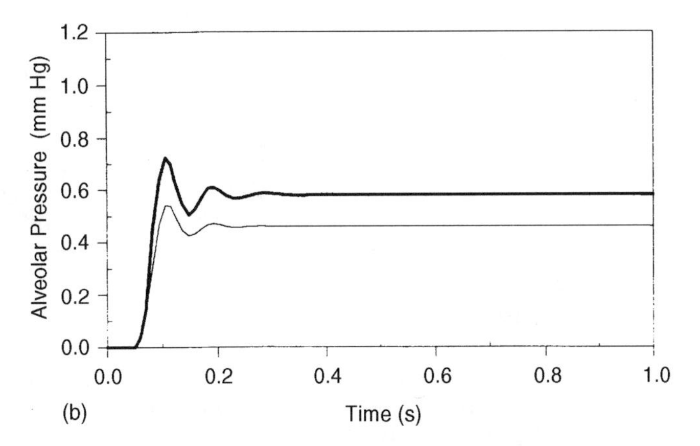


**Figure 9.2** (a) Responses of the linear lung mechanics model to a unit step (light tracing) and a step of magnitude 2 in $P_{ao}$. (b) Responses of the model with nonlinear feedback to the same inputs, showing that the principle of superposition is no longer valid in nonlinear systems.

example of the closed-loop nonlinear lung mechanics model of Figure 9.1b. When the nonlinear model is stimulated by a sinusoidal perturbation in $P_{ao}$ of frequency 0.16 Hz and amplitude 10 mm Hg, the response, as depicted by the time-series and phase-plane plots in the top panel of Figure 9.4, is essentially a very high frequency oscillation that rides on top of the (slower) fundamental frequency. When the forcing amplitude is decreased to 1 mm Hg (middle panel of Figure 9.4), the frequency of the "fast" oscillatory component is decreased and the response contains a mixture of periodic and aperiodic components. Finally, when the forcing amplitude is decreased to less than 0.01 mm Hg (bottom panel of Figure 9.4), the fluctuations in $P_{alv}$ become aperiodic and appear unpredictable. This type of dynamic behavior is known as *deterministic chaos*, since the seemingly random motion is generated by a perfectly deterministic model with no explicit noise input.

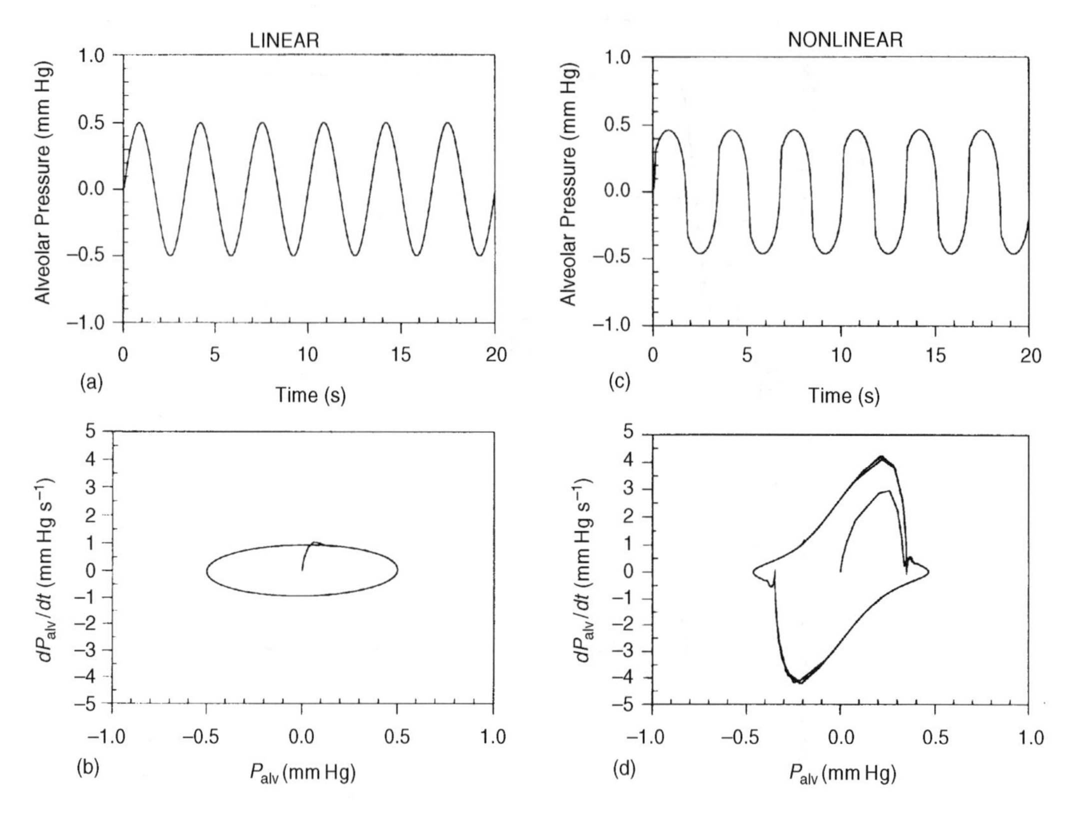


**Figure 9.3** Responses of the linear (a) and nonlinear (c) lung mechanics models to sinusoidal forcing in $P_{ao}$. The corresponding phase-plane plots are displayed in (b) and (d), respectively.

## 9.2 PHASE-PLANE ANALYSIS

Consider the motion of a simple linear spring–mass system that is characterized by the following second-order differential equation:

$$m\frac{d^2x}{dt^2} + kx = 0 \tag{9.1}$$

The steady-state solution to the above equation is given by:

$$x(t) = A\sin\left(\sqrt{\frac{k}{m}}\,t + \phi\right) \tag{9.2}$$

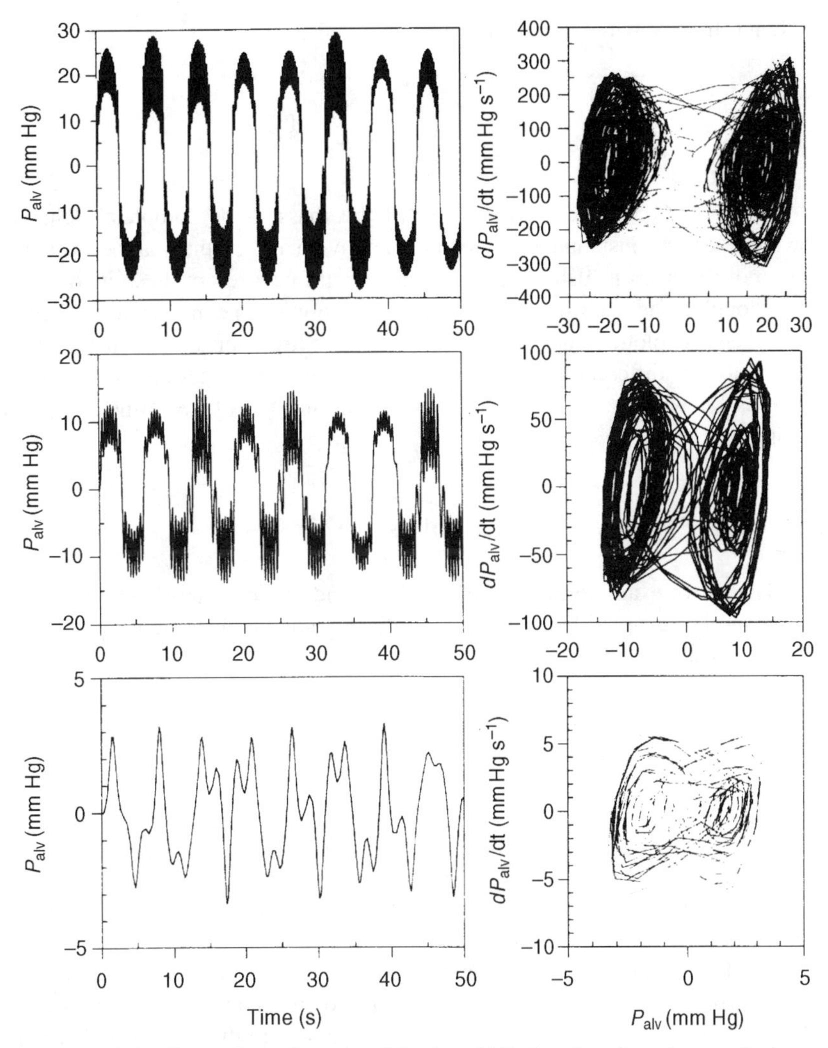


**Figure 9.4** Changes in the dynamics of the sinusoidally forced nonlinear lung mechanics model from almost periodic to chaotic as the forcing amplitude is reduced (top to bottom panels). Forcing frequency is 0.16 Hz.

where the constants $A$ and $\phi$ are determined by the initial conditions, i.e., the initial position and velocity of the mass. If we differentiate Equation (9.2) with respect to time, we obtain the velocity, $y(t)$, of the mass:

$$y(t) = x(t) = \sqrt{\frac{k}{m}}\, A \cos\left(\sqrt{\frac{k}{m}}\, t + \phi\right) \tag{9.3}$$

Then, normalizing Equation (9.2) by $A$ and Equation (9.3) by $(k/m)^{1/2}A$, and using the trigonometric equality

$$\sin^2\left(\sqrt{\frac{k}{m}}\, t + \phi\right) + \cos^2\left(\sqrt{\frac{k}{m}}\, t + \phi\right) = 1 \tag{9.4}$$

we obtain the following relation between $x(t)$ and $y(t)$:

$$x^2 + \frac{y^2}{k/m} = A^2 \tag{9.5}$$

Equation (9.5) allows the motion of the mass to be completely characterized by a knowledge of the instantaneous position and velocity of the mass, given that the initial position of the mass is also known. As such, $x$ and $y$ represent the *state* of the system. Note that, although $x$ and $y$ are functions of time, Equation (9.5) contains no explicit terms in time. Thus, when $y$ is plotted against $x$, a "stationary" ellipse appears, similar to that displayed in Figure 9.3b. For different values of $A$, ellipses of different sizes are generated. Each of these ellipses is known as a *trajectory* of the system, and the plane formed by the position and velocity axes is the *phase plane*.

### 9.2.1 Local Stability: Singular Points

The above example of a system with second-order dynamics can also be expressed in terms of a set of coupled first-order differential equations involving the position variable, $x(t)$, and the velocity variable, $y(t)$:

$$\frac{dx}{dt} = y \tag{9.6}$$

$$\frac{dy}{dt} = -\frac{k}{m}x \tag{9.7}$$

On the phase plane, the locus of points in which $dx/dt$ (velocity) or $dy/dt$ (acceleration) becomes zero is known as a *nullcline*. In general, the locus of points in the phase plane through which phase trajectories pass with constant slope is termed an *isocline*. The $x$-nullcline ($dx/dt = 0$) is a special case of an isocline that has infinite slope, while the $y$-nullcline ($dy/dt = 0$) is an isocline that has zero slope. In the linear oscillator, the $x$-nullcline coincides with the $x$-axis of the phase plane, while the $y$-nullcline lies on the $y$-axis. At the point where both nullclines intersect (i.e., at the origin), $dx/dt$ and $dy/dt$ are both simultaneously zero. This corresponds to an *equilibrium point*, a point at which there is no motion. Equations (9.6) and (9.7) can also be represented in the form of the following differential equation, in which there is no longer any explicit dependence on time:

$$\frac{dy}{dx} = \frac{-\frac{k}{m}x}{y} \tag{9.8}$$

At the equilibrium point, the numerator and denominator on the right-hand side of Equation (9.8) each becomes zero. As such, equilibrium points are also referred to as *singular points* in phase-plane terminology. Although there is "no motion" at the singular points, they do not necessarily represent stable points of equilibrium. The type of stability in the vicinity of each singular point can yield useful information about overall system dynamics.

Consider a second-order system that can be characterized by the following phase-plane equations (in which $y = dx/dt$):

$$\frac{dx}{dt} = F(x, y) \tag{9.9}$$

$$\frac{dy}{dt} = G(x, y) \tag{9.10}$$

where $F$ and $G$ can be nonlinear functions of $x$ and $y$. Suppose one of the singular points is at $(x_0, y_0)$. Consider the dynamics of motion at a point $(x, y)$ located in the proximity of the singular point, where

$$x = x_0 + u \tag{9.11}$$

$$y = y_0 + v \tag{9.12}$$

If we use Equations (9.11) and (9.12) to substitute for $x$ and $y$, respectively, in Equations (9.9) and (9.10) and perform a Taylor expansion about $(x_0, y_0)$, we obtain, after ignoring terms higher than first order, the following expressions for the local dynamics around $(x_0, y_0)$:

$$\frac{du}{dt} = F_x u + F_y v = \frac{\partial F}{\partial x} u + \frac{\partial F}{\partial y} v \tag{9.13}$$

$$\frac{dv}{dt} = G_x u + G_y v = \frac{\partial G}{\partial x} u + \frac{\partial G}{\partial y} v \tag{9.14}$$

where the partial derivative terms $(F_x, F_y, G_x, G_y)$ are all evaluated at the singular point $(x_0, y_0)$. Equations (9.13) and (9.14) can be combined in order to eliminate $v$, resulting in the following linear second-order differential equation:

$$\frac{d^2u}{dt^2} - (F_x + G_y)\frac{du}{dt} + (F_x G_y - G_x F_y)u = 0 \tag{9.15}$$

The solution to Equation (9.15) is given by

$$u = A_1 e^{\alpha_1 t} + A_2 e^{\alpha_2 t} \tag{9.16}$$

where the constants $A_1$ and $A_2$ depend on the initial conditions, and $\alpha_1$ and $\alpha_2$ are given by the roots of the following quadratic equation:

$$\alpha^2 - (F_x + G_y)\alpha + (F_x G_y - G_x F_y) = 0 \tag{9.17}$$

The solution for $v$ takes a form similar to that of Equation (9.16), except that the coefficient of each exponential term will be different from the corresponding term in Equation (9.16).

From Equation (9.16), it can be seen that the singular point in question will be *stable* only if the real parts of $\alpha_1$ and $\alpha_2$ are both *negative*. For this to be the case, two conditions must hold:

(A) The sum of the roots must be negative, i.e.,

$$F_x + G_y < 0 \tag{9.18}$$

(B) The product of the roots must be positive, i.e.,

$$F_x G_y - G_x F_y > 0 \tag{9.19}$$

Even if the singular point is stable, there is the additional question of whether the associated dynamics is oscillatory. For nonoscillatory dynamics, both roots must be real (i.e., have no imaginary parts):

$$\text{(C)} \qquad (F_x + G_y)^2 - 4(F_x G_y - G_x F_y) > 0 \tag{9.20}$$

Thus, depending on the values of the roots of Equation (9.17), one can have singular points with a variety of dynamics:

1. *Both roots real and negative*: Here, conditions (A), (B), and (C) are all satisfied. This singular point represents a *stable node*: the decay towards this equilibrium point is nonoscillatory.
2. *Both roots complex with negative real parts*: Conditions (A) and (B) are satisfied but not condition (C). The equilibrium point is stable but the decay towards it is oscillatory. This kind of singularity is known as a *stable focus*.
3. *Both roots real and positive*: Here, conditions (B) and (C) are satisfied but not condition (A). This produces an *unstable node*: any infinitesimal perturbation will cause the state point to move away from the singularity but the motion will not be oscillatory.
4. *Both roots complex with positive real parts*: Only condition (B) is satisfied. This produces an *unstable focus*: any infinitesimal perturbation will cause the state point to move away from the singularity with oscillatory dynamics.
5. *Both roots imaginary (zero real parts)*: Since the roots must be conjugate, their sum in this case is zero but their product is a positive real value. Thus, only condition (B) is satisfied. This leads to a *center*, which is considered neutrally stable. The singular point associated with the linear spring–mass system in Equation (9.1) is an example of a center.
6. *Both roots real, one positive and one negative*: Here, condition (C) is satisfied but not condition (B). However, condition (A) may or may not hold, depending on the magnitude of the negative root relative to that of the positive root. This gives rise to a peculiar type of unstable equilibrium point known as a *saddle point*.

### 9.2.2 Method of Isoclines

While the complete phase portrait of any given second-order system can be arrived at by simply solving the set of coupled first-order differential equations (Equations (9.9) and (9.10)), a good understanding of the dynamics of the system can often be obtained by applying an approximate, semigraphical analysis known as the *method of isoclines*. We illustrate the application of this method here by considering the dynamics of a simple nonlinear system: the pendulum. We assume that this pendulum consists of a heavy steel disk linked by a weightless rigid rod to a vertical fixture (Figure 9.5). If we apply Newton's Second Law to the motion of the bob in the direction tangential to the rod, we obtain the following second-order differential equation:

$$mL \frac{d^2\theta}{dt^2} = -mg \sin\theta \tag{9.21a}$$

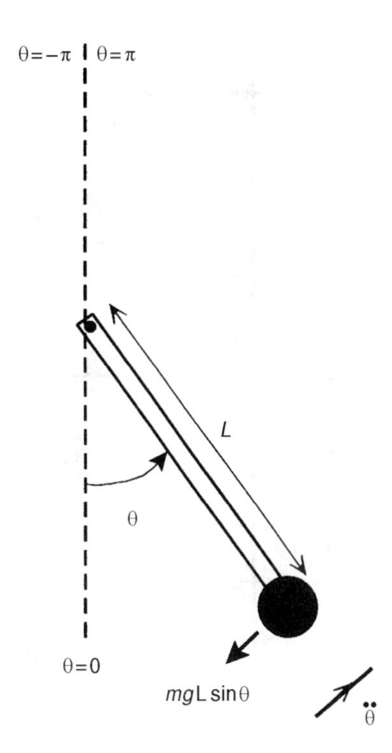


**Figure 9.5** Idealized rigid pendulum, showing tangential acceleration and tangential component of gravitational force.

where $\theta$, the angular displacement of the pendulum, is as shown in Figure 9.5. The above equation simplifies to

$$\frac{d^2\theta}{dt^2} + K\sin\theta = 0 \tag{9.21b}$$

where $K\ (= g/L)$ is a constant.

To apply the method of isoclines, we express Equation (9.21b) in the form of the equivalent phase-plane equations. Thus, we have

$$\frac{d\theta}{dt} = \varphi \tag{9.22}$$

$$\frac{d\varphi}{dt} = -K\sin\theta \tag{9.23}$$

These can also be combined to give

$$\frac{d\varphi}{d\theta} = \frac{-K\sin\theta}{\varphi} \tag{9.24}$$

The $\theta$-nullcline (i.e., points along which $d\theta/dt = 0$) is defined by the line $\varphi = 0$, which corresponds to the $\theta$-axis. The $\varphi$-nullcline (i.e., along which $d\varphi/dt = 0$) is given by $\sin\theta = 0$. Since $\theta$ can take on values between $-\pi$ and $\pi$ radians, there are three possible solutions for $\sin\theta = 0$ in this range and therefore three $\varphi$-nullclines: $\theta = 0$, $\theta = -\pi$ and $\theta = \pi$. However, it should be noted that $\theta = -\pi$ and $\theta = \pi$ correspond to the same physical configuration for the pendulum, i.e., when the bob is vertically above the hinge (see Figure 9.5). Intersection of the $\theta$-nullcline with the three $\varphi$-nullclines yields three singular points.

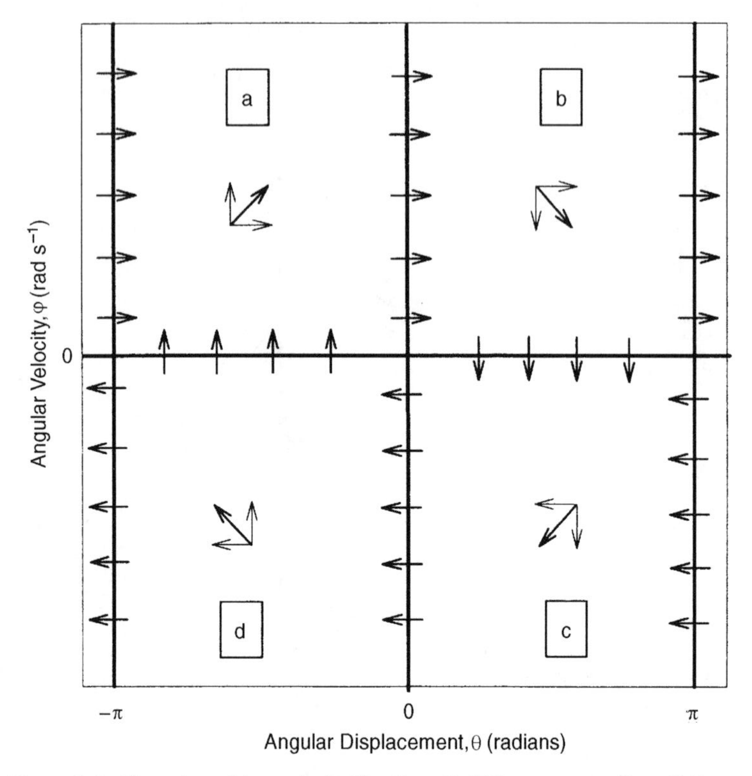


**Figure 9.6** Illustration of the method of isoclines. Bold lines represent the nullclines of the system. Arrows indicate the direction of the phase-plane trajectories. a, b, c, and d represent four regions of the phase space in which the general flow directions are different.

Treating $\theta$ as the horizontal axis and $\varphi$ as the vertical axis, these singular points are located at coordinates (0, 0), $(-\pi, 0)$ and $(\pi, 0)$, as indicated by the filled circles in Figure 9.7. The $\theta$- and $\varphi$-nullclines also divide up the phase-plane into four regions, labelled a, b, c and d in Figure 9.6. In region a, $\theta$ is negative and $\varphi$ is positive, so that by Equations (9.22) and (9.23), $d\theta/dt > 0$ and $d\varphi/dt > 0$; thus, the trajectories in this region will generally be directed upward and to the right. In region b, both $\theta$ and $\varphi$ are positive, so that from Equations (9.22) and (9.23), $d\theta/dt > 0$ and $d\varphi/dt < 0$; therefore, the flow is now directed downward and to the right. In region c, $\theta$ is positive and $\varphi$ is negative so that $d\theta/dt < 0$ and $d\varphi/dt < 0$. Finally, using similar considerations, it may be shown that in region d, the flow is directed upward and to the left. In Figure 9.6, we have also included arrows to indicate the directions of the flows on the nullclines. Thus, the overall "picture" we obtain from this approximate analysis is that for $\varphi > 0$, the trajectories generally flow from left to right, while for $\varphi < 0$, they flow from right to left. Furthermore, there is also a tendency for the flow to rotate in a clockwise manner around the origin.

Figure 9.7 shows a set of phase trajectories for the pendulum system, computed by solving Equation (9.21b) numerically for $K = 39.5$. Each phase trajectory is obtained by assigning different values to the initial conditions for $\theta$ and $\varphi$ before computing the numerical

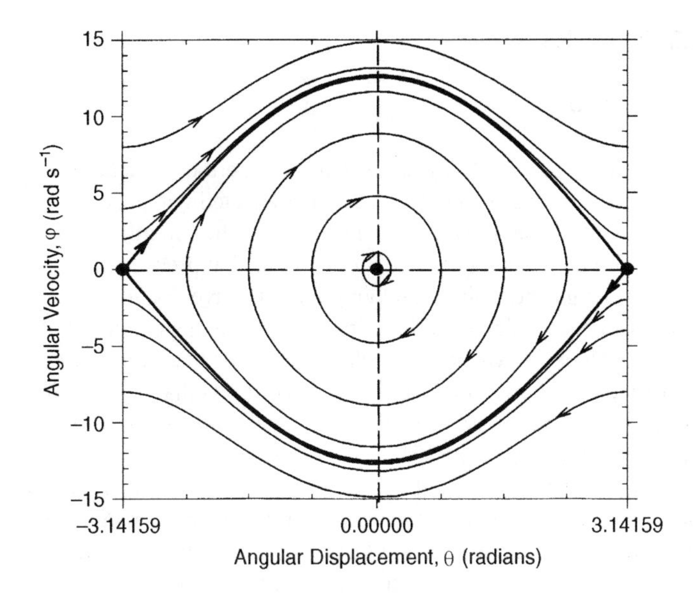


**Figure 9.7** Phase-plane portrait of the dynamics of the rigid pendulum. Filled circles represent the singular points, while the bold trajectories represent the separatrices that divide the oscillatory mode (inside) of the pendulum from the rotating mode (outside).

solution. It is clear that the trajectories shown here are consistent with the inferences made using the method of isoclines. For small starting values of $\theta$ and $\varphi$, the motion of the pendulum is a sinusoidal function of time, oscillating about $\theta = 0$ radians. Applying Equations (9.18), (9.19), and (9.20) to this example, we find that the singular point (0,0) corresponds to a center (imaginary roots). As the starting value for $\theta$ approaches $-\pi$ or $\pi$, the oscillations become less sinusoidal in character. Finally, if the initial condition for $\theta$ is set equal to $-\pi$ or $\pi$, the pendulum in principle would achieve an equilibrium position with its bob directly above its hinge. Theoretically, one can imagine that a virtually imperceptible nudge in either direction would send the pendulum swinging in that direction, making a $2\pi$ rotation until it comes to rest again with its bob balanced directly above its hinge. Thus, the singular points $(-\pi, 0)$ and $(\pi, 0)$ correspond to saddle points that are attracting when the phase trajectories approach them from one direction but repelling for trajectories in the orthogonal direction. Application of Equations (9.18), (9.19), and (9.20) will allow a verification that $(-\pi, 0)$ and $(\pi, 0)$ are saddle points.

For phase trajectories that begin at $\theta = -\pi$ or $\theta = \pi$ with nonzero velocity (i.e., $\varphi \neq 0$), $\varphi$ remains uniformly positive or negative over the whole range of $\theta$ values. This implies that the motion of the pendulum now is no longer oscillatory, but instead the pendulum simply rotates either in clockwise or anticlockwise manner around its hinge. In Figure 9.7, it is clear that the phase trajectories that lead into or away from the singular points $(-\pi, 0)$ and $(\pi, 0)$ define the boundaries that separate the oscillatory type of motion from the rotational type of motion. These trajectories, which divide the phase plane into regions of differing dynamic modes, are known as *separatrices*.

## 9.3 NONLINEAR OSCILLATORS

### 9.3.1 Limit Cycles

The only type of singularity associated with periodic oscillations that was discussed in the previous section is the *center*. The motion associated with a center takes the form of phase trajectories that close on themselves and enclose the singularity. Which particular phase trajectory is taken depends on the initial conditions that preceded the dynamics. For example, in the case of the rigid pendulum, applying an impulsive disturbance to the bob at the end of its swing can increase or decrease the swing amplitude, depending on the relative direction of the disturbance. On the phase portrait, this corresponds to a sudden change in phase trajectory to another of the concentric elliptical (or circular) orbits that enclose the center singularity. In Figure 9.8a, the pendulum bob is given a knock directed towards the equilibrium point ($\theta = 0$) at the end of its swing. This allows it to pass the $\theta = 0$ position with increased velocity, and consequently allows it to achieve an oscillation of larger amplitude. In the phase-plane diagram; this is represented by a change in trajectory from a to b (Figure 9.8a). Since no damping is present, the state-point will not return to the original phase trajectory unless another externally imposed disturbance forces it to do so.

Many physiological oscillators exhibit a behavior that is quite different from that displayed in Figure 9.8a. On the phase plane, these oscillations assume the form of a stable, closed trajectory called a *limit cycle*. What distinguishes the limit cycle from the type of oscillation discussed previously is that, although external perturbations can move the state point away from the limit cycle trajectory, it eventually always rejoins the original trajectory. This is illustrated in Figure 9.8b. Whether the state point is moved to a location outside the limit cycle (point a in Figure 9.8b) or a location inside the limit cycle (point b in Figure 9.8b), the original oscillatory behavior is always reestablished after some time.

### 9.3.2 The van der Pol Oscillator

In 1928, van der Pol and van der Mark proposed the first dynamic model of oscillatory activity in the heart. Their model consisted of the following second-order nonlinear differential equation:

$$\frac{d^2x}{dt^2} - c(1 - x^2)\frac{dx}{dt} + x = 0 \tag{9.25}$$

where the constant $c > 0$. The phase-plane properties of the van der Pol equation are most conveniently explored by applying *Lienard's transformation*, i.e.,

$$y = \frac{1}{c}\frac{dx}{dt} + \frac{x^3}{3} - x \tag{9.26a}$$

Differentiating Equation (9.26a) with respect to time, and substituting the result into Equation (9.25), we obtain

$$\frac{dy}{dt} = -\frac{x}{c} \tag{9.27}$$

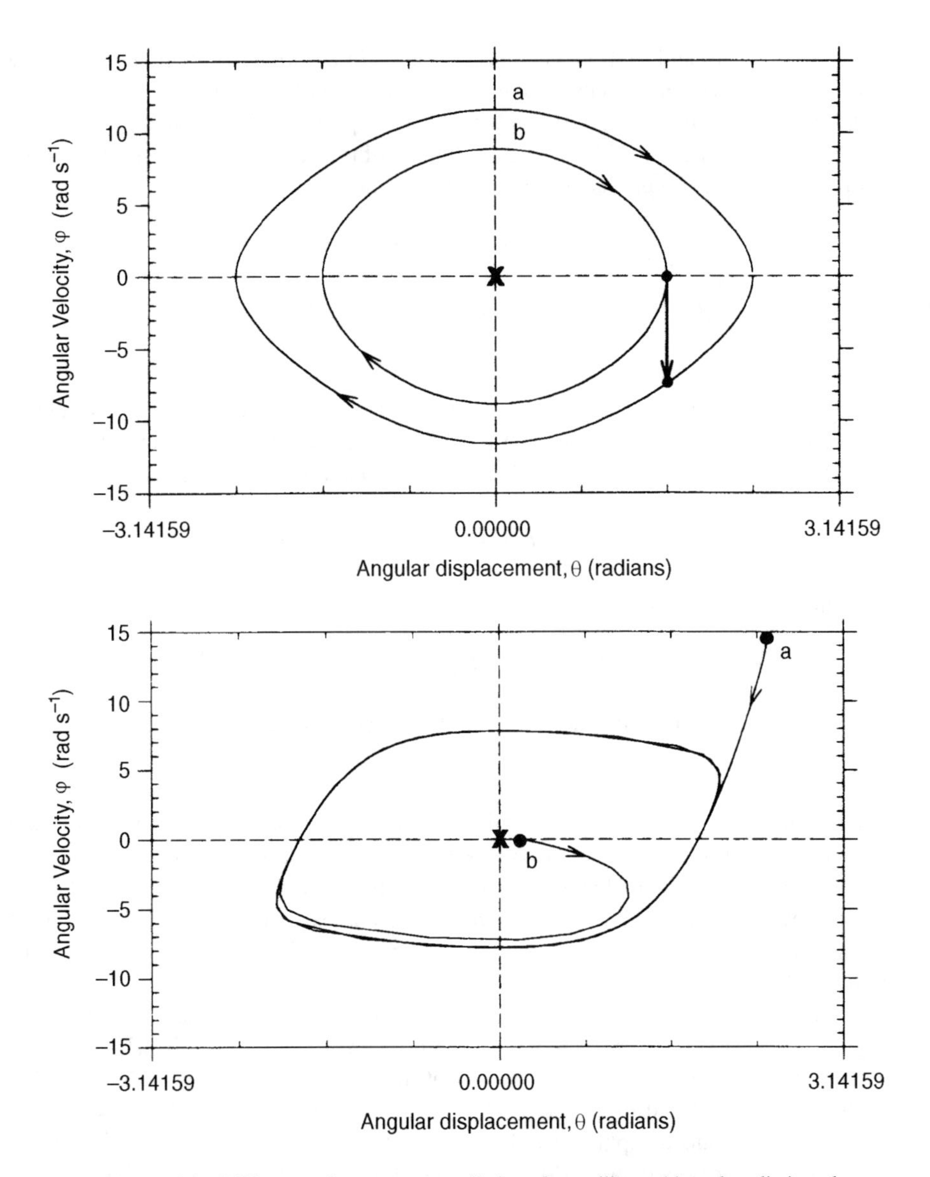


**Figure 9.8** Differences between a non-limit-cycle oscillator (a) and a limit-cycle oscillator (b). In the former, external disturbance (bold arrow) moves the phase trajectory to a different orbit (b vs. a) around the center. In the stable limit cycle, the state point always returns to its original trajectory even after an external disturbance moves it to a different location.

Rearranging Equation (9.26a), we have:

$$\frac{dx}{dt} = c\left(y - \frac{x^3}{3} + x\right) \tag{9.26b}$$

Equations (9.26b) and (9.27) form a set of coupled first-order differential equations which do not have a closed-form analytic solution. However, the techniques of Section 9.2 can be employed to provide a rough picture of the phase portrait of this dynamic system.

First, we deduce the phase-plane locations of the nullclines. The $x$-nullcline ($dx/dt = 0$) corresponds to the locus defined by the cubic function:

$$y = \frac{x^3}{3} - x \tag{9.28}$$

The $y$-nullcline ($dy/dt = 0$) is given by the vertical axis, or

$$x = 0 \tag{9.29}$$

The $x$- and $y$-nullclines intersect at only one point, i.e., at the origin (0, 0). We determine the nature of the singular point found at (0,0) by evaluating the coefficients of Equation (9.17) and subsequently the roots of the characteristic equation. Assuming that

$$F(x, y) = c\left(y - \frac{x^3}{3} + x\right) \tag{9.30}$$

and that

$$G(x, y) = -\frac{x}{c} \tag{9.31}$$

the characteristic equation describing the dynamics in the vicinity of (0, 0) takes the form:

$$\alpha^2 - c\alpha + 1 = 0 \tag{9.32}$$

Thus, condition (A) is clearly not satisfied while condition (B) is valid for all values of $c$. Whether condition (C) is satisfied depends on the value of $c$. When $c \geq 2$, the roots of Equation (9.32) will be real and positive, in which case, the singular point will be an unstable node. However, when $c < 2$, the roots become complex with positive real parts; in this case, the singular point is an unstable focus. Therefore, for all feasible values of $c$, the equilibrium point at the origin will be an unstable one.

The nullclines divide the phase plane into four major regions, as shown in Figure 9.9. In region a, $x > 0$ and $y > x^3/3 - x$. Therefore, from Equation (9.26b), $dx/dt > 0$ and from Equation (9.27), $dy/dt < 0$. This means that the phase trajectories here in general would be directed downward and to the right. In region b, $x > 0$ and $y < x^3/3 - x$, so that $dx/dt < 0$ and $dy/dt < 0$, and the phase trajectories would tend to point downward and to the left. Applying similar considerations, the pattern of flow is upward and to the left in region c, and upward and to the right in region d. For consistency, the directions of flow on the nullclines must be as displayed in Figure 9.9. Thus, in general, there is a clockwise flow of phase trajectories around the origin; however, at the same time, because of the unstable node or focus, the trajectories are also directed away from the origin.

To complete the picture, we numerically integrate Equations (9.26b) and (9.27) for given values of $c$, but with several different initial conditions. This can be achieved quite easily by implementing the system defined by Equations (9.26b) and (9.27) as a SIMULINK model. This model, the source code for which may be found in the SIMULINK file "`vdpmod.mdl`," is displayed in Figure 9.10. Figure 9.11a shows an example of the oscillatory dynamics generated by the van der Pol model for $c = 3$. This saw-toothed type of waveform is commonly referred to as a *relaxation oscillation*. A more complete representation of van der Pol dynamics is displayed in Figure 9.11b, which shows the

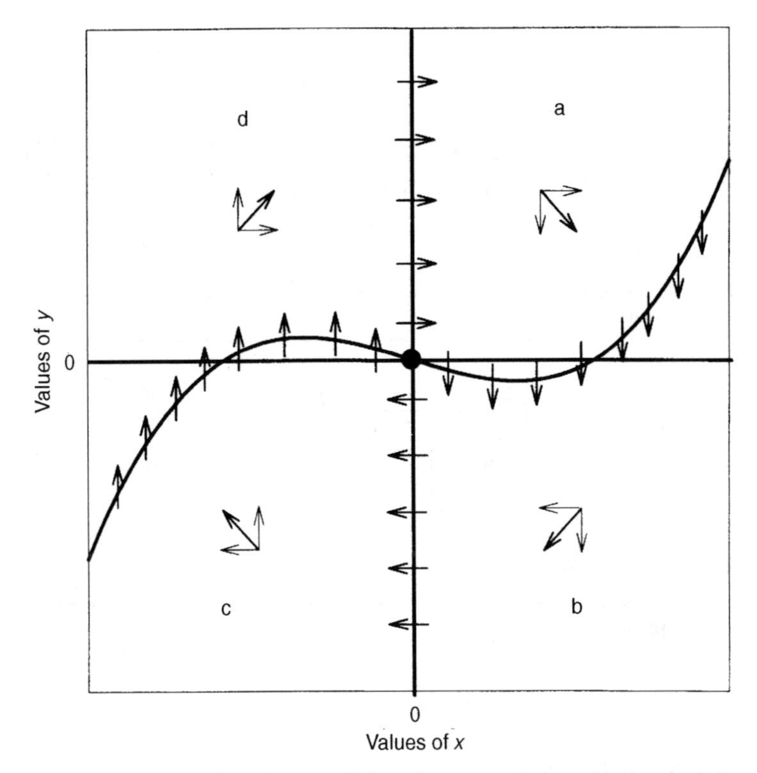


**Figure 9.9** Method of isoclines applied to the van der Pol model. Arrows indicate direction of phase-plane trajectories. a, b, c, and d represent four regions in which the general flow directions are different. The nullclines ($y = x^3/3 - x$ and $x = 0$) are shown as bold lines. The filled circle at the origin represents the only singular point for this system.

portraits for six phase trajectories that originate from different starting points (or initial conditions, shown as filled circles) in the phase space. The $x$-nullcline (shown as the dotted curve) is displayed for reference. Note that phase trajectories that originate from inside the limit cycle move away from the singular point toward the limit cycle, whereas the phase trajectories from outside the limit cycle move inward toward the limit cycle. All trajectories tend to circulate around the origin in a clockwise pattern, as predicted in Figure 9.9.

One feature that distinguishes a nonlinear oscillator, such as the van der Pol system, from a linear oscillatory system is that the former can exhibit the phenomenon of *entrainment* or *phase-locking*. The coupling of two linear systems with different natural oscillatory frequencies leads to *beating*, in which the combined output shows the original two frequencies of oscillation plus a new oscillation that corresponds to the difference between the two frequencies. However, when a nonlinear oscillator is driven by an external periodic stimulus whose frequency is quite different from the former, the output of the oscillatory system will contain a mixture of components that result from the interaction of the driving periodicity and the natural oscillation. As the driving frequency approaches the natural frequency of the nonlinear oscillator, there will be a range of frequencies over which the

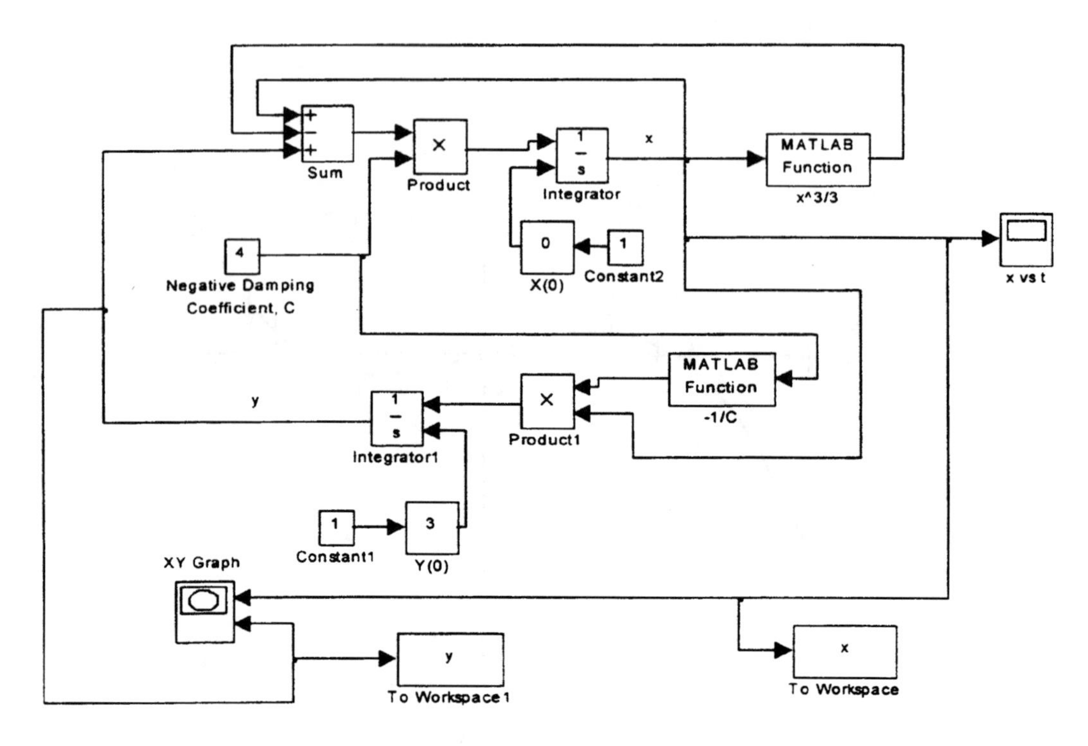


**Figure 9.10** SIMULINK implementation of the van der Pol oscillator. This model is contained in SIMULINK model file "vdpmod.mdl."

nonlinear system will adopt the frequency of the driving stimulus. This is illustrated for the case of the van der Pol oscillator in Figure 9.12. Here, the system of equations becomes

$$\frac{dx}{dt} = c\left(y - \frac{x^3}{3} + x\right) \tag{9.33}$$

and

$$\frac{dy}{dt} = -\frac{x}{c} + \frac{B}{c}\sin 2\pi ft \tag{9.34}$$

where $B$ and $f$ represent the amplitude and frequency, respectively, of the periodic stimulus. As shown in Figure 9.12a, the natural frequency of the van der Pol oscillator is 0.113 Hz. When this system is stimulated periodically at frequencies that are substantially lower (e.g., Figure 9.12b) or higher (e.g., Figure 9.12e) than 0.113 Hz, the result is a mixture of the forcing and natural frequencies. However, at frequencies close to 0.113 Hz, the van der Pol system adopts the frequencies of the driving stimulus (Figures 9.12c and 9.12d). Frequency entrainment is an important phenomenon from a practical standpoint, since it forms the basis on which heart pacemakers work. Frequency entrainment also explains the synchronization of many biological rhythms to the light–dark cycle, the coupling between respiration and blood pressure, as well as the synchronization of central pattern generators during walking and running.

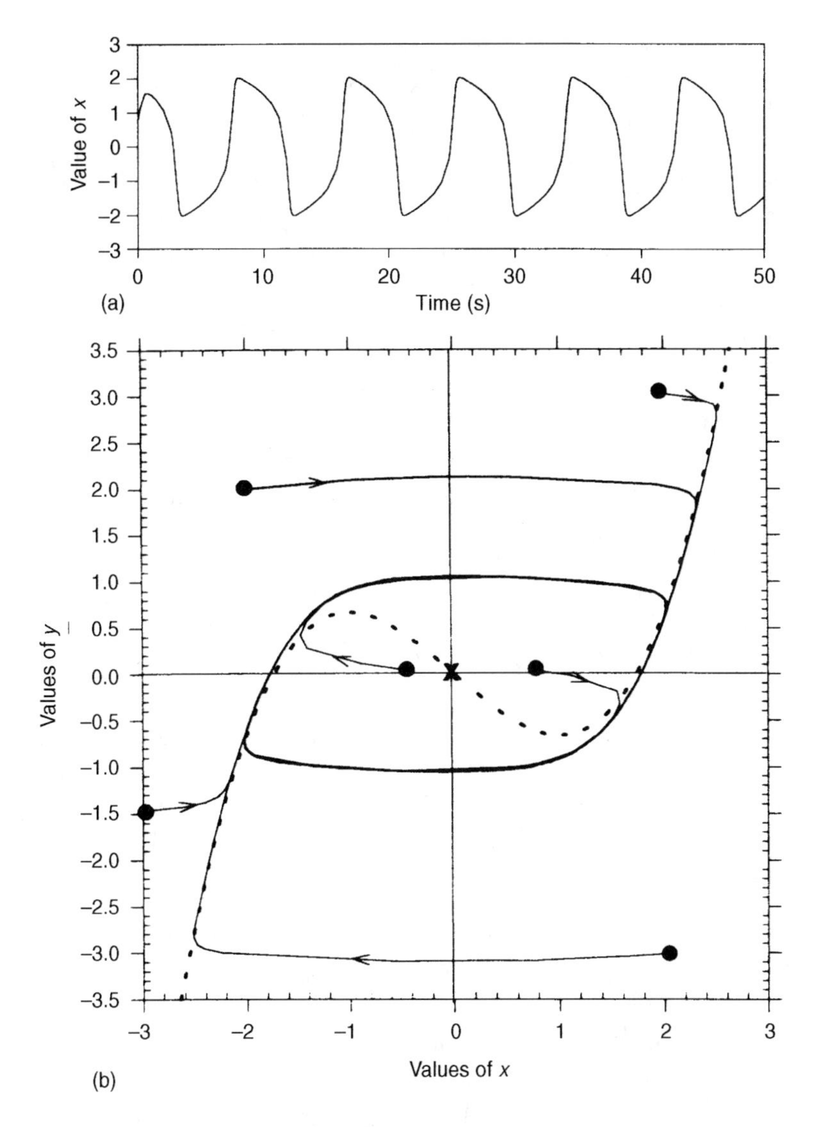


**Figure 9.11** (a) Time-course of oscillatory activity generated by the van der Pol model. (b) Phase portrait showing limit cycle formed by several trajectories initiated at different starting points (shown as filled circles). The $x$-nullcline is shown as the dotted curve, while the singular point at the origin is marked as a cross. Note consistency of flows with Figure 9.9.

### 9.3.3 Modeling Cardiac Dysrhythmias

Under normal circumstances, the cardiac cycle originates as electrical activity generated by the sinoatrial node. This impulse spreads through the atrial musculature, the atrioventricular node and finally through the Purkinje network of conducting fibers to elicit ventricular contraction. A common class of disorders, known as *atrioventricular heart block*, can occur in which the relative timing between atrial and ventricular contractions becomes impaired.

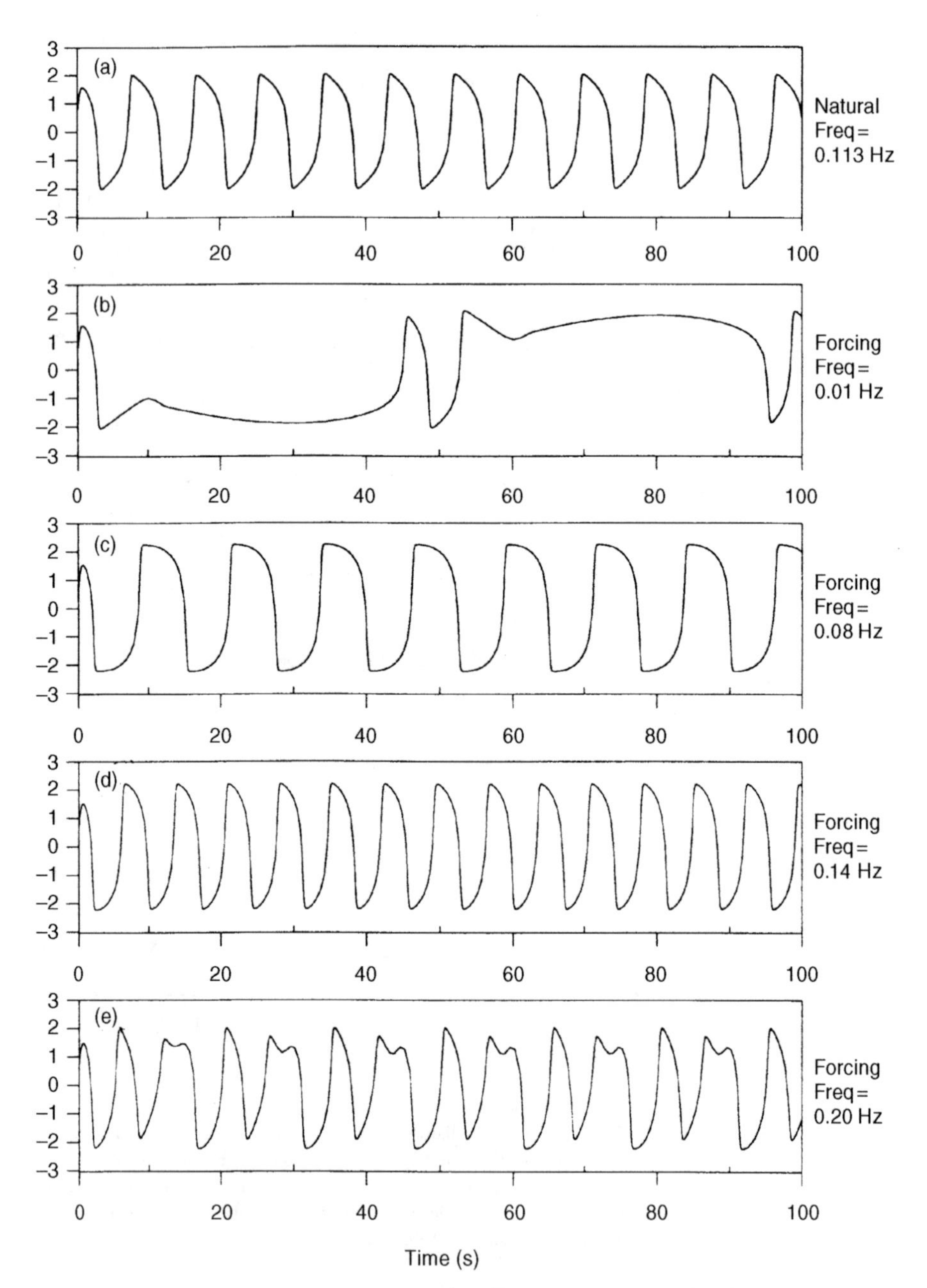


**Figure 9.12** Responses of the van der Pol oscillator to external sinusoidal forcing. The top panel (a) shows spontaneous oscillations at 0.113 Hz. The system becomes entrained to the frequency of the external forcing when the latter is close to the natural frequency of the oscillator.

One line of thought postulates that these dysrhythmias are the result of dynamic interaction among two or more coupled nonlinear oscillators in heart tissue. One of the simplest oscillator models that can demonstrate this type of phenomena is the *Poincaré oscillator*. This dynamic system is characterized by the following set of differential equations:

$$\frac{dr}{dt} = ar(1 - r) \tag{9.35}$$

and

$$\frac{d\Phi}{dt} = 2\pi \tag{9.36}$$

where $r$ represents the radial coordinate and $\Phi(-\infty < \Phi < \infty)$ represents the angular coordinate (in radians) of the state point in the phase plane. These dynamics give rise to a limit cycle that rotates anticlockwise on the unit circle (Figure 9.13). As such, it is more convenient to define the new angular coordinate, $\phi$, as follows:

$$\phi = \frac{\Phi}{2\pi} (\text{mod } 1) \tag{9.37}$$

so that $0 \leq \phi < 1$. $\phi$ is also known as the (normalized) *phase* of the oscillation.

Guevara and Glass (1982) considered what would occur if this oscillator were perturbed by an isolated, brief stimulus in the limit where $a \to \infty$. This is illustrated in Figure 9.13. The stimulus is represented by the heavy arrow that shifts the state point from its prestimulus location, corresponding to phase $\phi$ (shown in black), to its poststimulus location, corresponding to phase $\theta$ (shown in grey). The length of the arrow represents $b$, the magnitude of the stimulus. Because $a$ is infinite, the new state point moves instantaneously along the radial direction back to the limit cycle. It can be seen that when $0 < \phi < 0.5$, the state point is pushed back to a location that it had previously traversed; thus, the perturbation causes a *phase delay*. On the other hand, when $0.5 < \phi < 1$, the same stimulus would push the state point to a location further along the limit cycle; in this case, the perturbation causes a *phase advance*. This type of phenomenon is known as *phase resetting*. By careful consideration of

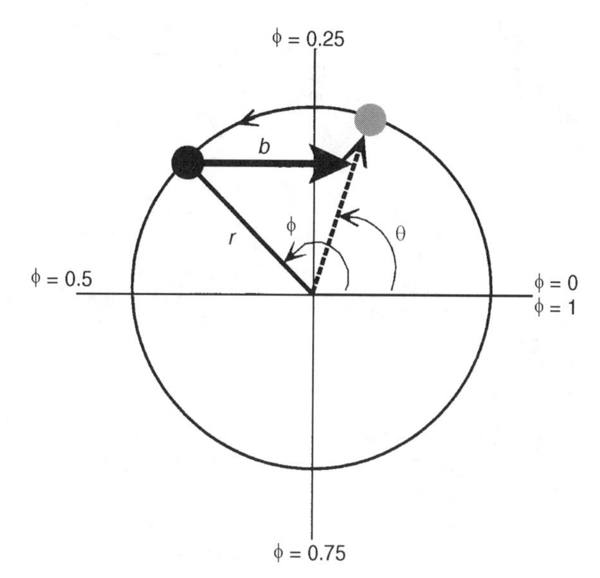


**Figure 9.13** The Poincaré oscillator. Application of a brief stimulus of magnitude $b$ leads to a resetting of the phase from $\phi$ to $\theta$ (old state point shown as black circle, new state point shown as grey circle). When $0 < \phi < 0.5$, the perturbation produces a delay in phase, but when $0.5 < \phi < 1$, the same stimulus causes an advance in phase.

the geometrical details of Figure 9.13, it can be shown that the new phase $\theta$ is related to the old phase $\phi$ through the following relationship:

$$\cos 2\pi\theta = \frac{b + \cos 2\pi\phi}{\sqrt{1 + 2b\cos 2\pi\phi + b^2}} \tag{9.38}$$

However, depending on the magnitude of the stimulus, $b$, Equation (9.38) can yield very different-looking functions that relate $\theta$ to $\phi$; these functions are termed *phase transition curves*. As illustrated in Figure 9.14, when $b < 1$ (*weak resetting*), there is a phase delay (i.e., $\theta < \phi$) for the range $0 < \phi < 0.5$ and a phase advance (i.e., $\theta > \phi$) for $0.5 < \phi < 1$, as noted earlier. Since the average slope of the phase transition curve is unity, this type of phase resetting is also commonly referred to as *type 1* resetting. However, when $b > 1$ (*strong resetting*), the effect of the perturbation on the trajectory of the state point becomes interesting and somewhat surprising at first glance. As $\phi$ increases from zero toward 0.5, $\theta$ initially increases but subsequently decreases so that, when $\phi$ attains the value of 0.5, $\theta$ becomes zero. The reason for this form of relationship may be better understood if one considers what happens for the case when $\phi$ equals 0.5: the perturbation forces the state point to a location on the horizontal axis that is past the center of the circle. The closest point on the limit cycle to this new state location is at $\phi = 0$. When $\phi$ increases beyond 0.5, perturbation of the state point leads to a resetting of phase to points that begin at $\phi = 1$, decrease below 1, but eventually increase back toward 1. As a result, the phase transiiton curve shows an apparent discontinuity and the average slope becomes zero (Figure 9.14, bold curve). This kind of resetting is known as *type 0* resetting.

Equation (9.38) characterizes the effect on the Poincaré oscillator of a single, isolated stimulus, delivered when the phase of the oscillation is $\phi$. This can be extended to produce a corresponding formula that characterizes how a periodic train of impulses would affect the

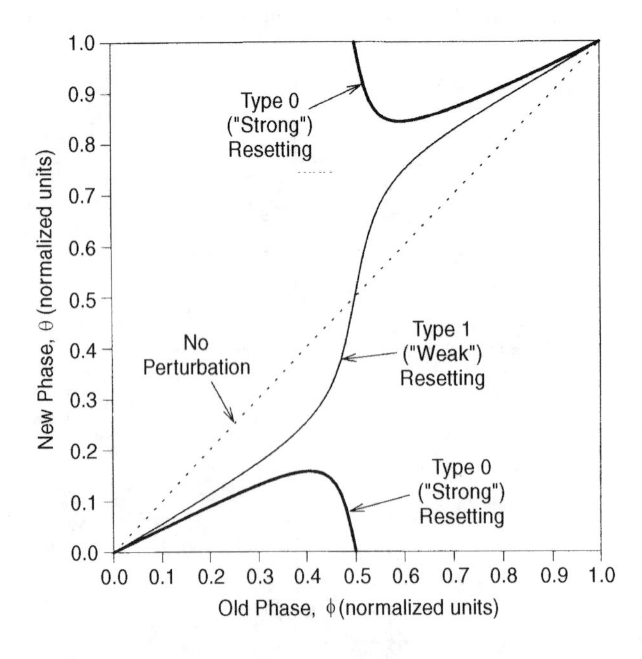


**Figure 9.14** Phase transition curves for the Poincaré oscillator. The dotted (identity) line represents the case in which there is no perturbation. When the system is perturbed by a brief stimulus of magnitude $b$ ($b < 1$), type 1 or weak resetting occurs (light solid curve). However, when $b > 1$, type 0 or strong resetting occurs (bold solid curve) in which there is an apparent discontinuity at $\phi = 0.5$.

behavior of the oscillator. If $\phi_i$ is the phase of the oscillator immediately prior to the $i$th stimulus, then the phase just before the next stimulus occurs is given by

$$\phi_{i+1} = \frac{1}{2\pi}\cos^{-1}\left(\frac{b + \cos 2\pi\phi_i}{1 + 2b\cos 2\pi\phi_i + b^2}\right) + \frac{T_s}{T_0} \tag{9.39}$$

where $T_0$ is the period of the limit cycle and $T_s$ is the interval between successive stimuli. Guevara and Glass showed that the nonlinear finite difference equation represented by Equation (9.39) can give rise to dynamics that are qualitatively similar to the dysrhythmias that have been observed in the electrocardiogram. To reproduce their simulations, we have developed the SIMULINK implementation (labeled "`poincare.mdl`") of the Poincaré model, as shown in Figure 9.15. Since the constant $a$ in Equation (9.35) is taken to be infinitely large, the state point is assumed to instantaneously return to the limit cycle after each perturbation by the external stimulus. As such, $r$ is assumed to be always equal to unity, and the radial dynamics in Equation (9.35) are neglected. In this model, we have also assumed that, whenever the rotating arm of the oscillator passes through $\phi = 0$, the system will generate a unit impulse (simulating a neural spike). The "stimulus period" ("`tau`" in Figure

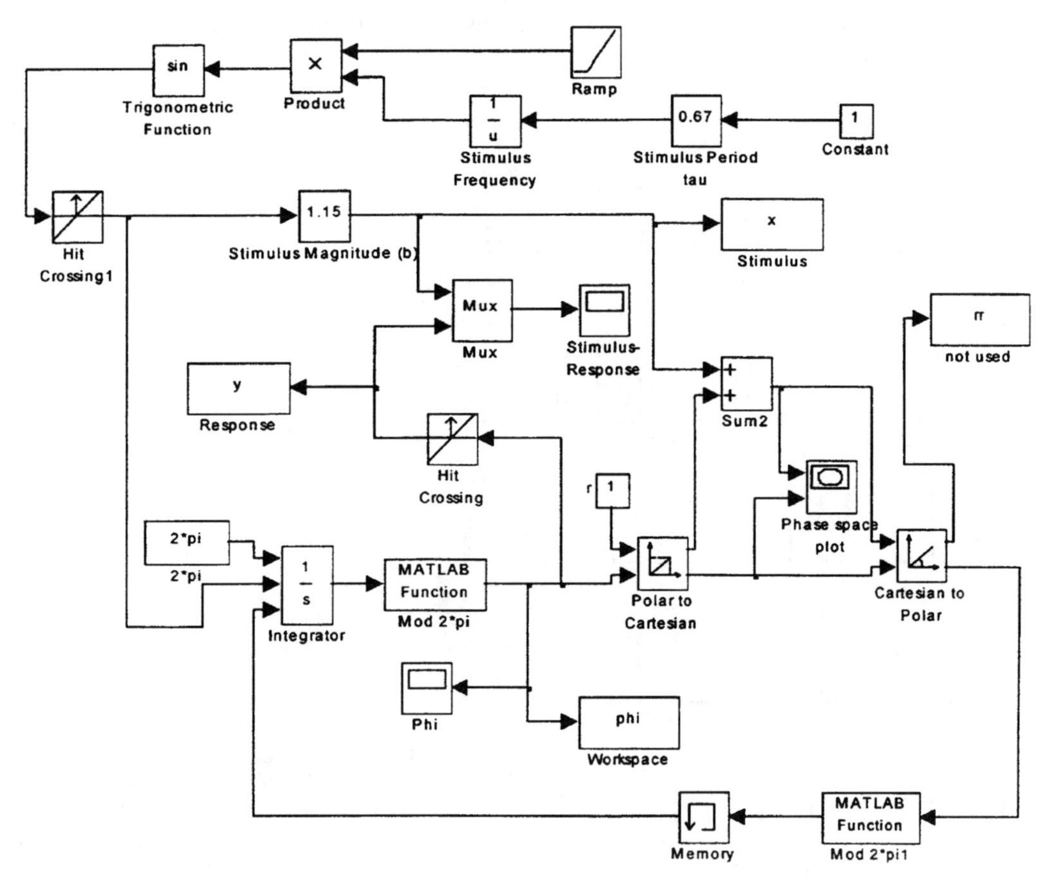


**Figure 9.15** SIMULINK implementation of the Poincaré oscillator model. This model may be found in the file "`poincare.mdl`."

9.15) that has to be specified prior to running the simulation is normalized with respect to the natural period of the limit cycle, i.e., `tau` $= T_s/T_0$.

Figure 9.16 displays some examples of the response of the Poincaré oscillator to a periodic stimulation of magnitude $b = 1.13$. In Figure 9.16a, the period of the stimulation is 75% of the length of the natural oscillatory cycle of the model. Entrainment occurs so that the Poincaré oscillator "fires" at approximately the same frequency as the external periodic stimulus. This kind of entrainment is also called *1:1 phase-locking*. When the normalized stimulation period is reduced to 0.69 (Figure 9.16b), the Poincaré oscillator now alternates between a long interspike interval and a short interspike interval. For every two stimulus spikes, the system responds with one long interval and one short interval; then the pattern repeats itself. This type of phenomenon is known as *2:2 phase-locking*. With further decrease of $T_s/T_0$ to 0.68 (Figure 9.16c), four stimulus spikes give rise to four response impulses, but the time relationship between each stimulus spike and its corresponding response is different for the four pairs, producing *4:4 phase-locking*. In Figure 9.16d, where $T_s/T_0$ is decreased to 0.65, the periodicity in the response disappears. Now the response spikes appear in an unpredictable fashion, giving rise also to skipped beats (e.g., at $t \sim 2.5$ s and $t \sim 4.5$ s in Figure 9.16d). Guevara and Glass have argued that this pattern reflects chaotic dynamics arising in this highly nonlinear system. Finally, in Figure 9.16e, decreasing $T_s/T_0$ to 0.6 leads to a reemergence of periodicity in the response. However, under these conditions, four stimulus spikes correspond to only two response spikes, producing what is known as *4:2 phase-locking*.

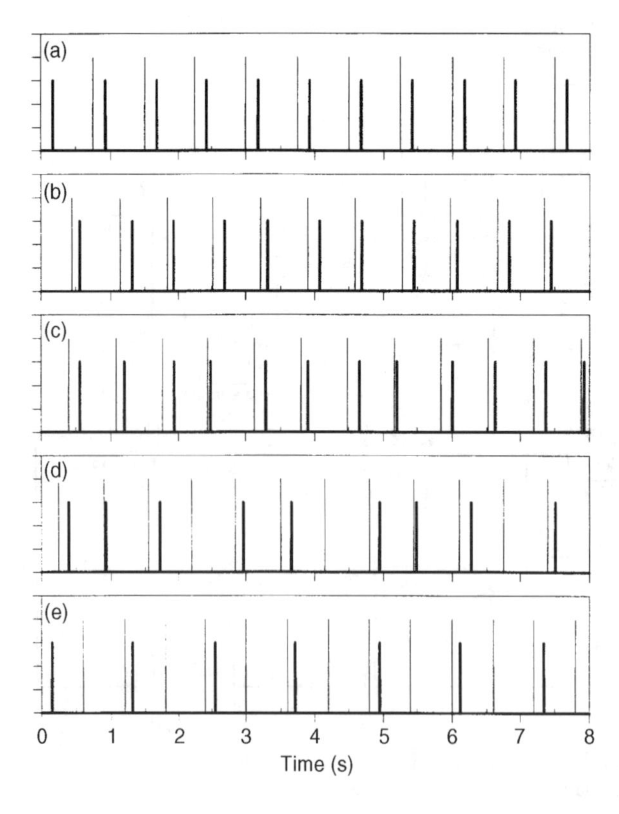


**Figure 9.16** Responses (heavy bars) of the Poincaré oscillator to periodic stimulation (light bars) of magnitude $b = 1.13$. The different panels represent responses to different stimulation periods: $T_s/T_0 = 0.85$, 0.69, 0.68, 0.65, and 0.60 in panels (a), (b), (c), (d), and (e), respectively. See text for further details.

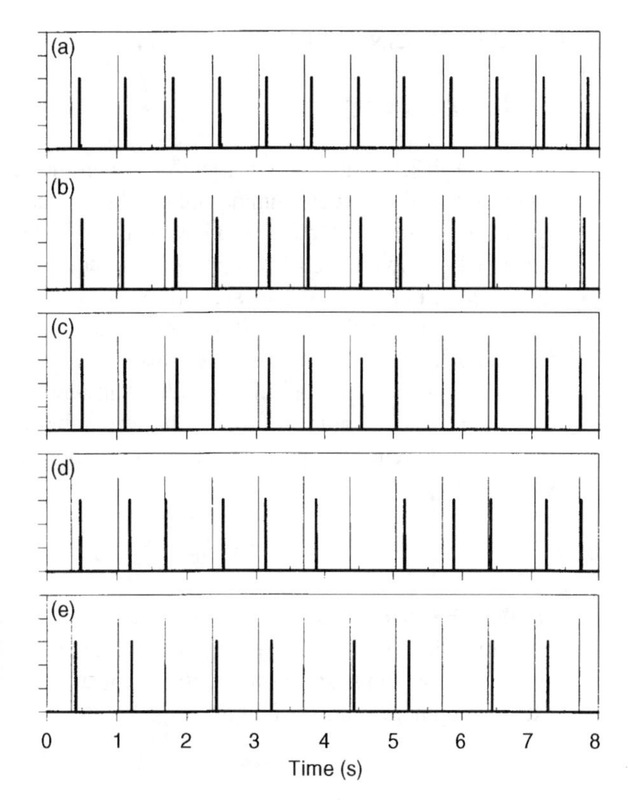


**Figure 9.17** Responses (heavy bars) of the Poincaré oscillator to periodic stimulation (light bars) of period $T_s/T_0 = 0.67$. The different panels represent responses to different stimulation magnitudes: $b = 1.40$, 1.22, 1.15, 1.12, and 1.02 in panels (a), (b), (c), (d), and (e), respectively. See text for further details.

The changes in dynamics exhibited by the Poincaré oscillator or any nonlinear system that occur abruptly as a system parameter is decreased or increased are commonly referred to as *bifurcations*. In the examples considered above, the bifurcations occurred at the points where changes in the value of the stimulus period led to sudden changes from one type of phase-locking to another mode. Bifurcations also occur when the magnitude of the stimulus ($b$) is continually varied. Figure 9.17 shows examples of the model response to periodic stimulation when $T_s/T_0$ is kept constant at 0.67 but the stimulus magnitude is varied. In Figure 9.17a, when $b = 1.4$, there is 1:1 phase-locking. Decreasing $b$ to 1.22 leads to 2:2 phase-locking (Figure 9.17b), and subsequently, 4:4 phase-locking when $b = 1.15$ (Figure 9.17c). Decreasing $b$ a little further to 1.12 produces chaotic dynamics (Figure 9.17d). Finally, with $b$ decreased to 1.02, the system once again exhibits periodic dynamics (Figure 9.17e). There is now 3:2 phase-locking of the type where the stimulus–response spike interval becomes progressively longer until the Poincaré oscillator misses a beat (e.g., at $t \sim 2$ s and $t \sim 4$ s in Figure 9.17e). This kind of pattern is similar to the clinically observed electrocardiographic phenomenon known as "second-degree AV (atrioventricular node) block with Wenckebach periodicity."

## 9.4 THE DESCRIBING FUNCTION METHOD

### 9.4.1 Methodology

The describing function method, sometimes also known as the method of *harmonic balance*, is useful in determining the conditions that produce limit cycles in relatively simple nonlinear systems. It may be viewed as an extension of the Nyquist stability criterion discussed in Chapter 6. The method assumes a closed-loop nonlinear model of the type displayed in Figure 9.18. This system can be decomposed into two parts: a linear portion that contains dynamic features, $G(s)$, and a static nonlinear component, characterized by the function $F(\epsilon)$. It is also assumed that this system is oscillating at some fundamental frequency $\omega$ (given in radians per unit time) without any input perturbation (i.e., in Figure 9.18, $u = 0$). In general, the output $x$ of the nonlinear subsystem will be a periodic oscillation with fundamental frequency plus its harmonic components. This can be expressed as a Fourier series:

$$x(t) = X_0 + \sum_{n=1}^{\infty}(a_n \sin n\omega t + b_n \cos n\omega t) \tag{9.40a}$$

On the other hand, if we assume the linear "plant" subsystem to be low-pass in nature, the harmonics in $x$ will be filtered out and the output, $y$, of the linear subsystem is likely to be approximately sinusoidal in form. Since $\epsilon$ is equal to the negative of $y$ in the absence of any external input, we can assume that

$$\epsilon(t) = E \sin \omega t \tag{9.40b}$$

And for purposes of assessing stability of the closed-loop system, we focus only on the fundamental component of $x(t)$, so that from Equation (9.40) we obtain

$$x(t) \approx X_0 + a_1 \sin \omega t + b_1 \cos \omega t \tag{9.41}$$

The *describing function*, *DF*, of the nonlinearity $F(\cdot)$ is defined as the complex coefficient for the fundamental frequency output divided by the input signal amplitude. The mathematical definition is as follows:

$$DF(E) = \frac{a_1 + jb_1}{E} \tag{9.42}$$

In Equation (9.42), *DF* is shown explicitly to be a function of the input amplitude, $E$. Although, in principle, *DF* can also be a function of frequency, this dependence is rare under

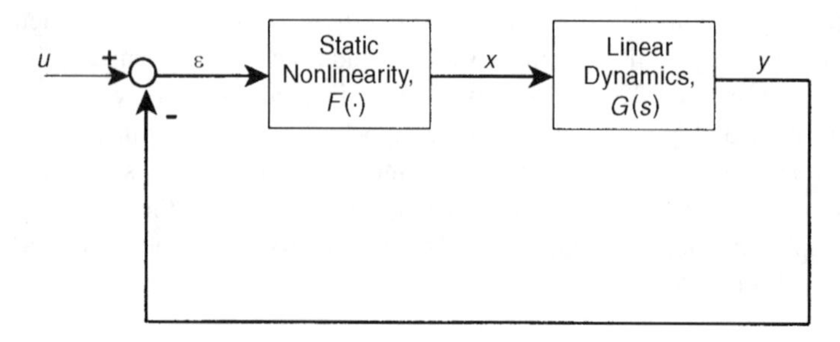


**Figure 9.18** Closed-loop nonlinear control system with static nonlinearity and linear dynamic components.

most practical circumstances. In general, the nonlinearity $F(\cdot)$ is assumed to be static, and therefore, independent of frequency.

If we let $\theta = \omega t$, then by making use of the property of orthogonality for $\sin\theta$, we can deduce $a_1$ by multiplying both sides of Equation (9.41) by $\sin\theta$ and then integrating over the range of 0 to $2\pi$. After simplification and rearrangement of terms, we obtain

$$a_1 = \frac{1}{\pi}\int_0^{2\pi} x(\theta)\sin\theta\, d\theta \tag{9.43}$$

Similarly, it can be shown that

$$b_1 = \frac{1}{\pi}\int_0^{2\pi} x(\theta)\cos\theta\, d\theta \tag{9.44}$$

Since $x(\theta)$ is periodic, changing the range of integration in Equations (9.43) and (9.44) will not alter the values of $a_1$ and $b_1$. For reasons that will become self-evident as our discussion proceeds, we choose to change the integral limits in Equation (9.43) and (9.44) to the range $-\pi/2$ to $3\pi/2$. Also, we introduce the following change of variable:

$$z = \sin\theta \tag{9.45}$$

so that the differentials $dx$ and $d\theta$ are related by

$$dz = \cos\theta\, d\theta = \pm\sqrt{1-z^2}\, d\theta \tag{9.46}$$

Note that the square-root term in Equation (9.46) will take on positive values when $-\pi/2 < \theta < \pi/2$ and negative values when $\pi/2 < \theta < 3\pi/2$. Thus, Equation (9.44) becomes

$$\begin{aligned} b_1 &= \frac{1}{\pi}\left(\int_{-\pi/2}^{\pi/2} x(\theta)\cos\theta\, d\theta + \int_{\pi/2}^{3\pi/2} x(\theta)\cos\theta\, d\theta\right) \\ &= \frac{1}{\pi}\left(\int_{-1}^{1} x(z)\, dz + \int_{1}^{-1} x(z)\, dz\right) = 0 \end{aligned} \tag{9.47}$$

The result derived from Equation (9.47) is important, as it implies that the imaginary part of $DF$ will be negative as long the nonlinear function $F(\cdot)$ is *single-valued*. If there is hysteresis in the nonlinearity, the two integrals in Equation (9.47) (second line) would not be equal in magnitude and opposite in sign, and, as a consequence, $b_1$ would not be zero.

The same analysis applied to Equation (9.43) yields the following result for $a_1$:

$$\begin{aligned} a_1 &= \frac{1}{\pi}\left(\int_{-\pi/2}^{\pi/2} x(\theta)\sin\theta\, d\theta + \int_{\pi/2}^{3\pi/2} x(\theta)\sin\theta\, d\theta\right) \\ &= \frac{1}{\pi}\left(\int_{-1}^{1} x(z)\frac{z}{\sqrt{1-x^2}}\, dz + \int_{1}^{-1} x(z)\frac{-z}{\sqrt{1-z^2}}\, dz\right) \\ &= \frac{2}{\pi}\int_{-1}^{1} x(z)\frac{z}{\sqrt{1-z^2}}\, dz \end{aligned} \tag{9.48}$$

The following expression, analogous to the Nyquist stability criterion, provides the conditions under which a limit cycle of amplitude $E$ and angular frequency $\omega$ might exist:

$$1 + DF(E)G(j\omega) = 0 \tag{9.49a}$$

Note the similarity in form between the above equation and Equation (6.8), which characterizes the condition in which any *linear* closed-loop system becomes unstable. One might consider Equation (9.49a) to be an extension of the Nyquist criterion (see Section 6.4) to a particular class of nonlinear closed-loop systems.

Equation (9.49a) can be rearranged into the form

$$G(j\omega) = \frac{-1}{DF(E)} \tag{9.49b}$$

Equation (9.49b) can be solved graphically by plotting $G(j\omega)$ on the Nyquist plane, and determining the values of $E$ and $\omega$ at which $G(j\omega)$ and $-1/DF(E)$, which is represented by a line running along part of the real axis, intersect.

### 9.4.2 Application: Periodic Breathing with Apnea

To illustrate a specific application of the describing function method, we turn to the model of Cheyne–Stokes breathing discussed in Section 6.7. As one might recall, this was a linearized model. However, for our present purposes, we will introduce a *thresholding nonlinearity* into the model by assuming that the controller output will become zero once the operating level of $P_{aCO_2}$ falls below a certain value, $B$. In other words, the simulated episodes of Cheyne–Stokes breathing would include periods of apnea. A schematic block diagram of this model is shown in Figure 9.19a, and examples of the waveforms in $P_{aCO_2}$ and $\dot{V}_E$ that one would expect to find are displayed in Figure 9.19b.

For simplicity, in the current example we will assume that there is only one chemoreflex loop in the system and that, unlike the example considered in Section 6.7, the controller responds instantaneously to changes in $P_{aCO_2}$. Suppose the controller response is given by

$$\begin{aligned} \dot{V}_E &= S_{CO_2}(P_{aCO_2} - B), \qquad && P_{aCO_2} > B \\ &= 0, && P_{aCO_2} \le B \end{aligned} \tag{9.50}$$

where $S_{CO_2}$ is the slope of the steady state ventilatory response to $CO_2$. We assume also that during periodic breathing, the $P_{aCO_2}$ waveform can be characterized by

$$P_{aCO_2} = A\sin\theta + P_M \tag{9.51}$$

where $\theta = \omega t$ and $P_M$ represents the mean level of the arterial $P_{CO_2}$ signal. Substituting Equation (9.51) into Equation (9.50), we obtain

$$\begin{aligned} \dot{V}_E &= S_{CO_2}(A\sin\theta + P_M - B), \qquad && 0 < \theta < \theta_1 \text{ or } \theta_2 < \theta < 2\pi \\ &= 0, && \theta_1 \le \theta \le \theta_2 \end{aligned} \tag{9.52}$$

where $\theta_1$ and $\theta_2$ represent the two points in the periodic breathing cycle at which $P_{aCO_2}$ crosses the apneic threshold, $B$ (see Figure 9.19b). Thus, $\theta_1$ and $\theta_2$ can be computed from Equation (9.51):

$$\theta_{1,2} = \sin^{-1}\left(\frac{B - P_M}{A}\right) \tag{9.53}$$

If we assume, as illustrated in Figure 9.19b, that $B < P_M$, then $\sin\theta_1 < 0$, implying that $\theta_1$ will be in the third quadrant ($\pi < \theta_1 < 3\pi/2$). If we define

$$\theta_0 = \sin^{-1}\left(\frac{|B - P_M|}{A}\right) \tag{9.54}$$

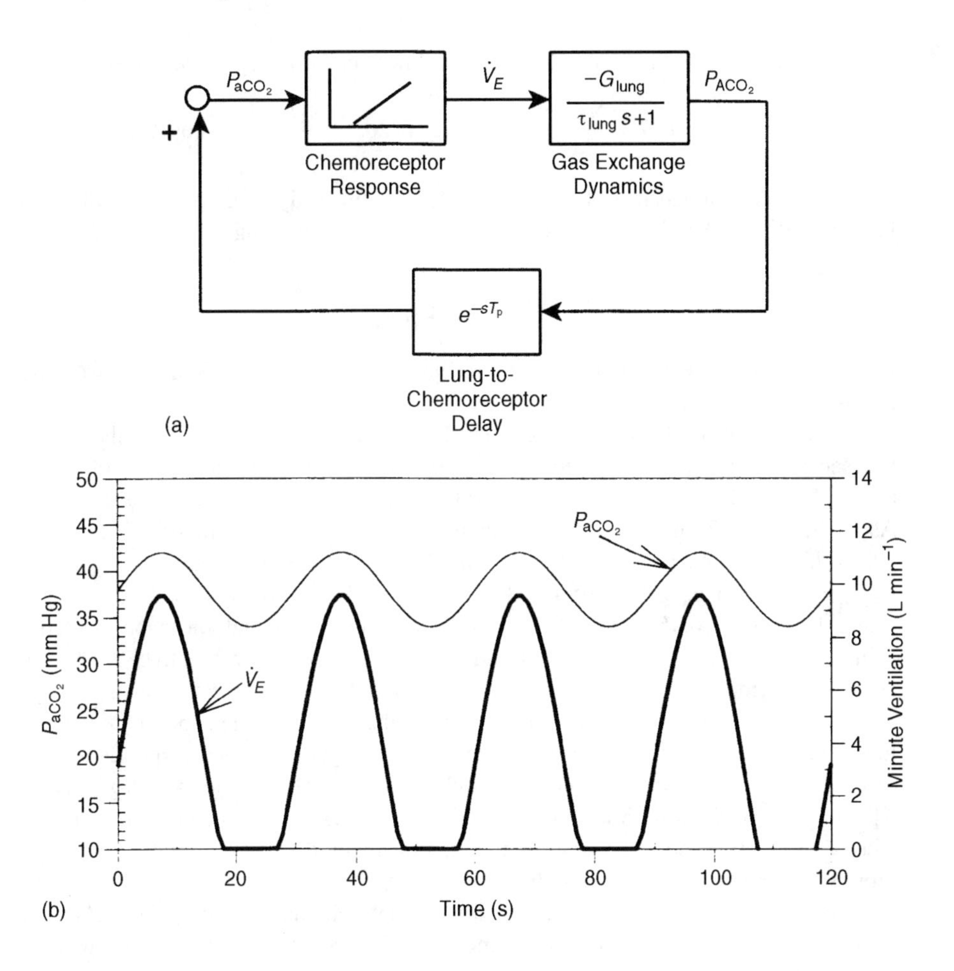


**Figure 9.19** (a) Model of periodic breathing with apnea. (b) Input ($P_{aCO_2}$) to and output ($V_E$) from the thresholding nonlinearity.

so that $0 < \theta_0 < \pi/2$, then

$$\theta_1 = \pi + \theta_0 \qquad \text{and} \qquad \theta_2 = 2\pi - \theta_0 \tag{9.55a,b}$$

We know from the previous section that, since Equation (9.50) is a single-valued function, the imaginary part ($b_1$) of *DF* must equal zero. In order to evaluate the real part, $a_1$, we substitute the expressions in Equation (9.52) and (9.55a,b) into Equation (9.43) to obtain

$$\begin{aligned} a_1 &= \frac{1}{\pi}\left(\int_0^{\theta_1} S_{CO_2}(A\sin\theta + P_M - B)\sin\theta\, d\theta + \int_{\theta_2}^{2\pi} S_{CO_2}(A\sin\theta + P_M - b)\sin\theta\, d\theta\right) \\ &= -\frac{S_{CO_2}A}{2\pi}\left(\pi + 2\theta_0 - \sin 2\theta_0 + \frac{4(P_M - B)}{A}\cos\theta_0\right) \end{aligned} \tag{9.56}$$

Therefore, the describing function of the nonlinear chemoreceptor characteristic is

$$DF(A) = \frac{S_{CO_2}}{2\pi}\left(\pi + 2\theta_0 - \sin 2\theta_0 = \frac{4(P_M - B)}{A}\cos\theta_0\right) \tag{9.57}$$

where $\theta_0$ is determined from Equation (9.54). The linear dynamic portion of this model (see Figure 9.19b and Section 6.7) is characterized by the following frequency response:

$$G(j\omega) = \frac{G_{lung}}{(1 + j\omega\tau_{lung})}e^{-j\omega T_d} \tag{9.58}$$

where $T_d$ represents the lung-to-chemoreceptor delay, and $G_{lung}$ and $\tau_{lung}$ were defined in Equations (6.45) and (6.46).

The existence of a limit cycle is predicted if the locus of $G(j\omega)$ on the Nyquist plane intersects with the locus defined by $-1/DF(A)$. Figure 9.20 shows the solution obtained using MATLAB m-file, `df_resp.m` for the case of a patient with congestive heart failure. As in Section 6.7, the parameter values used to represent this type of subject were $V_{lung} =$ 2.5 L, $K_{CO_2} = 0.0065$ mm Hg$^{-1}$, $\dot{V}_E = 0.12$ L s$^{-1}$, $\dot{V}_D = 0.03$ L s$^{-1}$, $P_{ICO_2} = 0$, $P_{aCO_2} = P_{ACO_2} = 40$ mm Hg, $Q = 0.05$ L s$^{-1}$, $B = 37$ mm Hg, and $S_{CO_2} = 0.02$ L s$^{-1}$ mm Hg$^{-1}$. Using Equations (6.45) and (6.46), it can be determined that these parameters produce values for $G_{lung}$ and $\tau_{lung}$ of 108 mm Hg s L$^{-1}$ and 6.75 s, respectively. The circulatory delay ($T_D$) employed was 14.2 s. Since the function $-1/DF(A)$ consists only of real values, its locus merely retraces a portion of the real axis. The point at which the two functions intersect (indicated by the open circle in Figure 9.20) corresponds to a frequency of 0.026 Hz, which translates into a periodicity of 38.5 s. This point also yields a value of 9.95 mm Hg for the amplitude of the oscillation in $P_{aCO_2}$. This calculation was undertaken merely to illustrate how the describing function method can be applied. As a predictor, it grossly underestimates the periodicity associated with Cheyne–Stokes breathing since it does not take into account the contributions of both chemoreflex loops and also ignores the response time associated with the chemoreflex. Incorporating these factors would produce

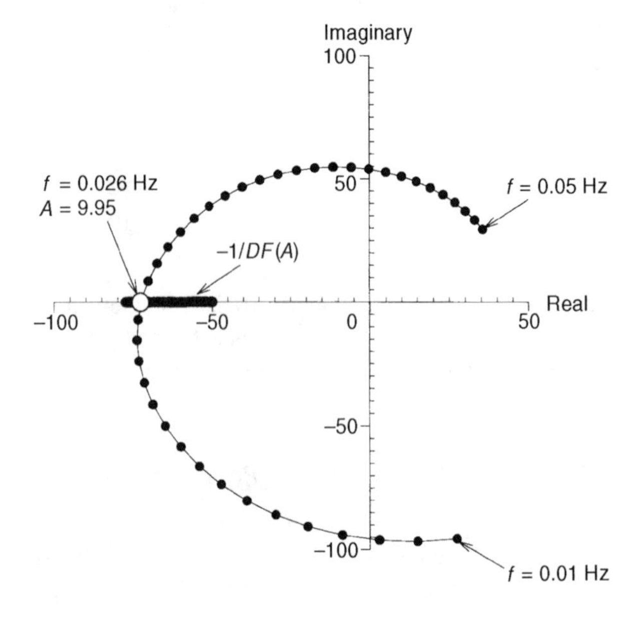


**Figure 9.20** Illustration of the describing function method for determining the periodicity and amplitude of Cheyne–Stokes breathing in congestive heart failure. Thick line on real axis represents the locus of the functions $-1/DF(A)$. Intersection of this locus with that of the linear transfer function $G(j\omega)$ yields solution for limit cycle.

more realistic predictions, but would make the expression for $G(j\omega)$ much more complicated.

## 9.5 MODELS OF NEURONAL DYNAMICS

We saw in Section 9.3 the utility of employing the van der Pol and Poincaré models as theoretical constructs for characterizing the dynamic behavior observed in physiological oscillators, such as cardiac and circadian pacemakers. However, they represent only the class of systems that are spontaneously oscillating. There is an even larger class of systems that do not spontaneously oscillate, but which can oscillate given sufficient stimulation. These systems provide a better description of the properties of nerve and muscle tissue. As the following discussion will show, these models may be viewed as closed-loop systems with both negative and positive feedback.

### 9.5.1 The Hodgkin–Huxley Model

The first relatively complete mathematical model of neuronal membrane dynamics was published by Hodgkin and Huxley in 1952. This work laid the foundation for further development of a quantitative approach to understanding the biophysical mechanism of action potential generation and was the seminal achievement that won them the Nobel Prize in 1963. Their model was based largely on empirical findings obtained through application of the voltage-clamp technique, which we discussed briefly in Section 7.4.4. This melding of physical intuition, modeling principles, and excellent experimental design is a classic example of first-class bioengineering research.

Under resting conditions, the intracellular space of the nerve cell is on the order of 60 mV more negative relative to the extracellular fluid. This net equilibrium potential is determined by the ionic concentration gradients across the slightly permeable membrane as well as by effect of active transport by the sodium–potassium pump. There is a higher concentration of potassium ions inside the cell versus a higher concentration of sodium and chloride ions on the outside. However, the membrane permeabilities to sodium and potassium are strongly dependent on the membrane potential. Depolarization of the membrane potential leads to rapid changes in sodium permeability and a somewhat slower time-course in potassium permeability. The Hodgkin–Huxley model postulates that it is the initial rapid influx of sodium ions and the subsequent outflow of potassium ions that account for the generation of the action potential that follows the depolarizing stimulus. The chloride ions do not play much of a role but account primarily for a small leakage current into the cell. The electrical circuit analog of this model is displayed in Figure 9.21a. $E_K$, $E_{Na}$ and $E_{Cl}$ represent the Nernst potentials for potassium, sodium, and chloride, respectively. Based on their measurements on the squid giant axon, Hodgkin and Huxley employed values of $-12$, 115 and 10.6 mV for $E_K$, $E_{Na}$, and $E_{Cl}$, respectively. (It should be noted that they assumed the membrane potential difference to be measured outside relative to inside; thus, their sign convention was opposite to what has generally been adopted since their early work.) $C$ represents the membrane capacitance, which is on the order of 1 μF cm$^{-2}$. $g_K$, $g_{Na}$, and $g_{Cl}$ represent the respective conductances for potassium, sodium, and chloride ions that correspond to the resistive elements displayed in Figure 9.21a; because of their voltage dependence, $g_K$ and $g_{Na}$ have been shown as variable resistors.

Application of Kirchhoff's law to the circuit in Figure 9.21a yields the following equation relating the total membrane current $I$ to the potential difference $V$ across the membrane:

$$I = C\frac{dV}{dt} + g_k(V - E_k) + g_{Na}(V - E_{Na}) + g_{Cl}(V - E_{Cl}) \tag{9.59}$$

The dependence of $g_{Na}$ on membrane voltage is characterized by the following expressions:

$$g_{Na} = G_{Na}m^3h \tag{9.60}$$

where $G_{Na}$ is a constant and assigned the value of 120 millimho cm$^{-2}$ by Hodgkin and Huxley. The time-course of $g_{Na}$ is assumed to be the result of interaction between two processes, one represented by the "activation" state variable, $m$, and the other by the "inactivation" state variable, $h$, where $m$ and $h$ each may vary from 0 to 1. These state variables each obey first-order dynamics:

$$\frac{dm}{dt} = \alpha_m(1 - m) - \beta_m m \tag{9.61}$$

and

$$\frac{dh}{dt} = \alpha_h(1 - h) - \beta_h h \tag{9.62}$$

where the rate "constants" are voltage-dependent quantities defined by the following:

$$\alpha_m = (0.1(25 - V)(e^{(25-V)/10} - 1)^{-1} \tag{9.63}$$

$$\beta_m = 0.125e^{-V/80} \tag{9.64}$$

$$\alpha_h = 0.07e^{-V/20} \tag{9.65}$$

$$\beta_h = 1/(e^{(30-V)/10} - 1) \tag{9.66}$$

The potassium conductance follows similar but somewhat simpler dynamics:

$$g_K = G_K n^4 \tag{9.67}$$

where $G_K$ is constant and given the value of 36 millimho cm$^{-2}$ in Hodgkin and Huxley's simulations. The single state variable, $n$, is assumed to obey the following first-order differential equation:

$$\frac{dn}{dt} = \alpha_n(1 - n) - \beta_n n \tag{9.68}$$

where

$$\alpha_n = 0.01(10 - V)(e^{(10-V)/10} - 1)^{-1} \tag{9.69}$$

and

$$\beta_n = 0.125e^{-V/80} \tag{9.70}$$

Finally, the membrane conductance for chloride ions is assumed to be constant and equal to 0.3 millimho cm$^{-2}$.

Equations (9.59) through (9.70) constitute the Hodgkin–Huxley model. Functionally, the dynamic behavior represented by this set of equations can also be modeled in terms of the closed-loop system shown in Figure 9.21b. A depolarizing stimulus that exceeds the threshold produces an increase in sodium conductance, which allows sodium ions to enter the intracellular space. This leads to further depolarization and greater increase in sodium conductance. This positive feedback effect is responsible for the rising phase of the action potential. However, fortunately, there is a built-in inactivation mechanism (represented by $h$) that now begins to reverse the depolarization process. This reversal is aided by the negative feedback effect of the increase in potassium conductance, which follows a time-course slower than that of the sodium conductance. The outflow of potassium ions leads to further repolarization of the membrane potential. Thus, the action potential is now in its declining phase. Because the potassium conductance remains above its resting level even after sodium conductance has returned to equilibrium, the nerve cell continues to be slightly hyperpolarized for a few more milliseconds following the end of the action potential. Figure 9.22, reproduced from the original Hodgkin–Huxley paper, shows the time-courses for $V$, $g_{Na}$, and $g_K$ as predicted by the model to occur during an action potential. The curve labeled $g$ represents the time-course of the overall membrane conductance. Multiplying this function with $V$ allows us to predict the time-course for the net membrane current during the action potential.

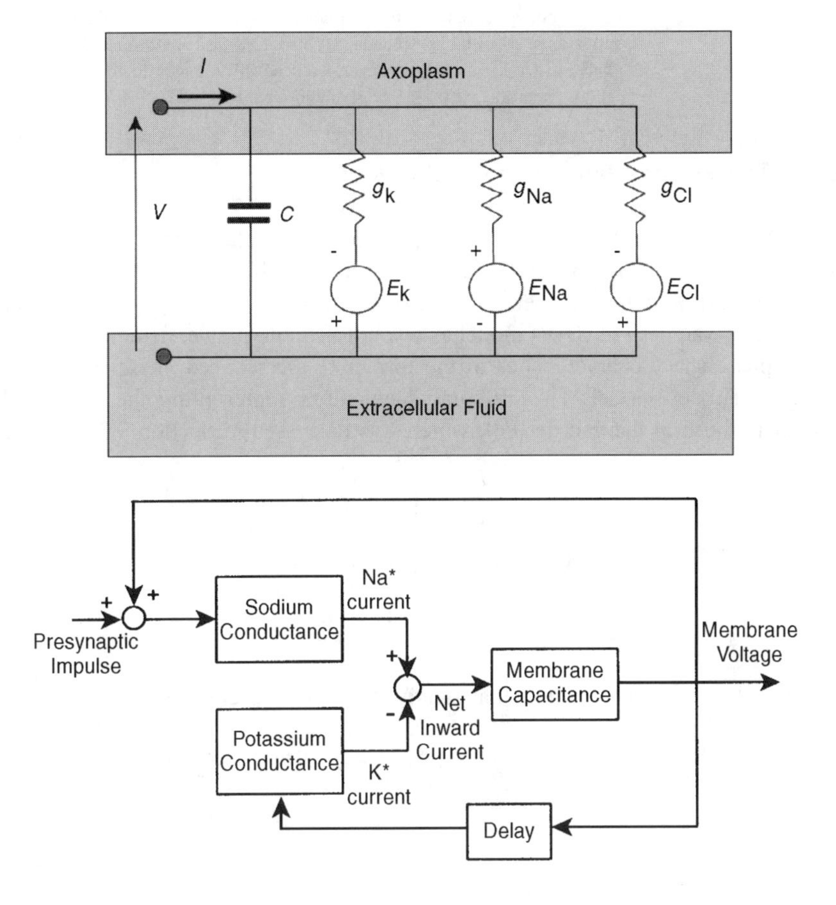


**Figure 9.21** (a) Circuit analog of the Hodgkin–Huxley nerve membrane model. (b) The Hodgkin–Huxley model as a closed-loop system with negative and positive feedback.

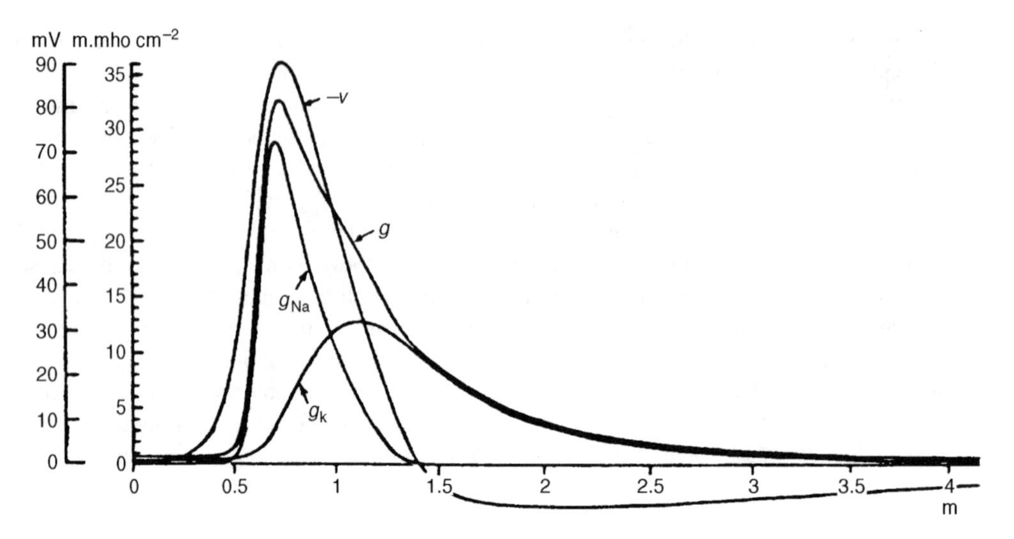


**Figure 9.22** Time-courses of ionic conductances and membrane potential during an action potential, as predicted by numerical solution of the Hodgkin–Huxley equations. Membrane voltage is displayed as predominantly negative because Hodgkin and Huxley referenced all voltages to the intracellular fluid. (Reproduced from Hodgkin and Huxley, 1952).

### 9.5.2 The Bonhoeffer–van der Pol Model

Although the Hodgkin–Huxley equation set is able to reproduce many features of neuronal dynamics, it constitutes a somewhat unwieldy model, containing several state variables and a large number of empirical constants. Fitzhugh (1961) considered the Bonhoeffer–van der Pol (BvP) model as a simplified alternative, demonstrating the similarity of the phase-space characteristics of the former to the reduced phase-space behavior of the Hodgkin–Huxley model. The differential equations representing the BvP model are very similar to those of the van der Pol, which was discussed in Section 9.3.2. These are

$$\frac{dx}{dt} = c\left(y - \frac{x^3}{3} + x + z\right) \tag{9.71}$$

and

$$\frac{dy}{dt} = -\frac{1}{c}(x - a + by) \tag{9.72}$$

where $a$, $b$, and $c$ are constants that satisfy the following constraints:

$$1 - \frac{2b}{3} < a < 1 \tag{9.73}$$

$$0 < b < 1 \tag{9.74}$$

$$b < c^2 \tag{9.75}$$

The variable $z$ in Equation (9.71) represents the magnitude of the stimulus applied to the model. This can consist of two components. The first is the steady-state level of the stimulus (i.e., $z =$ constant); as we will demonstrate below, since this enters into Equation (9.71) explicitly, it can change the dynamics of the model quite dramatically. The second component of $z$ is the transient contribution, which generally takes the form of a brief pulse of given magnitude.

By setting $dx/dt$ and $dy/dt$ in Equations (9.71) and (9.72) to zero, we can obtain expressions for the $x$- and $y$-nullclines, respectively. The $x$-nullcline is given by the cubic equation

$$y = \frac{x^3}{3} - x - z \tag{9.76}$$

which is the same as the $x$-nullcline for the van der Pol model except that, here, the vertical position (in phase space) of the cubic curve is controlled by the stimulus level ($z$). The $y$-nullcline is given by

$$y = \frac{(a - x)}{b} \tag{9.77}$$

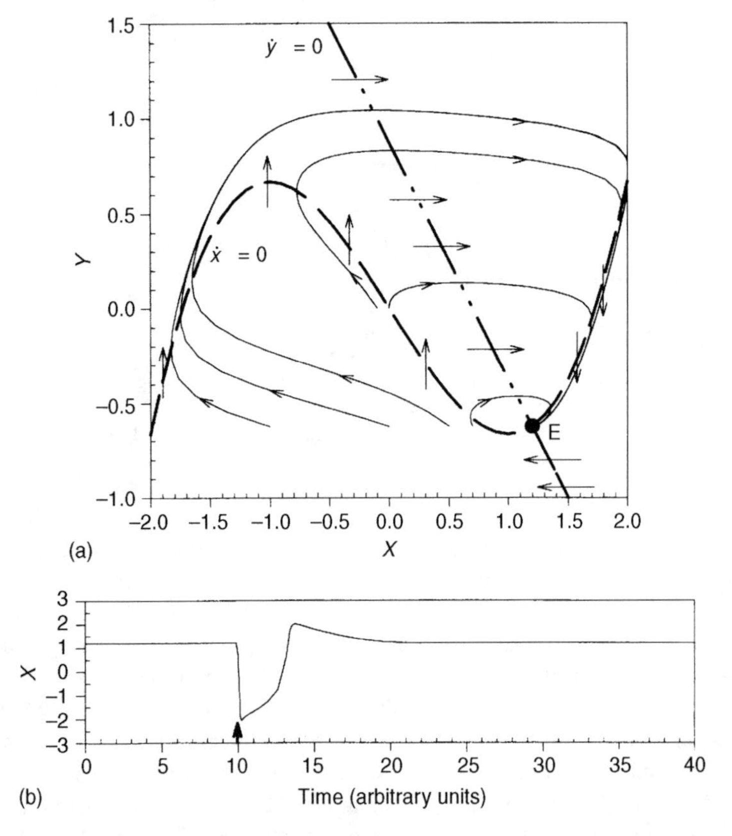


**Figure 9.23** (a) Phase-plane diagram of the Bonhoeffer–van der Pol model with steady-state stimulus level ($z$) set equal to zero. Bold dashed curve represents $x$-nullcline ($dx/dt = 0$, vertical arrows); bold chained line represents $y$-nullcline ($dy/dt = 0$, horizontal arrows). Other curves are sample phase trajectories. E is the stable singular point; $a = 0.7$, $b = 0.8$, and $c = 3$ (b) Time response of BvP model to impulsive disturbance delivered at $t = 10$ (bold arrow).

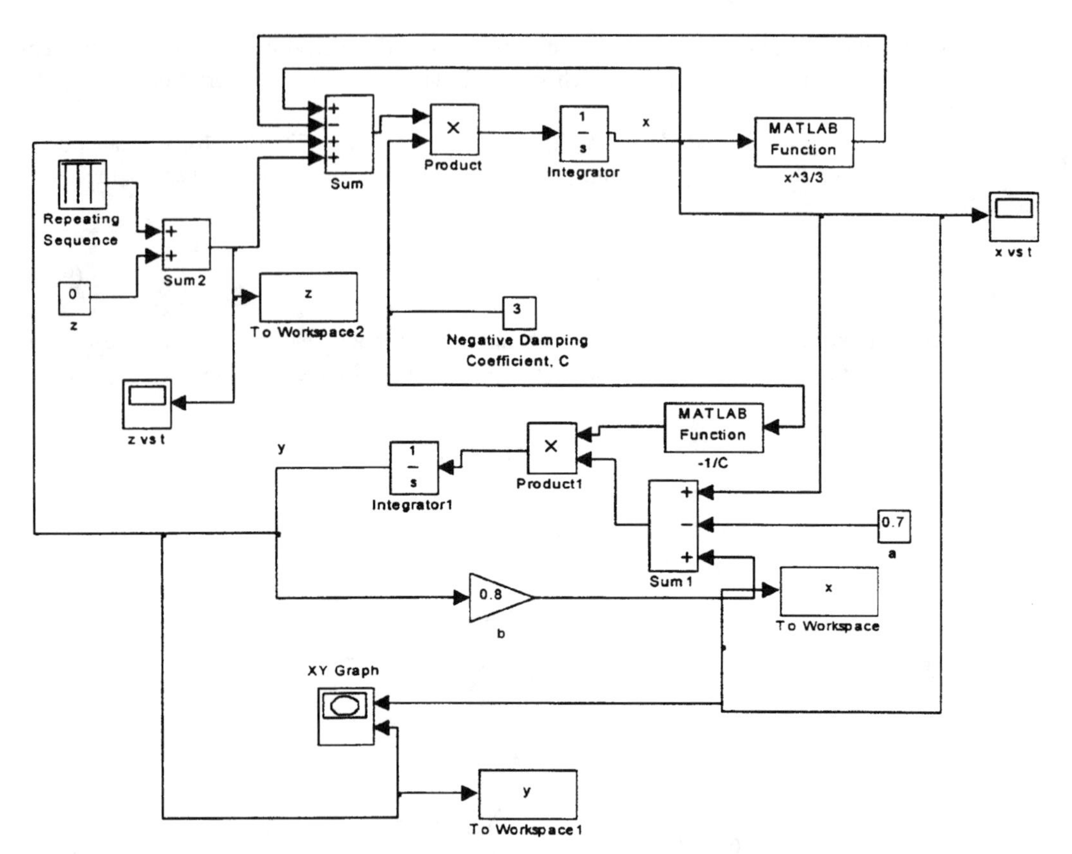


**Figure 9.24** SIMULINK implementation ("bvpmod.mdl") of the Bonhoeffer–van der Pol model for parameter values $c = 3$, $a = 0.7$, and $b = 0.8$. $x$ represents membrane voltage (with inverted sign), while $z$ represents the applied stimulus, which includes a constant level plus instantaneous impulses applied at times specified in the repeating sequence block.

which is a straight line with negative slope. The conditions specified in Equations (9.73) through (9.75) guarantee that there will be only one intersection between the two nullclines. An example of the phase plane diagram for $a = 0.7,\ b = 0.8,\ c = 3$, and $z = 0$ is displayed in Figure 9.23a. In this case, the single equilibrium point is located at coordinates $x = 1.2$ and $y = -0.625$ (point E in Figure 9.23a). This singular point may be shown, using the analysis technique of Section 9.2.1, to be a stable focus. To the left of the $y$-nullcline ($y < (a - x)/b$), evaluation of Equation (9.72) shows that $dy/dt$ must be positive. Thus, the trajectories that cross the $x$-nullcline in this region must be directed upward. Conversely, in the region to the right of the $y$-nullcline, the trajectories that cross the $x$-nullcline must point downward. In the region above the $x$-nullcline, evaluation of Equation (9.71) shows that $dx/dt$ is positive and therefore, the horizontal arrows on the $y$-nullcline in this region must be directed rightward. By similar reasoning, the trajectories that cross the $y$-nullcline below the $x$-nullcline must be directed leftward. Thus, the general flow is in a clockwise direction, as shown in Figure 9.23a.

The time-course of $x$, which simulates membrane voltage, for one of the sample phase trajectories is displayed in Figure 9.23b. Application of a brief pulse of magnitude ($\Delta z =$) $-$ 5

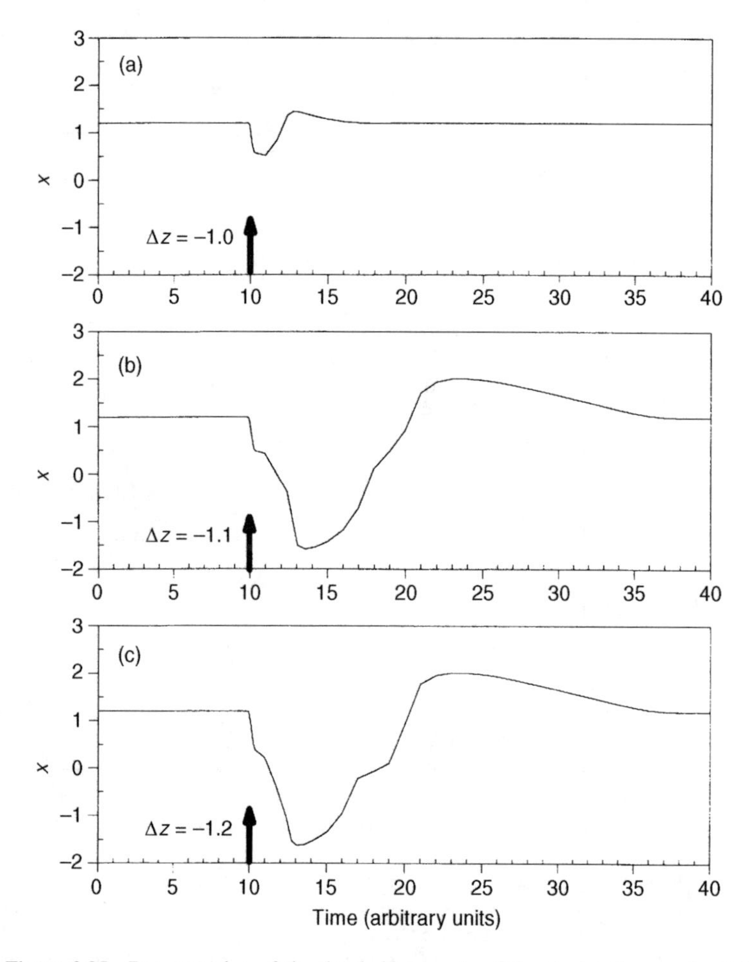


**Figure 9.25** Demonstration of the threshold property of the Bonhoeffer–van der Pol model. Brief pulses of magnitudes −1.0 (a), −1.1 (b) and −1.2 (c) are applied through the variable $z$ (baseline level of $z = 0$) at $t = 10$. Only the latter two cases are suprathreshold and produce "action potentials."

units at time $t = 10$ produces an abrupt decrease in $x$, followed later by a more gradual recovery, a small overshoot, and finally a slow return to baseline. Thus, this pattern of $x$ simulates an inverted action potential (compare this to Figure 9.22). For this reason, we will refer to negative changes in $x$ or $z$ as "depolarizing" and positive changes as "repolarizing." The simulation result shown in Figure 9.23b and others that are displayed in subsequent figures were produced by a SIMULINK implementation of the BvP model, named "`bvpmod.mdl`." The diagram of the SIMULINK configuration for this model is shown in Figure 9.24. A "repeating sequence" block is used to generate the brief "shocks" (pulses in $z$) that are applied to the model. It should be noted from Figure 9.23a that, in general, only negative pulses in $z$ (if sufficiently strong) would generate action potentials, since these displace the state point to the left of the equilibrium point.

One basic property of nerve and muscle tissue is the "all or none" phenomenon of thresholding. A pulse of insufficient magnitude, when applied to the nerve or muscle cell,

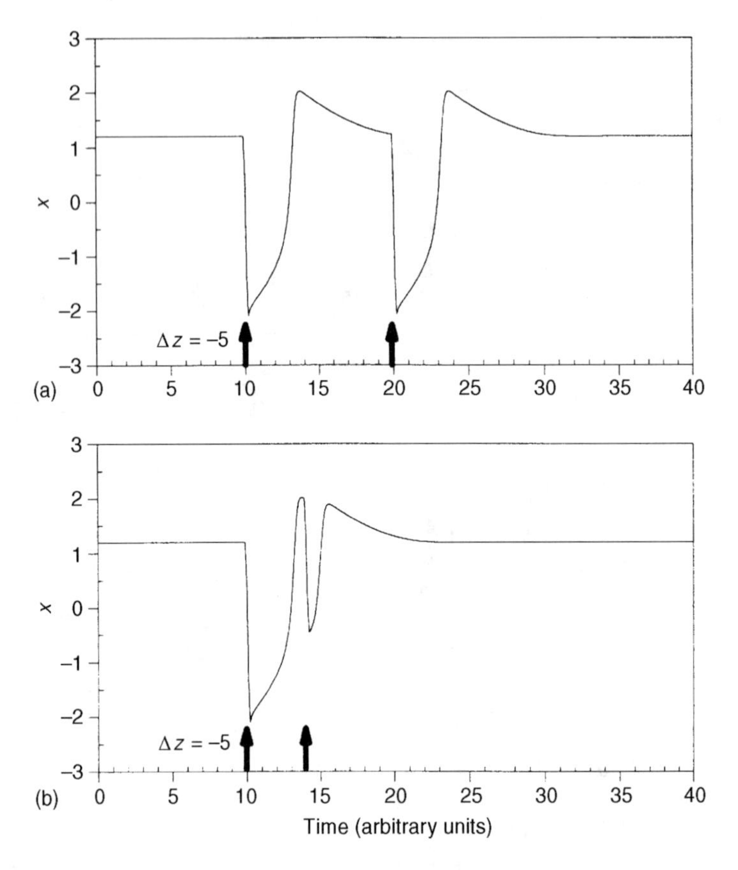


**Figure 9.26** Demonstration of the refractory property of the Bonhoeffer–van der Pol model. (a) Brief pulses of magnitude $-5$ applied at $t = 10$ and $t = 20$ produce separate "action potentials"; (b) The brief pulse applied at $t = 10$ produces an action potential, but the succeeding pulse at $t \sim 14$, during the "refractory interval," does not.

produces only a small "depolarization" but does not elicit a full-fledged action potential. However, if the stimulus magnitude is increased above threshold, the action potential becomes unstoppable. The BvP model shows this type of behavior, as illustrated in Figure 9.25. In Figure 9.25a, a brief pulse of magnitude $-1.0$ produces only subthreshold behavior— a small and brief depolarization before $x$ returns to baseline. In the phase-plane diagram, this is represented by a small displacement of the state point to the left of the equilibrium point, E. As long as the displaced state point falls to the right of the negatively-sloped portion of the $x$-nullcline or not too far to the left of it, the resulting phase trajectory will follow a small loop that leads back into E. However, if the pulse is large enough to push the state point sufficiently leftward of the $x$-nullcline, the subsequent phase trajectory will be one that moves leftward and upward, turns to the right, and then moves back toward E. The corresponding time-course of $x$ would be the action potential displayed in Figure 9.25b. Increasing the stimulus pulse magnitude does not alter the size of the action potential, as shown in Figure 9.25c and Figure 9.23a.

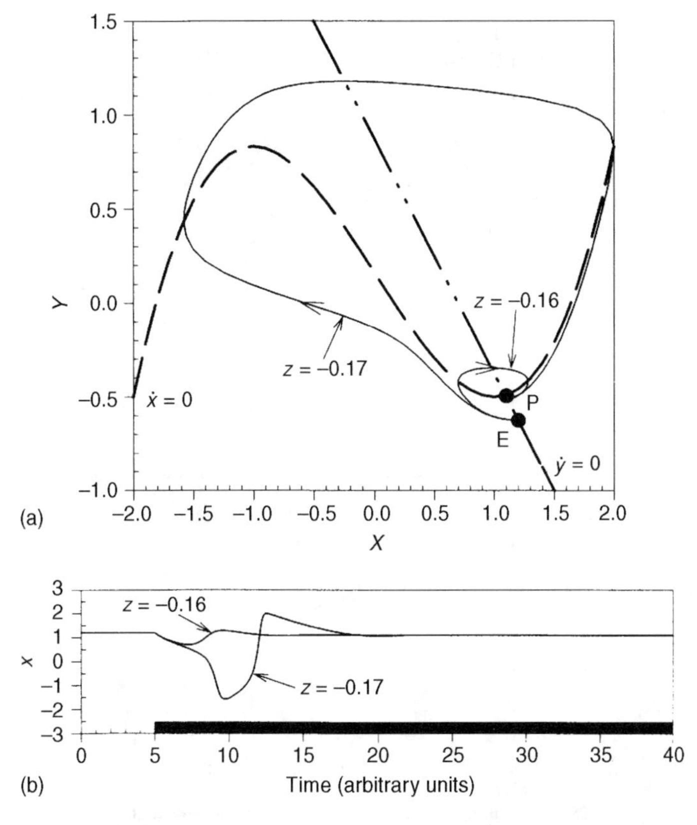


**Figure 9.27** Responses of the Bonhoeffer–van der Pol model to "step depolarizations." (a) Phase-plane diagram showing the change in singular point from E to P following application of step in $z$ of $-0.16$ units. The resulting response in $x$ is subthreshold. When the step is made slightly more negative ($z = -0.17$), an action potential is generated. The singular point corresponding to $z = -0.17$ is located very close to P and therefore is not shown separately. (b) Time-courses of $x$ following step depolarizations applied at $t = 5$ (indicated by black horizontal bar).

Another fundamental neuronal property is the presence of a refractory period. If a depolarizing stimulus is applied to a nerve cell too soon after the firing of an action potential, this stimulus will not elicit another action potential. The BvP model also exhibits this kind of behavior. Figure 9.26a shows the effect of stimulating the BvP model with two brief pulses of magnitude $-5$ units, spaced 10 time units apart. The second pulse occurs after much of the response to the first pulse has already taken place. Consequently, this second pulse leads to another action potential. In Figure 9.26b, the second pulse is applied only 4 time units after the application of the first pulse. This occurs during the early stages of the "repolarization." The net result is a small and brief depolarization, but a second action potential does not take place. This behavior can be better understood if one turns again to the phase-plane diagram in Figure 9.23. It can be seen that any stimulus that displaces the state point horizontally to the left (negative $\Delta z$) when the latter is on the final two portions of the phase trajectory, will not change its subsequent movement much, because the state point will basically follow its original course back toward E.

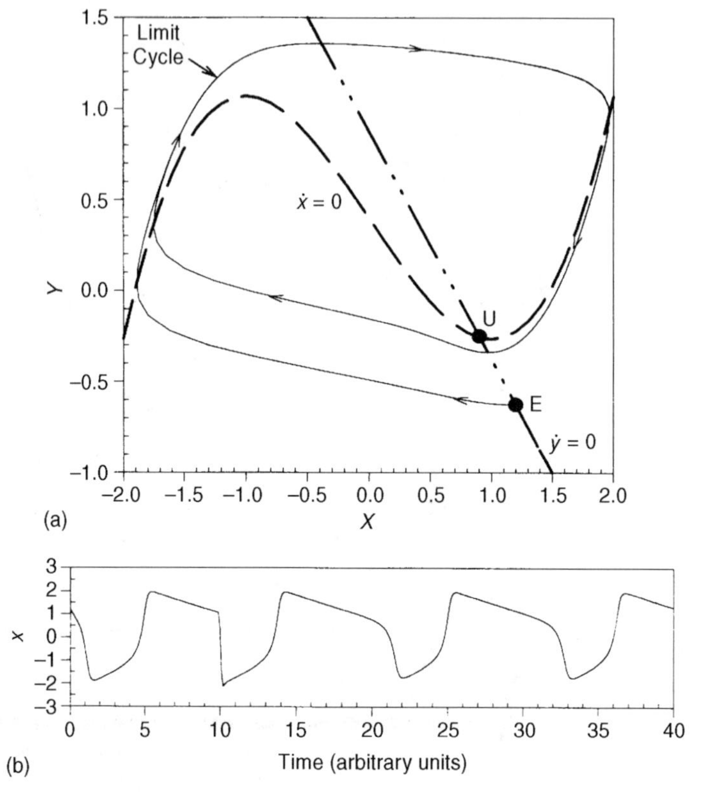


**Figure 9.28** Conditions that lead to a train of "action potentials" from the Bonhoeffer–van der Pol model. (a) Simulating a "step depolarization" by setting $z = -0.4$ shifts the $x$-nullcline upward and produces a new unstable singular point (U). The net result is the creation of a stable limit cycle that encloses the unstable focus. (b) Time-course of $x$ during application of depolarizing step.

Thus far, we have examined how the BvP model responds to brief pulses in $z$. What is the effect of changing the baseline level of $z$, which until now has been assumed to equal zero? From Equation (9.76), it is clear that giving $z$ a nonzero value would shift the vertical position of the $x$-nullcline: positive $z$-values would move the cubic curve downward, while negative $z$-values, corresponding to step depolarizations, would move it upward. Figure 9.27 shows how the model would respond when subjected to step changes in $z$ from zero to $-0.16$ and $-0.17$. In the phase-plane diagram (Figure 9.27a), the state-point starts off at E, prior to application of the step stimulus. When the step is applied, the $x$-nullcline moves upward so that the new singular point (labeled P) is now located at $x = 1.1$ and $y = -0.5$. The resulting response in $x$, however, is subthreshold (Figure 9.27b), and the corresponding phase trajectory is a small loop that begins at E and ends at P (Figure 9.27a). If the step in $z$ is made only slightly more negative ($z = -0.17$), the response becomes quite different. Now, the phase trajectory that describes the dynamics of $x$ between $z = 0$ and $z = -0.17$ takes the form of the large loop that corresponds to the generation of an action potential (Figure 9.27b). In both cases, the new singular points remain stable. Consequently, following the occurrence of the action potential, the state variable, $x$, settles down to a new steady level. However, when the applied steps in $z$ are made sufficiently negative, in addition to being displaced further upward, the new singular point (U) also becomes unstable (Figure 9.28). Instead of converging to the new "equilibrium" level, $x$ simply oscillates around it (Figure 9.28b). In the phase-plane diagram (Figure 9.28a), this corresponds to the phase trajectory that begins at E but eventually gets trapped in the limit cycle that encloses the unstable focus, U. Thus, the BvP model predicts that when sufficiently large step depolarizations are applied, an infinite periodic train of action potentials will be generated. However, both BvP and Hodgkin–Huxley models are not able to simulate finite trains of action potentials, a phenomenon commonly observed in experimental nerve preparations.

## BIBLIOGRAPHY

Bahill, A.T. *Bioengineering: Biomedical, Medical and Clinical Engineering*. Prentice-Hall, Englewood Cliffs, NJ, 1981.

Fitzhugh, R. Impulses and physiological states in theoretical models of nerve membrane. *Biophys. J.* **1**: 445–466, 1961.

Friedland, B. *Advanced Control System Design*. Prentice-Hall, New York, 1996.

Guevara, M.R., and L. Glass. Phase-locking, period doubling bifurcations and chaos in a mathematical model of a periodically driven oscillator: a theory for the entrainment of biological oscillators and the generation of cardiac dysrhythmias. *Math. Biology* **14**: 1–23, 1982.

Hodgkin, A.L., and A.F. Huxley. A quantitative description of membrane current and its application to conduction and excitation in nerve. *J. Physiol. (Lond.)* **117**: 500–544, 1952.

Kaplan, D., and L. Glass. *Understanding Nonlinear Dynamics*. Springer-Verlag, New York, 1995.

Korta, L.B., J.D. Horgan, and R.L. Lange. Stability analysis of the human respiratory system. *Proc. Natl. Electronics Conf.*, vol. 21, 1965; pp. 201–206.

Pavlidis, T. *Biological Oscillators: Their Mathematical Analysis*. Academic Press, New York, 1973.

Talbot, S.A., and U. Gessner. *Systems Physiology*. Wiley, New York, 1973.

Thompson, J.M.T., and H.B. Stewart. *Nonlinear Dynamics and Chaos*. Wiley, New York, 1986.

van der Pol, B., and J. van der Mark. The heartbeat considered as a relaxation oscillation, and an electrical model of the heart. *Phil. Mag.* **6**: 763–775, 1928.

## PROBLEMS

**P9.1.** Consider the dynamical system defined by the following pair of coupled first-order differential equations:

$$\frac{dx}{dt} = x - \frac{x^3}{3} + y - 1.5$$
$$\frac{dy}{dt} = \frac{1 - 0.6x - 0.48y}{5.4}$$

Sketch the phase-plane diagram for this system, showing the $x$- and $y$-nullclines, as well as the general directions of flow for the phase trajectories. Find out where the singular point in this system is located and determine its stability characteristics.

**P9.2** The regulation of the populations of two competing animal species, X and Y, can be modeled approximately by assuming the following pair of coupled differential equations:

$$\frac{dx}{dt} = (a - bx - cy)x$$

$$\frac{dy}{dt} = (e - fx - gy)y$$

where $x$ and $y$ represent the populations of X and Y, respectively, and $a$, $b$, $c$, $d$, $e$, $f$, and $g$ are all positive constants. Note from each equation that the rate of population growth of either species increases directly with the current population, but as the population becomes progressively larger relative to the availability of food, there will be a tendency for starvation to decrease the rate of growth (quadratic terms in $x$ or $y$). Furthermore, since the two species compete for the same limited food supply, a larger population of Y would inhibit the population growth of X and vice versa (cross-product terms). Depending on the values of the constants, stable populations of X and Y might coexist or one species might out-survive the other. Determine what would be the likely outcomes for these two species if:

(a) $\frac{a}{c} > \frac{e}{g}$ and $\frac{a}{b} > \frac{e}{f}$

(b) $\frac{a}{c} > \frac{e}{g}$ and $\frac{a}{b} < \frac{e}{f}$

(c) $\frac{a}{c} < \frac{e}{g}$ and $\frac{a}{b} < \frac{e}{f}$

(d) $\frac{a}{c} < \frac{e}{g}$ and $\frac{a}{b} > \frac{e}{f}$

**P9.3.** Develop the SIMULINK model of the system given in Problem P9.1. Determine the phase portrait of this dynamical system by computing the phase trajectories from several different starting locations on the phase plane.

**P9.4.** Use the describing function method to determine whether a limit cycle exists for the respiratory control model discussed in Section 9.4.2 if the controller response is characterized by the following equations:

$$\dot{V}_E = 0.02(P_{aCO_2} - 37), \qquad P_{aCO_2} > 39$$
$$= 0, \qquad P_{aCO_2} \leq 39$$

where $\dot{V}_E$ is given in units of L s$^{-1}$. This controller response function differs from that given in Equation (9.50) in that there is an abrupt silencing of chemoreceptor output when $P_{aCO_2}$ is decreased from values slightly greater than 39 mm Hg to values below 39 mm Hg. Assume in your computations the values for the other parameters as given in Section 9.4.2. If a limit cycle exists, determine the periodicity and amplitude of the oscillation.

**P9.5.** In the example of describing function analysis given in Section 9.4.2, the predicted oscillation period for Cheyne–Stokes breathing was on the order of 40 s. This is substantially shorter than the ~60 s cycle-time that is more frequently observed. Determine whether the inclusion of additional factors, such as circulatory mixing and chemoreceptor response time, can account for much of the difference. To do this, extend the overall transfer function of the linear component so that it takes on the following frequency response:

$$G(j\omega) = \frac{G_{lung}}{(1 + j\omega\tau_{lung})(1 + j\omega\tau_{circ})(1 + j\omega\tau_{chemo})} e^{-j\omega T_d}$$

Assume that $\tau_{circ} = 2$ s and $\tau_{chemo} = 15$ s. For other parameters, use the values employed in Section 9.4.2.

**P9.6**. By decreasing the normalized period ($T_s/T_0$) of the stimulation from 0.95 to 0.5 in the SIMULINK program "`poincare.mdl`," explore the changes in the phase relationship between the stimulus and response changes in the Poincaré oscillator during type 1 resetting with the magnitude of the stimulus, $b$, set equal to 0.9.

**P9.7**. Derive Equation (9.38), which relates the old phase $\phi$ to the new phase $\theta$ of the Poincaré oscillator when it is perturbed by a single, isolated shock of magnitude $b$.

**P9.8**. Modify the SIMULINK implementation ("`vdpmod.mdl`") of the van der Pol model so that it can be driven by an external periodic stimulus of magnitude $B$ and frequency $f$. Thus, modify this model so that its dynamics are described by Equations (9.33) and (9.34). First, set $B$ equal to 2 and vary $f$ in small increments from 0.01 to 0.2 Hz. Then, set $B$ equal to 0.5 and vary $f$ over the same range of frequencies. In each case, determine the band of frequencies over which entrainment occurs. Does the strength of the external forcing affect the entrainment band?

**P9.9**. Figure P9.1 shows the SIMULINK implementation ("`linhhmod.mdl`") of the linearized, closed-loop representation of the Hodgkin–Huxley model displayed in Figure 9.21b. Use this model to determine the membrane voltage response to a brief depolarization pulse of 60 mV. Does this model display the properties of thresholding and refractoriness? Modify the model to simulate voltage clamp experiments, in which the membrane voltage is constrained to follow a step change of +60 mV. Determine the time-courses of the sodium and potassium currents following the step depolarization.

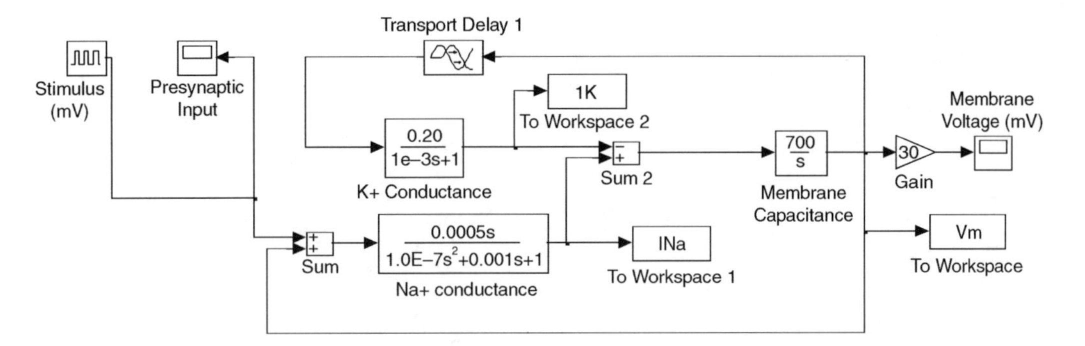


**Figure P9.1** Linearized SIMULINK representation ("`linhhmod.mdl`") of the Hodgkin–Huxley model.

# 10

# Complex Dynamics in Physiological Control Systems

## 10.1 SPONTANEOUS VARIABILITY

The examples presented in Chapter 9 showed that the presence of nonlinearity can dramatically increase the range of dynamic behavior exhibited by relatively simple open-loop or closed-loop systems. Linear systems are, by definition, constrained to obey the principle of superposition, which makes prediction of their future behavior relatively easy, unless the output measurements are heavily contaminated by random noise. By contrast, nonlinear systems can produce responses that are difficult to predict, even in the absence of noise. However, unless one subjects these signals to careful analysis, it is often difficult to distinguish one kind of "complexity" from the other. This is very much the case with the naturally occurring fluctuations, or *spontaneous variability*, exhibited by physiological control systems of all kinds. An important question that continues to stimulate bioengineering research is whether these spontaneous physiological fluctuations represent the effect of random perturbations on the underlying system or whether they are the result of the complex dynamics generated by a deterministic, nonlinear system. It is likely that both random and nonlinear influences contribute in various degrees to these natural variations. The purpose of this chapter is to examine in greater detail some of the possible mechanisms through which complex dynamics can arise.

The spontaneous variability exhibited by physiological systems ranges from highly periodic and regular waveforms to highly complex temporal structures. To illustrate the breadth of complex physiological patterns, we examine four very different time-series, displayed in Figure 10.1 (panels (a), (c), (e), and (g)). From visual inspection, all four contain different degrees of regularity and irregularity. To aid in distinguishing one signal from the other, it is useful to also examine the frequency content of these signals. Panels (b), (d), (f), and (h) display the Fourier transform magnitudes of the four signals; these were computed using the "`fft`" command in MATLAB and subsequently applying the "`abs`" command to each result.

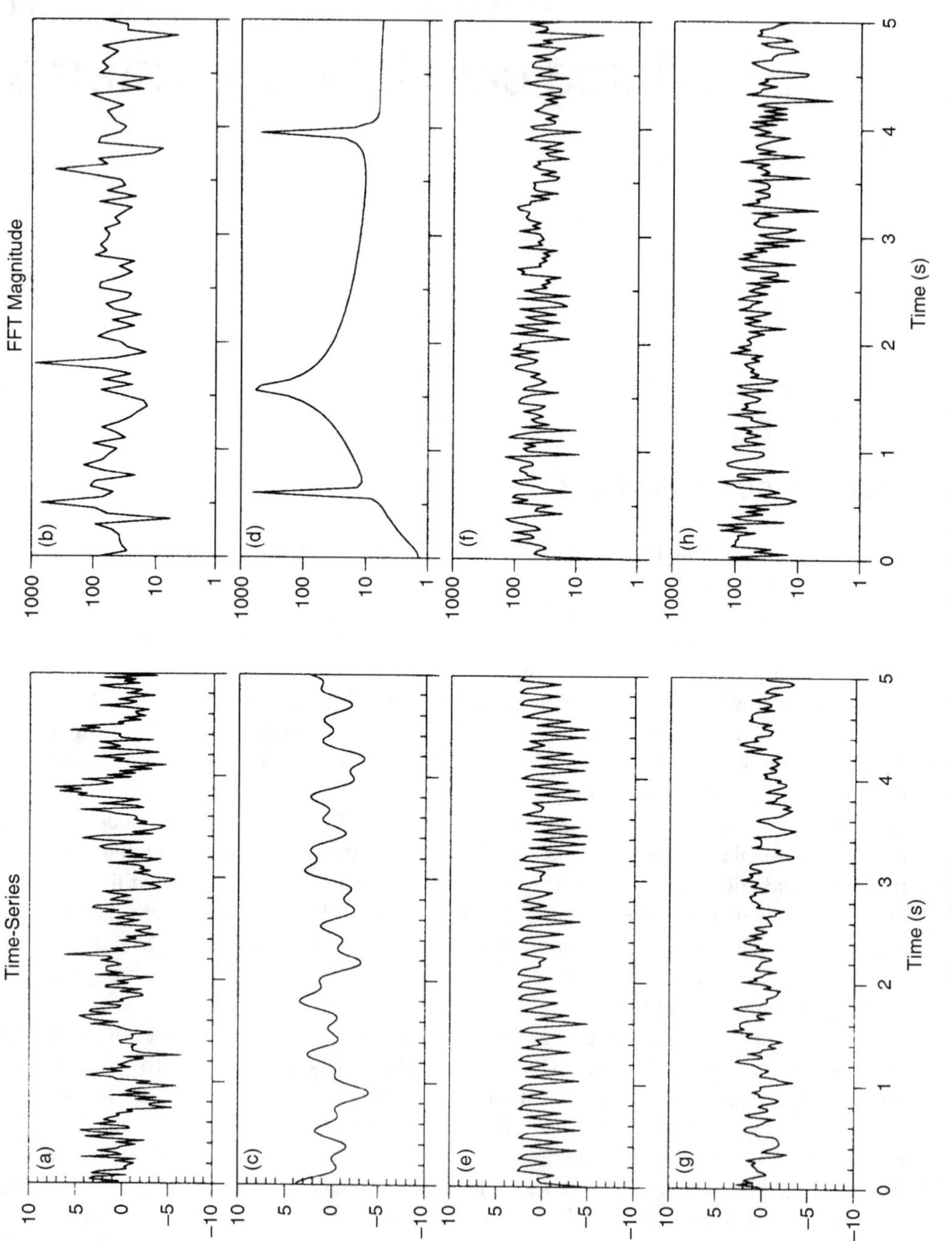


**Figure 10.1** Time-series plots and corresponding Fourier spectra of four types of system responses.

1. Although it is difficult to pick this out from the display in the time-domain, the signal displayed in Figure 10.1a actually contains three sinusoidal components with center frequencies of 1.2, 3.6, and 7.2 Hz, as the Fourier spectrum reveals in Figure 10.1b. However, the regularity of this signal is obscured by the substantial amount of random noise that accompanies it. On the other hand, linear filtering techniques can easily strip away the noise and reveal the underlying periodic nature of this signal.
2. The second signal, displayed in Figure 10.1c, does not exhibit the "noisiness" of the first signal and looks "almost" periodic, but is not. This is a *quasi-periodic* signal with no noise added. As its Fourier spectrum (Figure 10.1d) shows, this time-series, like the first in Figure 10.1a, is actually composed of three sinusoidal components. However, the three frequencies are *incommensurate* with one another, i.e., the ratio between any two of the frequencies is not a rational number. Thus, the signal exhibits *recurrence* (i.e., segments of the signal regularly "resemble" each other but are not completely reproducible) but not strict periodicity.
3. The third time-series, displayed in Figure 10.1e, appears quite "noisy," but the "noisiness" is different from that in Figure 10.1a. In fact, this signal was generated by a completely deterministic system and contains no random noise. At the same time, however, it is also quite different from the quasiperiodic signal in Figure 10.1c. The corresponding Fourier spectrum in Figure 10.1f shows no dominant periodicities, unlike the previous two signals. Instead, the power of the signal is spread over a broad band of frequencies. This is an example of a *chaotic* signal.
4. The final time-series, displayed in Figure 10.1g, looks like a hybrid between the first and second signals. However, its Fourier spectrum is broad-band and more similar to that of the chaotic signal (Figure 10.1f) than to the spectra of the other two time-series. In fact, this signal was generated by low-pass filtering white noise. This kind of signal is commonly referred to a *colored* or *correlated* noise.

The above discussion was aimed at highlighting the fact that it is not always easy to determine the source of spontaneous variability in any given system. Oscillatory systems that generate periodic signals are the easiest to recognize. However, measurement or system noise can "bury" the weaker oscillatory components (Figure 10.1a). Oscillations with incommensurate frequencies can lead to quasiperiodic outputs, that look "almost" but not exactly periodic (Figure 10.1b). However, such signals are easily distinguishable from the others by the presence of significant peaks in their Fourier spectra. Although the chaotic signal is very different from the periodic or quasiperiodic signal, a common feature is that the values taken by any of these signals are always bounded. In phase-space representation, one would observe that all the corresponding points fall only in certain defined regions, leaving other areas void. The multidimensional object formed by these filled regions is called the *attractor*. By contrast, the random signal would have a representation that filled the entire phase space.

In the rest of this chapter, we will highlight a number of control mechanisms that can give rise to the spontaneous variability that is so much a trademark of physiological signals. Nonlinear component properties and delays in the transmission of feedback information provide potent sources of periodic, quasiperiodic, and chaotic behavior. There are several oscillatory components in the body, and the interactions among these oscillators as well as the effects of external rhythms, such as the environmental light, can produce complex and unexpected consequences. Moreover, most physiological parameters are time-varying. Thus, one would expect these "nonstationarities" to exert an important influence on the physiological system in question. Finally, most people would agree on the existence of truly random processes at various hierarchical levels in physiology. The propagation of noise generated by these processes through the feedback loops of a given system can also lead to highly complex behavior.

## 10.2 NONLINEAR CONTROL SYSTEMS WITH DELAYED FEEDBACK

### 10.2.1 The Logistic Equation

The logistic equation, proposed by May (1976) as a model of population growth, is one of the best examples of how a simple nonlinear process can lead to highly complicated dynamics. In May's population model, $x_n$ represents the (normalized) population of the current generation of a given species. $x_n$ depends on two opposing influences. The first is a factor tending to increase the species population—this, of course, depends on the birth rate and the population of the previous generation of this species. The second factor represents the limiting influence to growth resulting from a finite supply of food or energy. This second factor is what imparts nonlinearity to the model. Thus, the model can be written as

$$x_n = \alpha x_{n-1} - \alpha x_{n-1}^2 = \alpha(1 - x_{n-1})x_{n-1} \tag{10.1}$$

This process can also be cast in the form of a nonlinear feedback control system, as illustrated in Figure 10.2. Note that, in this case, the parameter $\alpha$ can be thought of as a "gain factor" in the feedback control system.

The dynamic behavior of the logistic equation depends on the magnitude of $\alpha$. When $\alpha$ is less than 3, $x_n$ simply converges to a steady-state equilibrium value after a number of iterations. However, when $\alpha$ increases above 3 but is below 3.4495, this system exhibits a simple oscillation with a period of two time-steps; this is known as a *period-2* oscillation (Figure 10.3a). When $\alpha$ increases further and now falls into the range between 3.4495 and 3.5441, the behavior changes character and now becomes a *period-4* oscillation (Figure 10.3b). Subsequently, when $\alpha$ is increased to any value between 3.5441 and 3.5644, the cycling continues but now the oscillation is a *period-8* cycle (Figure 10.3c). With further increases of $\alpha$, there are successive doublings of the periodicity. However, when $\alpha$ increases beyond 3.57, the periodic behavior disappears and chaotic dynamics emerges instead (Figure 10.3d). This kind of change in dynamics of the system is a classic illustration of what has been termed *the period-doubling route to chaos*. The reader should be cautioned that the logistic system does not always behave chaotically when $\alpha$ is between 3.57 and 4. In certain subranges of $\alpha$, the system reverts to an orderly periodic behavior, such as the period-3 cycling shown in Figure 10.3e when $\alpha = 3.83$. Then, just as abruptly, in other regions, the behavior can become chaotic again (e.g., Figure 10.3f, when $\alpha = 3.87$).

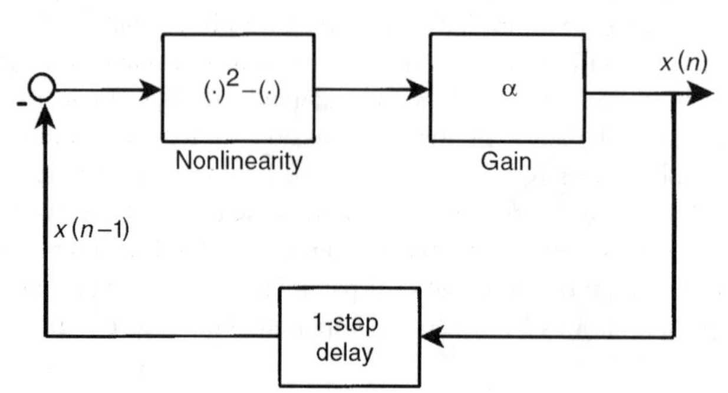


**Figure 10.2** The logistic equation as a nonlinear feedback control system.

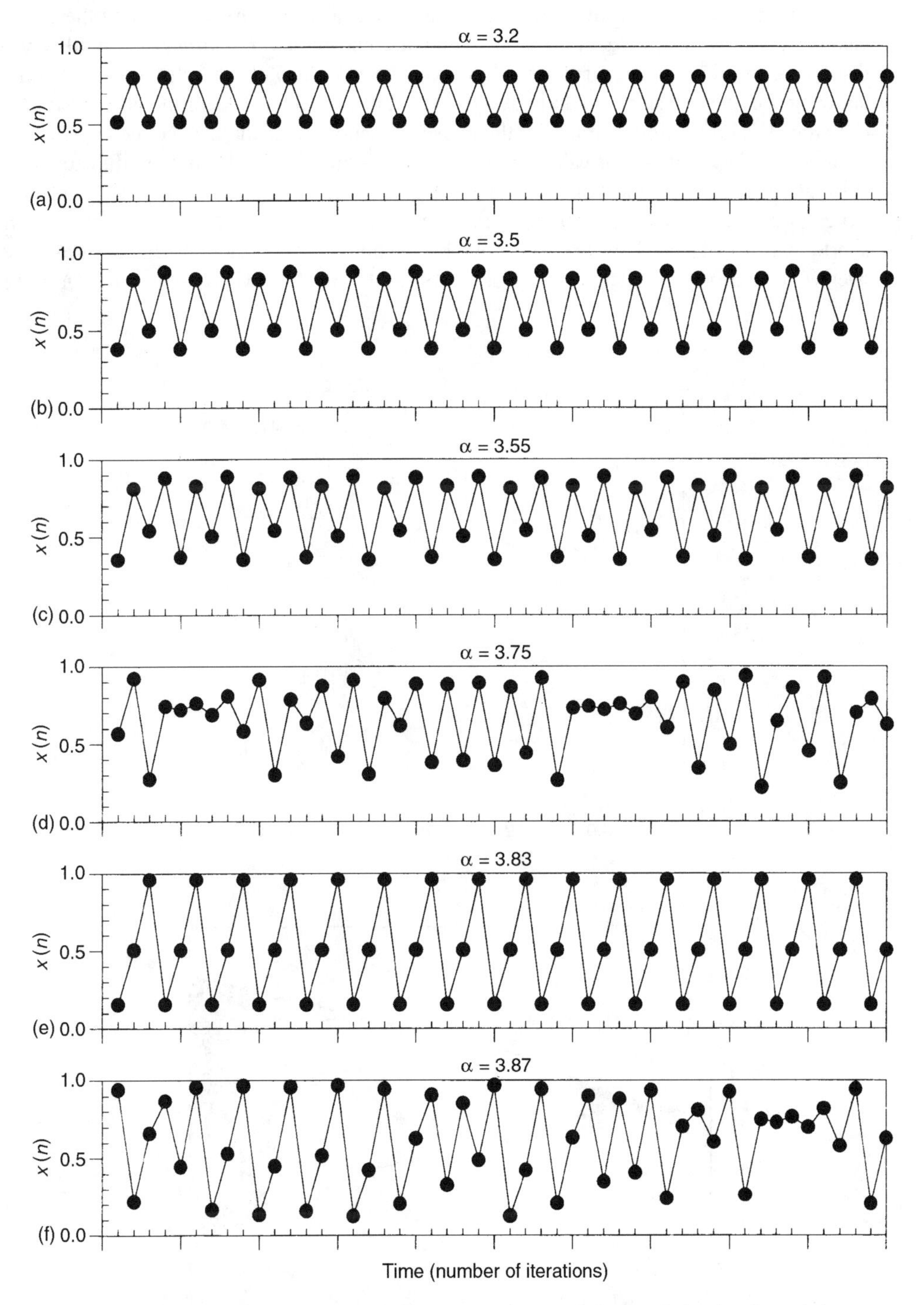


**Figure 10.3** Dynamic behavior of the logistic equation as $\alpha$ is increased, showing the period-doubling route to chaos.

As one may recall from Section 9.3.3, the points at which the dynamics of the system changes abruptly, as the "gain" $\alpha$ is increased, are known as *bifurcations*. A useful way of displaying the dependence of these bifurcations on $\alpha$ is to plot the solutions for $x_n$ (after allowing for the transients to fade away) as a function of $\alpha$. The result is a *bifurcation diagram*. The bifurcation diagram for the logistic equation, frequently called the *logistic map*, is shown in Figure 10.4a for values of $\alpha$ that range from 2.8 to 4. Here, the bifurcations are particularly evident as the points at which a given locus of points "pitchforks" into two daughter branches. One feature of the logistic map that is quite apparent to any observer is the shortening of distances between successive bifurcations. For instance, the range of $\alpha$-values over which period-2 oscillations occur is $3.4495 - 3 = 0.4495$, whereas the range in $\alpha$ over

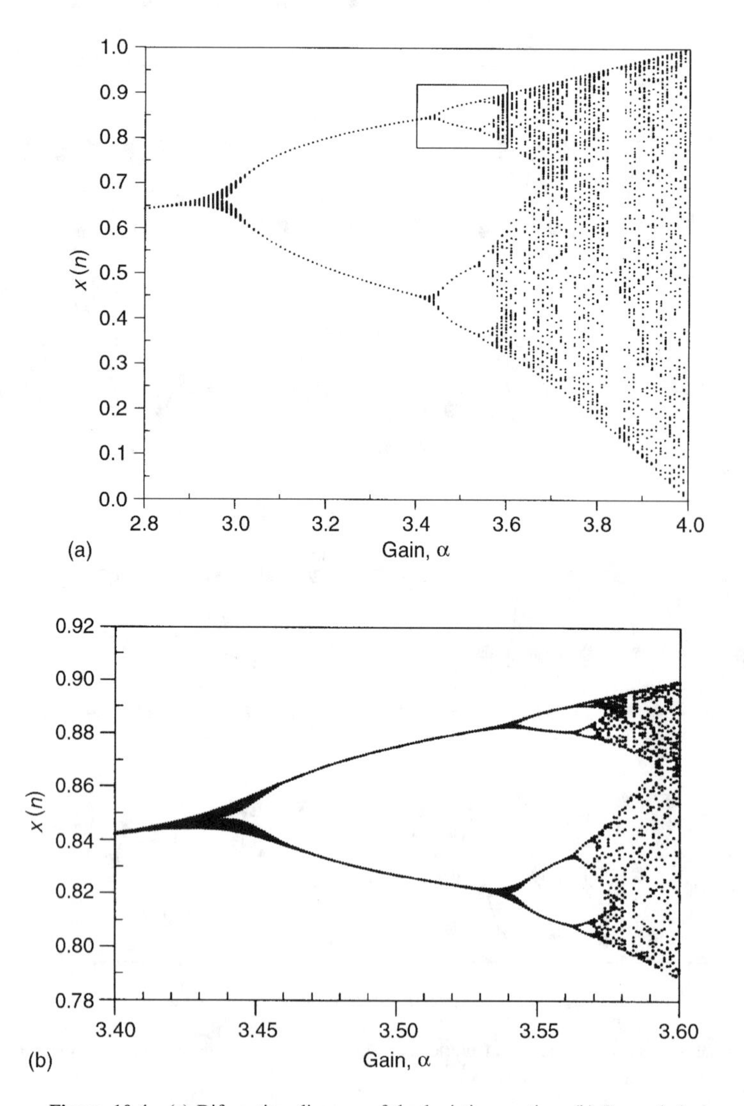


**Figure 10.4** (a) Bifurcation diagram of the logistic equation. (b) Expanded view of the logistic map within the rectangular region shown in (a).

which period-4 oscillations occur is 3.5441 − 3.4495 = 0.0946; subsequently, the range in $\alpha$ over which period-8 oscillations occur becomes 0.0203. Note that the ratio of the $\alpha$-range for period-2 oscillations to the $\alpha$-range for period-4 oscillations is approximately 4.75, whereas the ratio of the $\alpha$-range for period-4 oscillations to the corresponding range for period-8 oscillations is about 4.66. Feigenbaum (1980) showed theoretically that the ratio of the range in $\alpha$ for period-$n$ oscillations to the corresponding range for period-$2n$ oscillations approaches a value of approximately 4.6692 when $n$ approaches infinity. For obvious reasons, this limiting value is known as *Feigenbaum's number.* Another interesting feature of the logistic map is that it exhibits self-similarity. By this, we mean that if we were to "zoom" into a selected section of the bifurcation diagram (as illustrated by the small rectangle in Figure 10.4a), examination of the resulting view would yield a structure that is similar to the original diagram. This kind of self-similarity is characteristic of *fractal* structures. Further discussion of fractals falls beyond the scope of this volume, and the interested reader is encouraged to look up a number of excellent sources, such as Feder (1988), West (1990), and Bassingthwaighte, Liebovitch, and West (1994).

The key property that distinguishes the chaotic waveform from other periodic and quasiperiodic signals is its *sensitivity to initial conditions.* This property is illustrated in Figure 10.5, which shows two possible outputs (bold versus light tracing) generated by the logistic equation. Both were produced using exactly the same value of $\alpha$. The only difference is that in the bold tracing the initial value (i.e., $x_0$) was set equal to exactly 0.1, whereas for the light tracing the initial state was assigned the value of 0.1001. For the first few iterations, both trajectories are extremely close. However, after the tenth iteration, the light tracing begins to diverge further and further away from the bold tracing, so that eventually two totally distinct trajectories emerge. In the other types of deterministic signal, two state points that are initially close together will always remain close to each other. The logistic system is an example of a process that is both fractal and, when $\alpha \geq 3.57$, sensitive to initial conditions. However, it is important to remember that not all fractal processes are chaotic and not all chaotic processes are fractal.

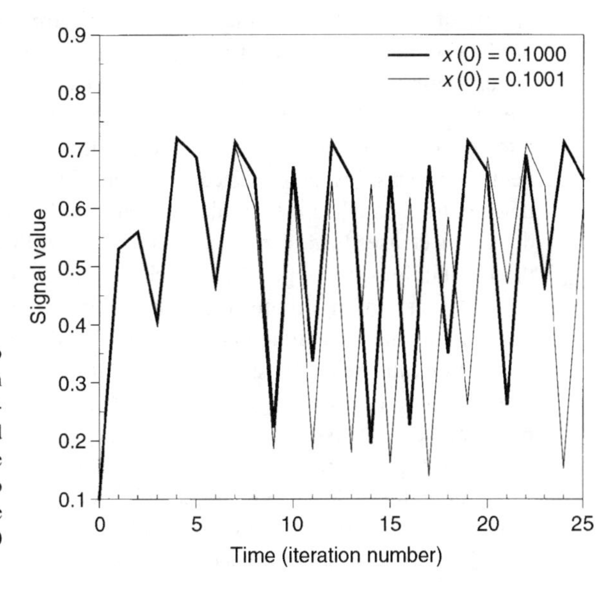


**Figure 10.5** Illustration of "sensitivity to initial conditions" in a chaotic signal. Both signals (bold and light tracings) were generated by the logistic model. In the bold tracing, initial value was set to 0.1; in the light tracing, initial value was set equal to 0.1001. The two time-courses diverge considerably from one another after $\sim 10$ iterations.

Sensitivity to initial conditions implies that it is impossible to predict the future values of a chaotic signal in the long run. Because we can only measure this signal with finite precision, the initially very small difference between our initial measurement and the true signal value at that point would, over time, grow exponentially. Referring again to Figure 10.5, if we take the light tracing to be our prediction of the "true" signal (bold tracing), it is clear that the predicted values after the tenth iteration are totally off the mark. The rate at which two initially nearby trajectories become increasingly separated from each other as time progresses is quantified by the dominant *Lyapunov exponent.* The presence of a positive Lyapunov exponent indicates that the underlying system is chaotic. However, the computation of the Lyapunov exponent itself is difficult and the statistical reliability of the solution depends on the quality and quantity of the available data. The reader is again referred to other excellent references that exclusively cover the topic of chaos, such as Thompson and Stewart (1986), Moon (1987), and West (1990).

### 10.2.2 Regulation of Neutrophil Density

The white blood cell counts of patients with *chronic myeloid leukemia* (CML) are known to fluctuate wildly about elevated levels. These fluctuations are roughly periodic with cycle durations that range from 30 to 70 days. Mackey and Glass (1977) speculated that these oscillations may be related to changes in the dynamic properties of the physiological control system that regulates the balance between production and destruction of the neutrophils that circulate throughout the body. They proposed the following differential-delay equation to account for the dynamics of this regulatory process:

$$\frac{dx}{dt} = \frac{\beta\theta^n x(t - T_d)}{\theta^n + x(t - T_d)^n} - \gamma x(t) \tag{10.2}$$

where $x(t)$ represents the neutrophil density in blood at time $t$, and $T_d$ is the "maturation time," i.e., the delay between the time the new neutrophils are produced by the stem cells in the marrow and the time the mature neutrophils are released into the circulation. The parameter $\gamma$, assumed constant, represents the rate at which the cells are destroyed due to a variety of factors. The parameters $\theta$ and $n$ determine the relationship between the neutrophil production rate and the past neutrophil density, while the parameter $\beta$ represents a scaling factor. Figure 10.6 shows this nonlinear function (bold curve) when $\theta$, $n$, and $\beta$ are assigned the values of 1, 10, and 0.2, respectively. Also shown is the linear function relating destruction rate to neutrophil density; here, a value of 0.1 is assumed for $\gamma$. Over a large range of neutrophil densities, poietin feedback control exerts its effects by reducing the production of new neutrophils when the circulating neutrophil density increases. However, as the neutrophil density decreases toward zero, it is assumed that the production rate also falls to zero. Under these conditions, the type of feedback therefore changes from negative to positive.

As shown in Figure 10.6, the intersection between the straight line representing destruction rate as a function of neutrophil density and the nonlinear curve representing production rate yields the steady-state (equilibrium) solutions for $x$. There is a trivial solution at the origin. The other solution (located at $x = 1, y = 0.1$) can be shown, using the method of phase-plane analysis (see Chapter 9), to be stable. To examine the dynamic behavior of this system in the presence of feedback time delay, we solve Equation (10.2) through computer simulation: the SIMULINK implementation (named "`hematop.mdl`") of the model is shown in Figure 10.7. Some simulation results are displayed in Figure 10.8. When $T_d$ is small (=2 days), as would be expected under normal circumstances, the system is stable and,

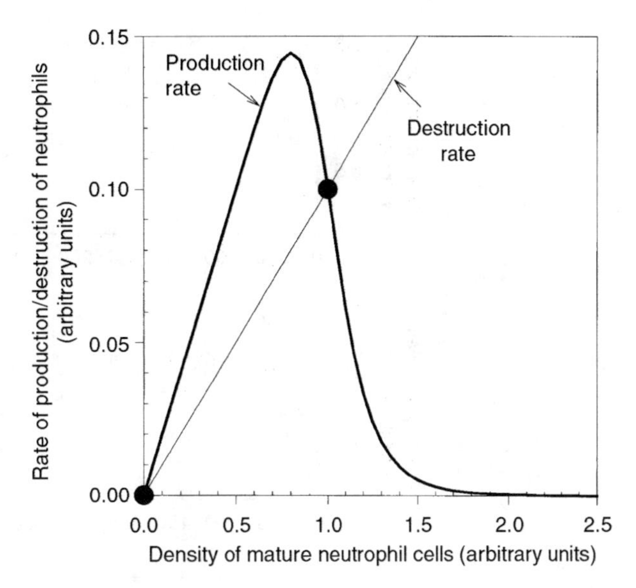


**Figure 10.6** The steady-state relationships characterizing the production (bold curve) and destruction (light line) of neutrophil cells as functions of the circulating neutrophil density in blood. There are two steady-state equilibrium points (filled circles).

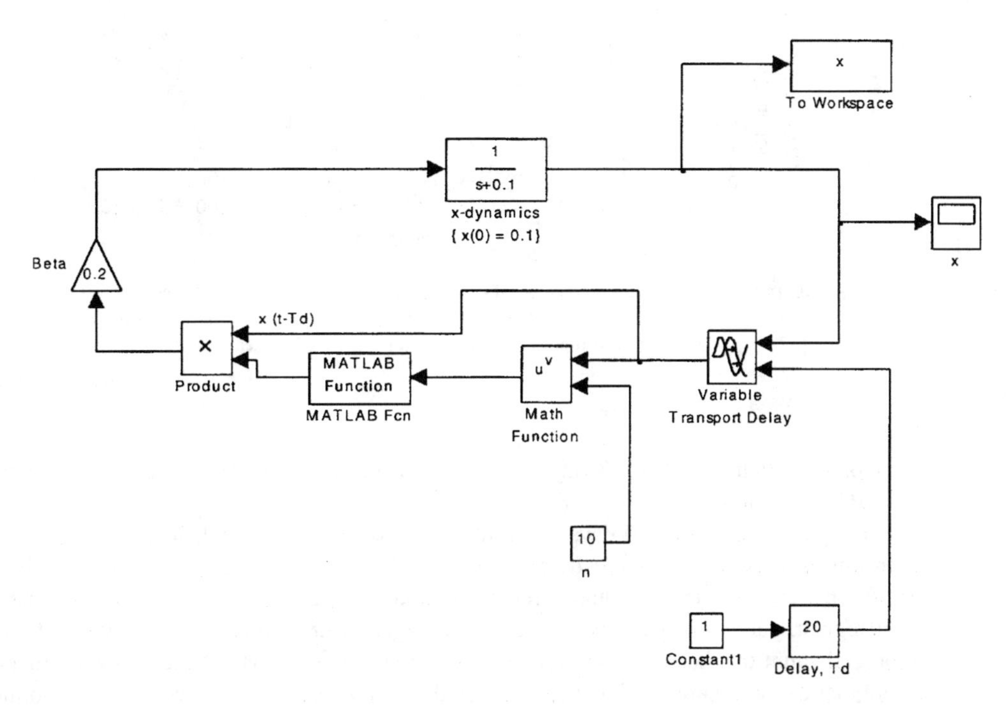


**Figure 10.7** SIMULINK implementation of the neutrophil density regulation model of Mackey and Glass (1977).

following any perturbation in neutrophil density (e.g., due to blood loss), there is a rapid return to the stable steady-state level (Figure 10.8a). However, if the maturation time is increased threefold to 6 days, following an initial perturbation, $x$ does not return to its previous stable steady state. Instead, it oscillates with a small amplitude around the original equilibrium level (Figure 10.8b). The period of this oscillation is 18 days, which falls in the

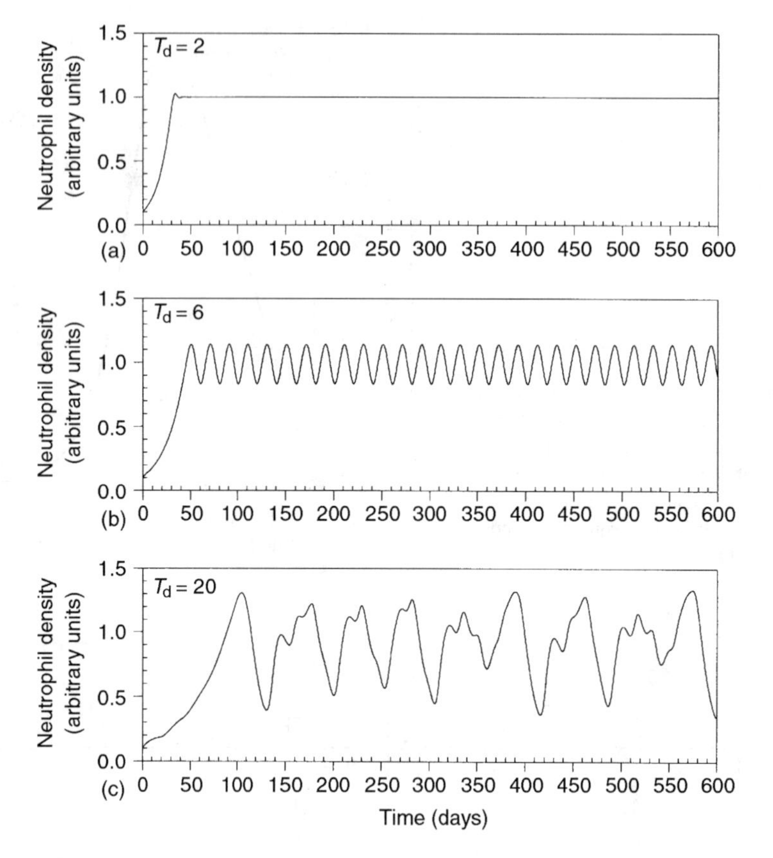


**Figure 10.8** Spontaneous dynamics of neutrophil density predicted by the Mackey–Glass model as the cell maturation time, $T_d$, increases from the "normal" value of 2 days (a), to 6 days (b), and in chronic myeloid leukemia to 20 days (c).

range of periodicities (17 to 28 days) observed in humans who have a disorder known as *cyclical neutropenia*.

If $T_d$ is increased to 20 days to simulate the conditions of CML, large fluctuations in neutrophil density occur, as illustrated in Figure 10.8c. These fluctuations look somewhat periodic but they are actually quite irregular, displaying a significant degree of "noisiness." The irregular, yet almost periodic, waveform of Figure 10.8c resembles the pattern of white blood cell count that has been observed in some patients with CML. Note, however, that the differential-delay equation (Equation (10.2)) that produced this time-course is absolutely deterministic—no random noise has been added to the simulation output. In other words, the neutrophil time-course in Figure 10.8c is chaotic. To confirm that this indeed reflects chaotic behavior, we need to determine that the dynamic evolution of the predicted neutrophil density is sensitive to small changes in initial conditions. Figure 10.9 shows the result of just such a determination. Solution of Equation (10.2) using the SIMULINK model in Figure 10.7 is performed using two sets of initial conditions. The first assumes the initial neutrophil density, $x(0)$, to be 1.22; in the second simulation, $x(0)$ is set equal to 1.21. As Figure 10.9 shows, after an initial period of remaining very close together, the two simulated time-courses begin

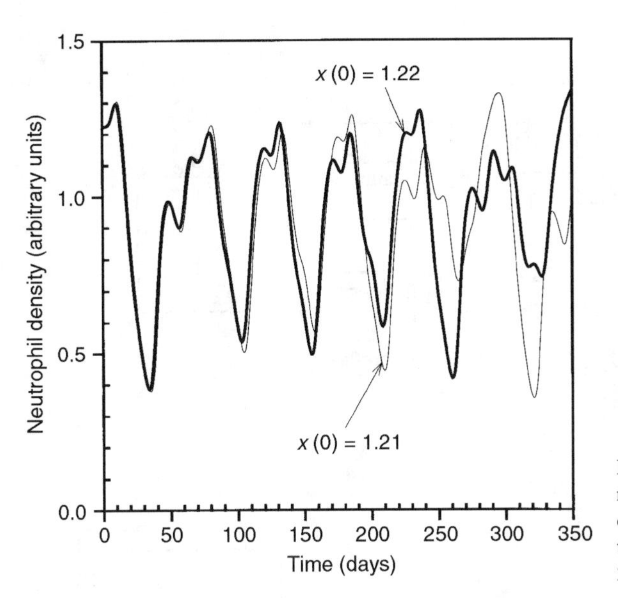


**Figure 10.9** Sensitivity of the neutrophil regulation model to initial conditions indicates that the model becomes chaotic when the cell maturation delay, $T_d$, is increased to 20 days.

to diverge after $t = 100$ days. Eventually, each trajectory appears to follow its own course, although both remain within a bounded range of values. This is the hallmark of chaotic behavior.

### 10.2.3 Model of Cardiovascular Variability

Beat-to-beat fluctuations in the duration of the cardiac cycle, arterial blood pressure, and cardiac output are well-known phenomena. Are these simply manifestations of intrinsically noisy processes? Recent applications of the tools of nonlinear dynamics to cardiovascular measurements suggest that deterministic chaos may be the underlying mechanism. In this section, we examine a model published by Cavalcanti and Belardinelli (1996), which postulates that the spontaneous variability in heart rate and blood pressure results from chaotic behavior occurring in the baroreflex control system. Functionally, the model contains two feedback loops: one representing the effect of the baroreflex on heart rate and the other representing the effect of the baroreflex on cardiac contractility, which in turn affects stroke volume.

To highlight the importance of interaction between the two feedback loops, we begin by discussing a simplified variant of this model. The schematic block diagram that represents this model is shown in Figure 10.10a, and the corresponding SIMULINK implementation of the model ("`cvvar1.mdl`") appears in Figure 10.10b. For a given cardiac output, circulatory mechanics determines the corresponding level of arterial blood pressure. In the model, circulatory mechanics is characterized by a 3-element *Windkessel* model (Figure 8.4). The latter consists of a resistance, representing the aortic characteristic impedance ($r$), placed in series with a parallel combination of the peripheral resistance ($R$) and the total arterial compliance ($C$). The differential equation describing the 3-element Windkessel model is given by

$$RC\frac{dP}{dt} + P = rRC\frac{dQ}{dt} + (R + r)Q \tag{10.3}$$

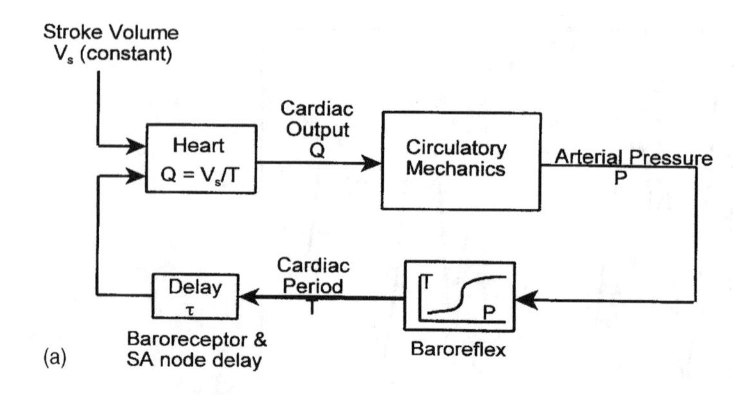


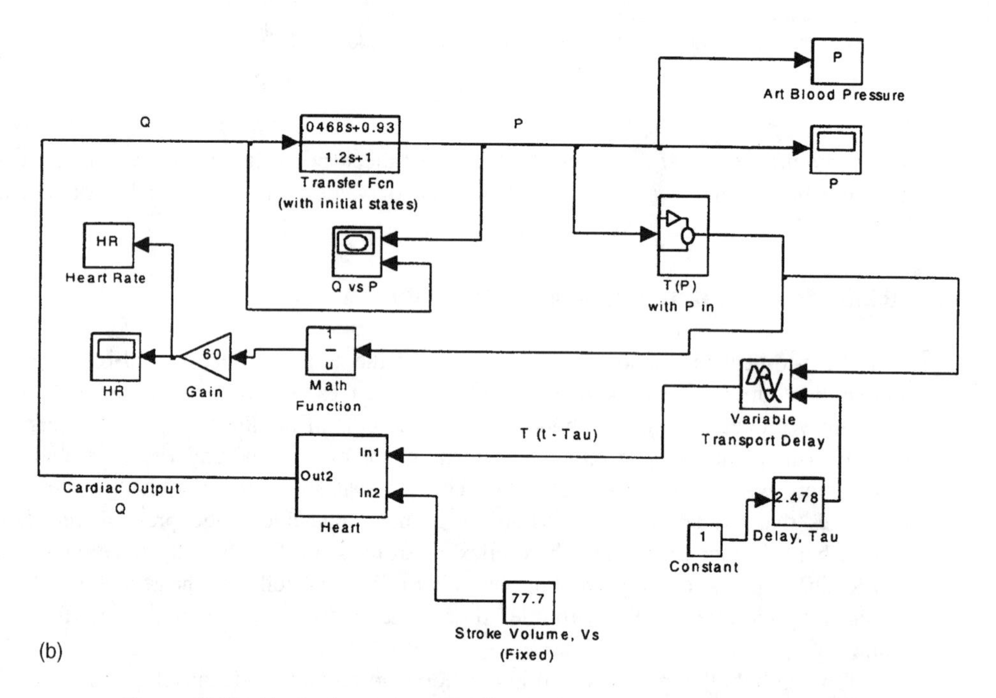


**Figure 10.10** (a) Cardiovascular variability model with single feedback loop. (b) SIMULINK implementation ("cvvar1.mdl") of the above model.

where $P$ and $Q$ represent arterial blood pressure and cardiac output, respectively. The resistances and compliance are assumed constant and given the values $C = 1.333$ ml mm $\text{Hg}^{-1}$, $R = 0.900$ mm Hg s $\text{ml}^{-1}$, and $r = 0.039$ mm Hg s $\text{ml}^{-1}$. Thus, the circulatory mechanics subsystem is linear.

Changes in $P$ are sensed by the baroreceptors, which relay this information back to the vasomotor center in the brain stem. The vasomotor center responds with changes in vagal and sympathetic nerve activity which, in turn, modulate the cardiac period, $T$ (and thus, heart rate $= 1/T$). In this simplified version of the Calvacanti model, we assume stroke volume ($V_s$) to be independent of $P$. The steady-state characteristic of this baroreflex response is shown in Figure 10.11a. As $P$ decreases, $T$ also decreases, meaning that cardiac output increases, since $Q = V_s/T$ and $V_s$ is constant. Conversely, as $P$ increases, $T$ also increases, decreasing $Q$. This accounts for the negative feedback effect of the baroreflex. However, when $P$ falls below 80 mm Hg or rises above 100 mm Hg, the sensitivity of $T$ to further changes in $P$ drops substantially. The baroreflex response therefore saturates at low and high levels of $P$. The dependence of $T$ on $P$ is given by

$$T(P) = T_{min} + \frac{T_{max} - T_{min}}{1 + \gamma e^{-\alpha P/P_e}} \tag{10.4}$$

where $T_{min}$ and $T_{max}$ are the lowest and highest possible values for $T$, assumed to be 0.66 and 1.2 s, respectively. $P_e$ is the equilibrium level of arterial pressure and is equal to 89 mm Hg. The constants $\alpha$ and $\gamma$ control the range and slope of the linear portion of the $T$–$P$ curve; they are assigned values of 31 and $6.7 \times 10^{13}$, respectively.

The model does not take into account the dynamics of the sinoatrial node, as in the model of Saul et al. (see Section 5.4). However, an overall delay, $\tau$, is incorporated. This delay represents the combined lag associated with the baroreceptor response time and the response times of the sinoatrial node to vagal and sympathetic stimulation. In the discussions that follow, we will examine how the dynamics of this baroreflex model are affected by $\tau$ as it is given different values that span the range of feasible delays.

Before proceeding with our exploration of the dynamics of this model, it is useful to apply the approach presented in Chapter 3 and to determine the steady-state operating point that results from the matching of the feedforward (circulatory mechanics) and feedback (baroreflex + delay + heart) portions of this closed-loop system. This is shown in terms of the variables $Q$ and $P$ in Figure 10.11b. Intersection of the curve representing the baroreflex and heart (bold curve) with the straight line representing linear circulatory mechanics yields the equilibrium point E1 at which $P = P_e = 89$ mm Hg and $Q = 78.8$ ml s$^{-1}$. Applying the phase-plane method of analysis (Section 9.2), it can be determined that this equilibrium point is stable. Since these considerations involve only the steady state, the delay plays no role in the determination of the equilibrium point.

Figure 10.12 shows some simulation results obtained with the SIMULINK model "`cvvar1.mdl`." The evolution of the time-course of instantaneous heart rate ($= 60/T$, expressed in beats per minute) is displayed in each of the left panels (a, c, e, and g) of Figure 10.12; the transient effects of starting the simulations with arbitrary initial conditions have been removed from these plots. The corresponding phase-space plots, with $Q$ plotted against $P$, are shown in the panels on the right. When $\tau$ is small, e.g., 0.5 s (as in Figure 10.12a), the response rapidly converges to the equilibrium level; this is represented as a single dot in the phase space (Figure 10.12b). As $\tau$ is increased, the system becomes oscillatory, as demonstrated by the periodic waveform (Figure 10.12c) and limit cycle behavior in the phase-space plot (Figure 10.12d). This periodic behavior persists with further increases in $\tau$ (Figures 10.12e through h). The cycle duration of the oscillation, however, increases as $\tau$ is increased. Again, this example demonstrates that prolongation of the time delays inherent in a closed-loop system constitutes a highly destabilizing effect.

A schematic block diagram representing the complete version of the model is displayed in Figure 10.13a. The SIMULINK implementation of this model, "`cvvar2.mdl`," is shown in Figure 10.13b. Here, stroke volume, $V_s$, is assumed to be a function of arterial pressure.

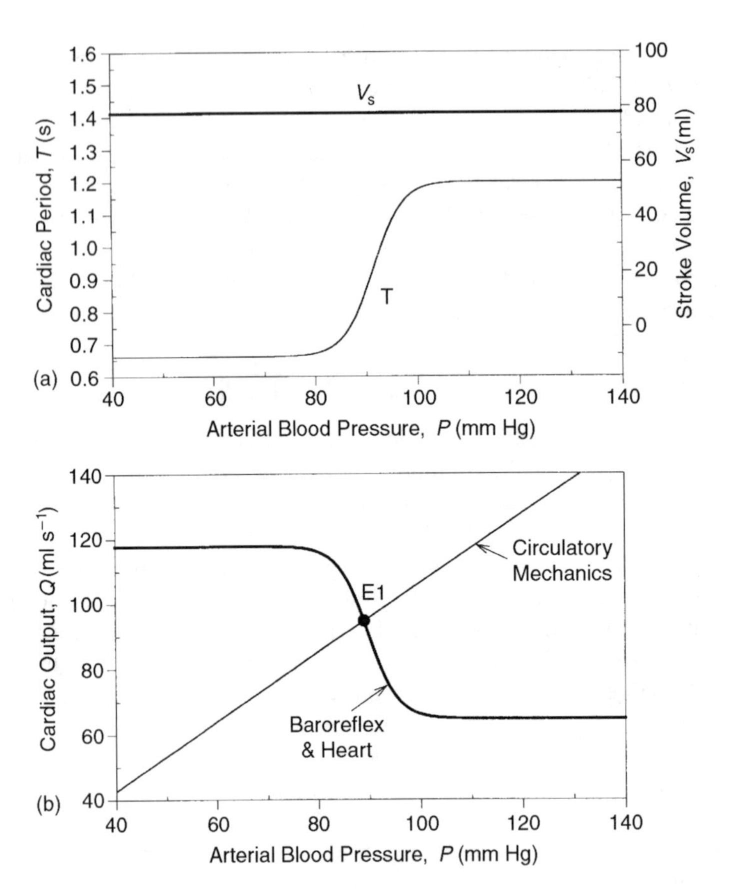


**Figure 10.11** (a) Steady-state characteristics of the baroreflex; stroke volume is assumed to be fixed and independent of arterial blood pressure. (b) Steady-state properties of the baroreflex and heart combined (bold curve), shown together with the arterial pressure–cardiac output dependence determined by circulatory mechanics. Intersection of these two functions yields a single equilibrium point, E1.

Figure 10.14a illustrates the dependence of $V_s$ on $P$, which is also basically sigmoidal in form as in the relation between $T$ and $P$. Above 90 mm Hg, $V_s$ remains relatively constant. Below 80 mm Hg, $V_s$ decreases steeply with decreases in $P$. This is due to the concomitant increase in heart rate, which reduces the time for ventricular filling. As $P$ decreases even further, $V_s$ decreases toward zero, as the heart begins to fail. Incorporation of the dependence of $V_s$ on $P$ produces a dual feedback-loop model. Furthermore, since cardiac output is the ratio of $V_s$ to $T$, this introduces a nonlinear interaction between the two feedback loops.

Figure 10.14b displays the steady-state relations between cardiac output and $P$ for both the circulatory mechanics subsystem (light line) and the subsystem representing the baroreflex and heart (bold curve). In the latter subsystem, notice that as $P$ decreases from physiological levels, $Q$ attains a peak and subsequently decreases monotonically toward zero. Consequently, intersection of the two curves yields two equilibrium points, E1 and E2, unlike the single feedback-loop case. Stability analysis shows that E1 is stable and corresponds to

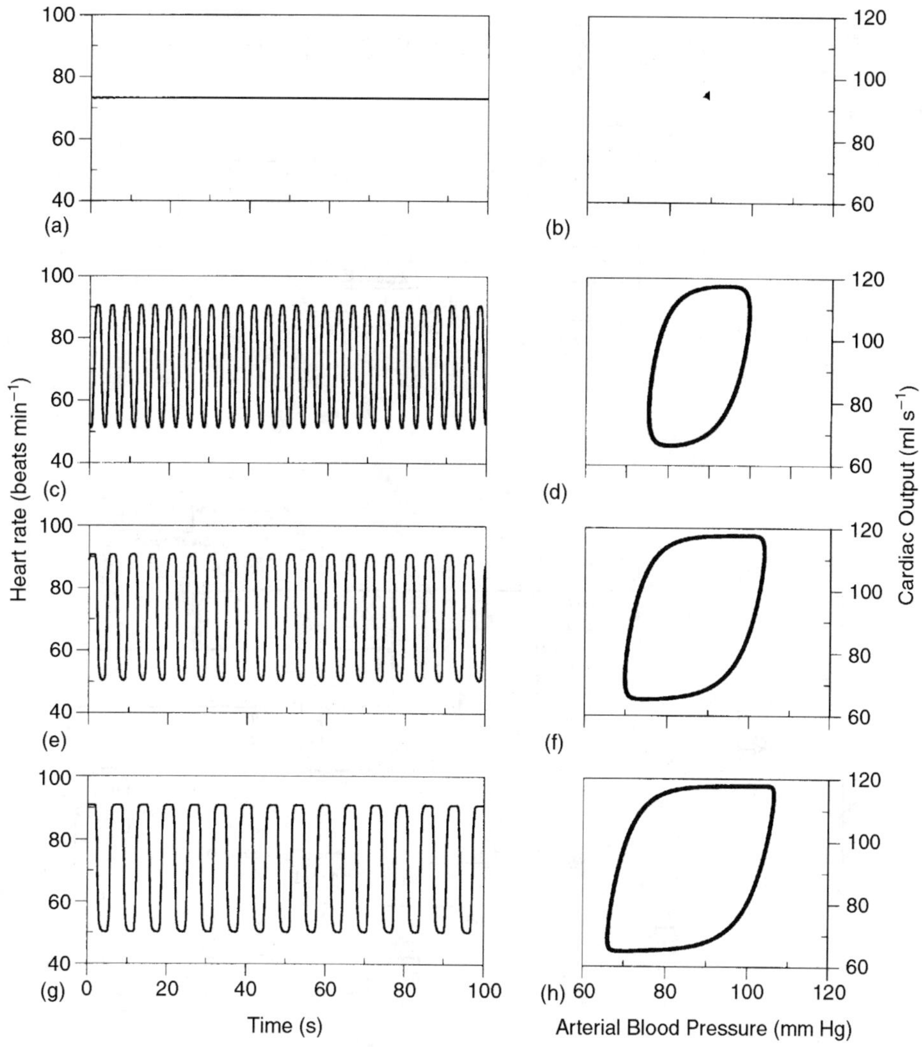


**Figure 10.12** Simulation results produced by "cvvar1.mdl." Left panels show predicted time-courses of heart rate; right panels show $x-y$ plots of cardiac output (=heart rate × stroke volume) versus arterial blood pressure (a, b) $\tau = 0.5$ s; (c, d) $\tau = 1.2$ s; (e, f) $\tau = 1.8$ s; and (g, h) $\tau = 2.5$ s. As the delay increases, the system becomes oscillatory with increasing cycle duration.

the same equilibrium point that we found for the simpler version of the model. However, it can be shown that E2 is an unstable fixed point. This is due basically to the fact that E2 falls on the portion of the baroreflex–heart *Q–NP* curve where the slope now has become positive, instead of negative as for the case of E1. In other words, the system contains negative feedback over the range of high $P$ values, but becomes one with positive feedback when $P$ falls below 80 mm Hg.

What is the dynamic behavior of this model when the time delay is taken into account? Figure 10.15 shows some examples of simulations performed with the SIMULINK program "cvvar2.mdl." When $\tau$ is kept small at, say, 0.5 s, the system trajectory converges rapidly

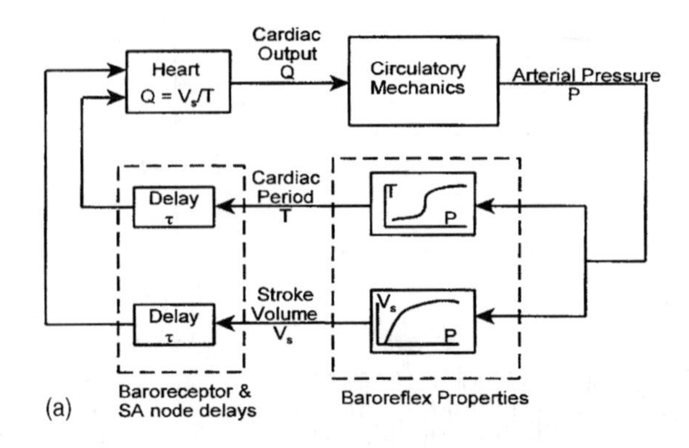


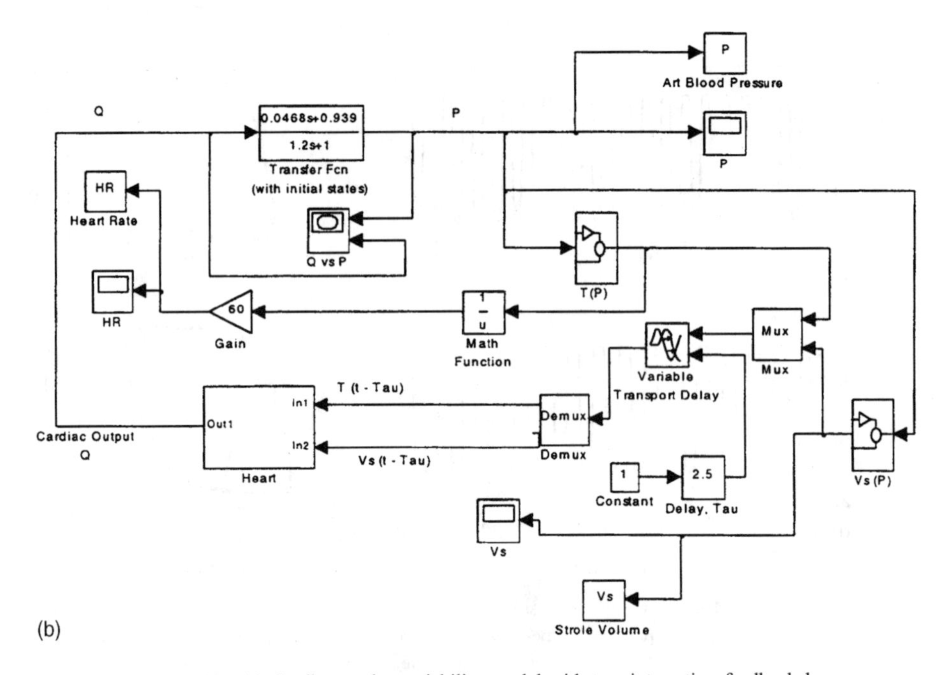


**Figure 10.13** (a) Cardiovascular variability model with two interacting feedback loops. (b) SIMULINK implementation ("cvvar2.mdl") of the above model.

to the steady state represented by E1, after the initial transient that depends on starting conditions (Figures 10.15a and b). As $\tau$ is increased, a periodic oscillation develops (Figure 10.15c and d for $\tau = 1.2$ s) as in the single feedback-loop case; in this case, the frequency of the oscillation is 0.28 Hz. However, with further increases in $\tau$, *period-doubling* occurs. At $\tau = 1.8$ s, the system remains periodic; however, the oscillations now contain multiple frequencies at a subharmonic and superharmonics of 0.19 Hz (Figure 10.15e). The corresponding phase-space plot shows a complicated double loop (Figure 10.15f). When $\tau$ is

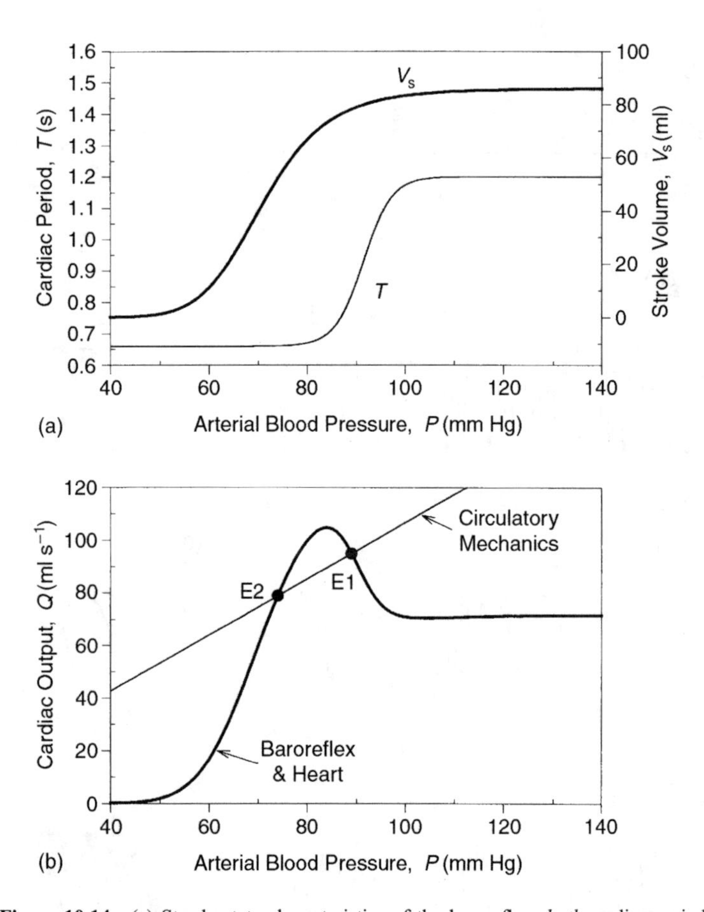


**Figure 10.14** (a) Steady-state characteristics of the baroreflex; *both* cardiac period and stroke volume are assumed to be dependent on arterial blood pressure. (b) Steady-state properties of the baroreflex and heart combined (bold curve), shown together with the arterial pressure–cardiac output dependence determined by circulatory mechanics. Intersection of these two functions yields two nontrivial equilibrium points, E1 and E2.

increased beyond 2 s, the oscillations turn into chaos, as illustrated in the example in Figure 10.15g and h; in this case, $\tau = 2.5$ s. Figure 10.16a provides a closer look at the relative time-courses of the system variables $P$, heart rate (or equivalently, $T$), and $V_s$ when the system is in chaotic mode, with $\tau$ set equal to 2.5 s. When these three state variables are plotted in 3-dimensional format, the picture of the chaotic attractor emerges (Figure 10.16b).

As in the neutrophil regulation model of Section 10.2.2, the preceding discussion demonstrates that chaotic behavior can be produced in systems with relatively low-order dynamics when the important ingredients of time delay and mixed feedback are present. However, the predictions made by this model are not so consistent with empirical observations, which tend to show more chaotic behavior in heart rate variability in normals (where $\tau$ would be expected to be small) and less chaotic, more periodic behavior in patients who suffer from myocardial infarction (where we would expect $\tau$ to be increased due to the dominant influence of the sympathetic nervous system). One reason could be that this model is highly simplistic in not explicitly incorporating the dynamics of the sinoatrial node, the baroreflex,

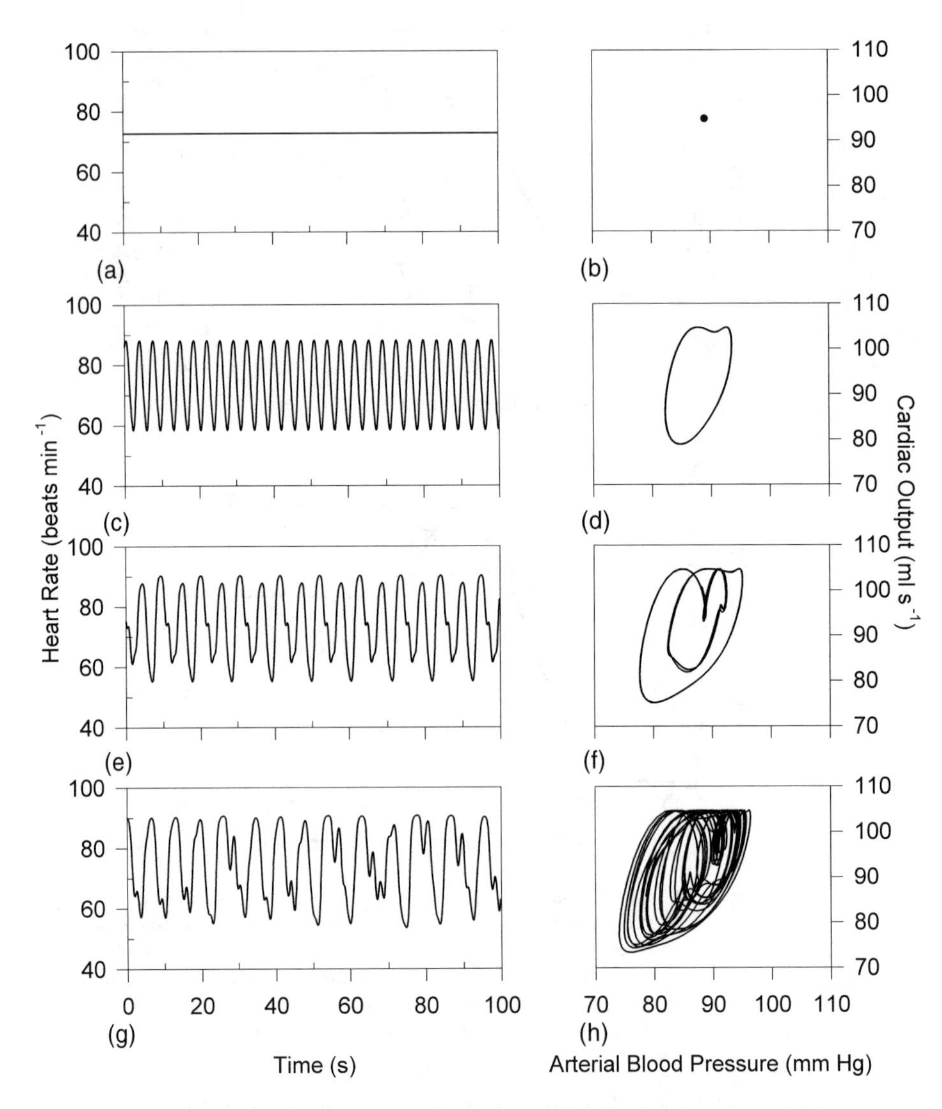


**Figure 10.15** Simulation results produced by "ccvar2.mdl." Left panels show predicted time-courses of heart rate; right panels show $x - y$, plots of cardiac output versus arterial blood pressure. (a, b) $\tau = 0.5$ s; (c, d) $\tau = 1.2$ s; (e, f) $\tau = 1.8$ s; and (g, h) $\tau$=2.5 s. As the delay is increased, the system becomes periodic; then period-doubling occurs and, finally, chaos sets in.

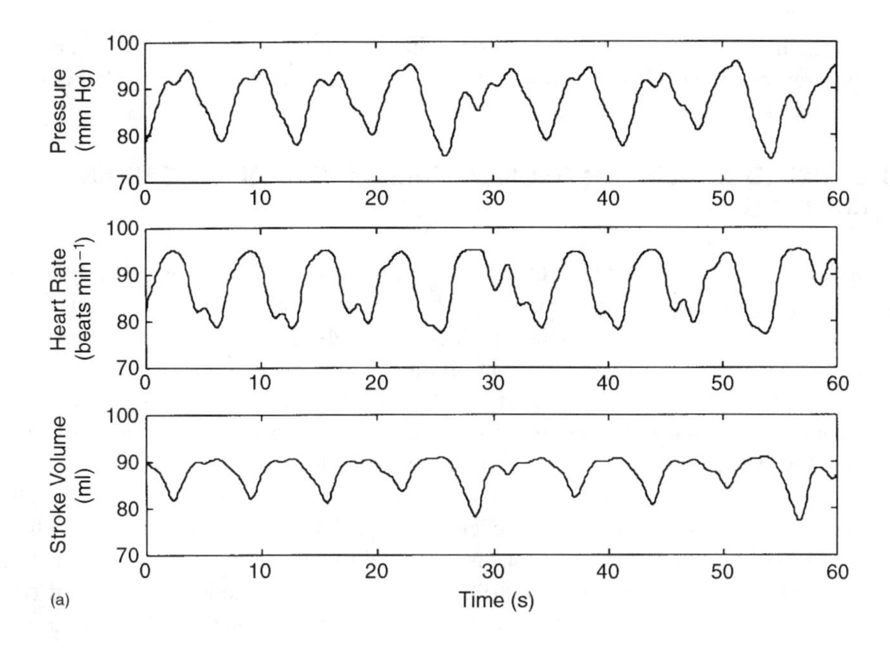


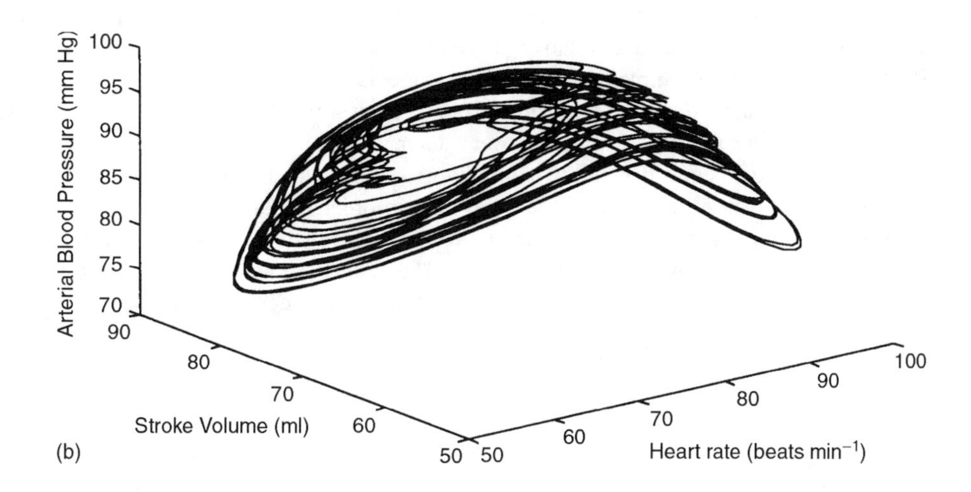


**Figure 10.16** (a) Time-courses of arterial blood pressure, heart rate, and stroke volume when the system is in chaotic mode ($\tau = 2.5\,\text{s}$). (b) Three-dimensional view of the chaotic attractor.

and the heart. Another important factor is that the influence of breathing on arterial pressure and heart rate fluctuations is ignored. Finally, the assumed dependences relating $P$ to $T$ and $V_s$ may be too simplistic.

## 10.3 COUPLED NONLINEAR OSCILLATORS: MODEL OF CIRCADIAN RHYTHMS

In Section 9.3.2, we pointed to the phenomenon of "entrainment" as a distinguishing feature of nonlinear oscillators that are coupled together. Since the body contains a number of pacemakers and we have evolved to adapt to the 24-hour rhythm of the light–dark cycle, it follows that several of these circadian oscillators must be entrained to the external *zeitgeber* (or, if translated literally from German, "time giver"). There is a large body of evidence to support this notion. For example, in subjects who have been isolated from all external time cues for over 2 months, the sleep–wake and body temperature rhythms become *internally desynchronized*, with the temperature oscillator assuming a periodicity that is slightly longer than 24 hours and the sleep–wake cycle being prolonged to approximately 30 hours. Other physiological rhythms in these subjects then tend to be entrained to either the temperature or sleep–wake cycle. Kronauer et al. (1982) have proposed a model consisting of two coupled van der Pol oscillators to represent the temperature and sleep–wake pacemakers. Under normal circumstances, both oscillators are entrained to the external 24-hour zeitgeber. However, under conditions that simulate temporal isolation, the model exhibits the complex variations in periodicities and relative phasing between the temperature and sleep–wake rhythms that closely resemble empirical measurements.

The schematic diagram in Figure 10.17a shows the temperature and sleep–wake oscillators and the mutual coupling between them. Kronauer and coworkers found that it was necessary to assume that the synchronizing zeitgeber is applied directly to the sleep–wake oscillator instead of the temperature system in order to obtain realistic phase relations between them during zeitgeber entrainment. We represent the outputs of the temperature and sleep–wake oscillators by $x$ and $y$, respectively. The zeitgeber output is represented by $z$. The model is characterized by the following pair of coupled van der Pol equations:

$$k^2\ddot{x} + k\mu_x(x^2 - 1)\dot{x} + \omega_x^2 x + F_{yx}k\dot{y} = 0 \tag{10.5}$$

and

$$k^2\ddot{y} + k\mu_y(y^2 - 1)\dot{y} + \omega_y^2 y + F_{xy}k\dot{x} = F_{zy}\sin\left(\frac{\omega_z t}{k}\right) \tag{10.6}$$

In the above equations, the *scaling factor* $k$ $(= 24/2\pi)$ is introduced so that the intrinsic periods of the (uncoupled) temperature and sleep–wake oscillators would equal 24 hours if the respective angular frequencies $\omega_x$ and $\omega_y$, were each set equal to unity. Similarly, $\omega_z$ is set equal to unity so that the zeitgeber period is 24 hours. The parameters $\mu_x$ and $\mu_y$ represent the "stiffness" of the temperature and sleep–wake oscillators, respectively. They determine the time constants of the transient duration of adjustment in phase of each oscillator to that of the zeitgeber following release from entrainment or after reentrainment. In our simulations, $\mu_x$ and $\mu_y$ are each assigned the value of 0.1. $F_{yx}$ and $F_{xy}$ represent the strengths of the coupling between the temperature and sleep–wake oscillators. $F_{yx}$ and $F_{xy}$ are assigned the values of $-0.04$ and $-0.16$, respectively. The relative magnitudes of these values imply that the temperature oscillator has a stronger influence on the sleep–wake oscillator than vice versa. $F_{zy}$ is assigned the value of unity. The SIMULINK implementation of this model

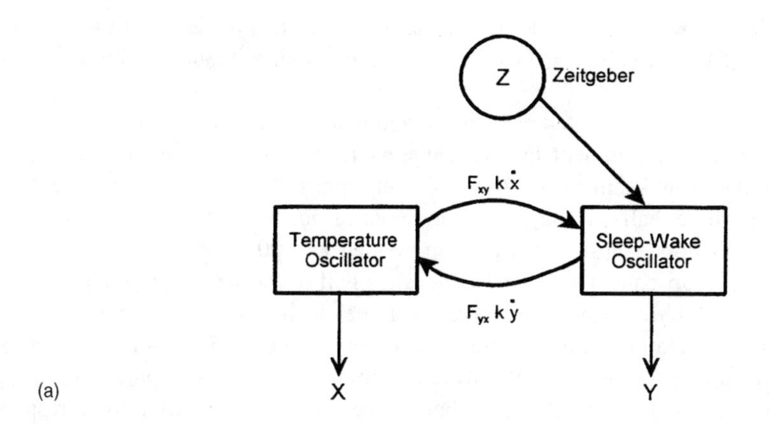


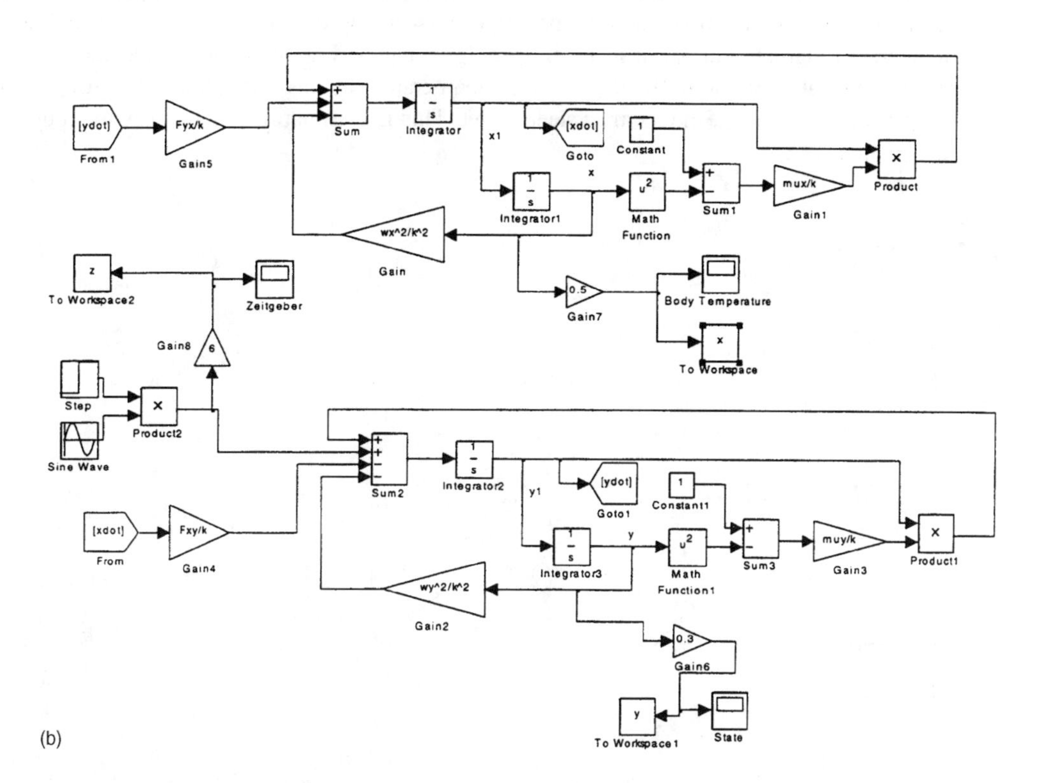


**Figure 10.17** (a) Schematic representation of the circadian model of Kronauer et al. (1982). (b) SIMULINK implementation ("circad.mdl") of the circadian model.

("`circad.mdl`") is shown in Figure 10.17b. Note the use in this model file of `Go to` and `From` blocks in order to couple the two oscillators to each other without creating a mess in signal lines.

Figure 10.18 shows the simulated behavior of temperature ($x$, middle panel) and activity ($y$, bottom panel) during entrainment by the zeitgeber ($z$, top panel). The selected model parameters allow the establishment of the following phase relations among the three oscillations. The positive half of the $z$-oscillation corresponds to the 12-hour duration between midnight and noon; these times are represented as "00" and "12", respectively, in Figure 10.18 and all subsequent graphs. The negative half of the $x$-oscillation corresponds to the duration over which core body temperature is below its mean level. Similarly, the negative half of the $y$-oscillation represents the duration over which activity is below average. Kronauer used the middle two-thirds of this duration to represent the period of sleep. However, for simplicity, we will take the entire below-average activity duration to correspond to sleep. The model simulation shows that the start of the "sleep period" (represented by the dark bars in Figure 10.18, bottom panel) occurs at approximately 9:30 p.m. Core body temperature starts to fall below its mean level (gray bars in Figure 10.18, middle panel) some 40 minutes later, at about 10:20 p.m. Thus, temperature attains its lowest value at approximately 4 a.m., some 6.5 hours after sleep onset. This is reasonably consistent with empirical observations.

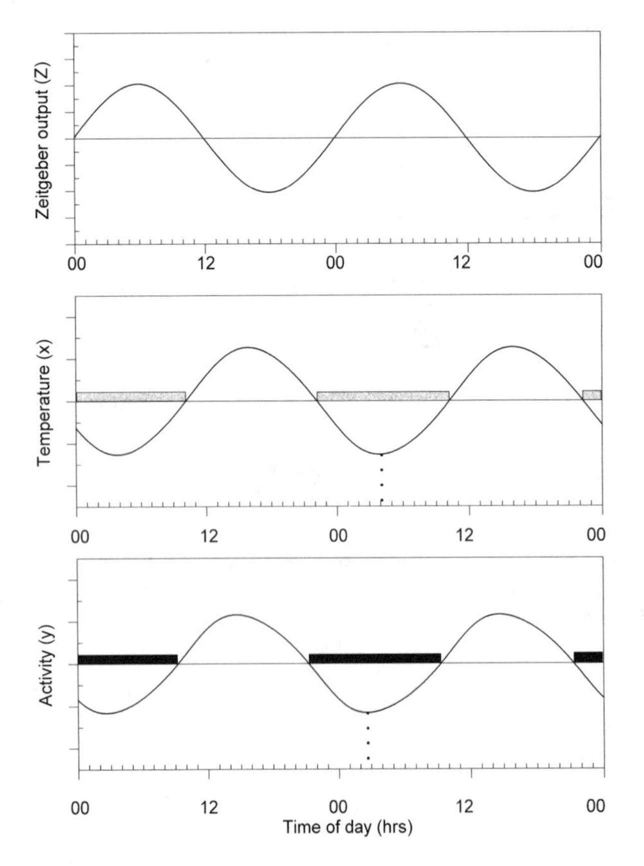


**Figure 10.18** Phase relations between the oscillations in core body temperature ($x$) and activity ($y$) during entrainment by the external 24-hour zeitgeber ($z$). Time is shown in hours starting at midnight ("00"); "12" represents noon. The gray and dark horizontal bars indicate the durations over which temperature and activity, respectively, are below their mean levels. Dotted lines indicate times at which temperature and activity are at their lowest levels.

In the simulation shown in Figure 10.18, the following intrinsic frequencies of the temperature and sleep-wake oscillators were assigned: $\omega_x = 0.99$ and $\omega_y = 0.92$. These frequencies correspond to periodicities of 24.37 hours for temperature and 26.09 hours for activity. To simulate the "free run" condition (i.e., release from zeitgeber entrainment), $F_{zy}$ was set equal to zero. In addition, $\omega_y$ was assumed to decrease linearly so that by the end of the 100th day after the start of free run, $\omega_y$ would become 0.78, which corresponds to a period of 30.8 hours. Kronauer found this latter assumption to be necessary to produce a better match between the simulation results and empirical data.

Figure 10.19 shows the relative phasing between $x$ (light waveforms) and $y$ (bold waveforms) during selected segments of the free run duration. In the first 5 days following release from zeitgeber drive, the time of sleep onset, which previously preceded the drop in temperature below its mean level, can be seen to be delayed progressively (top panel, Figure 10.19). On day 5 after release into free run, the point of lowest activity (*mid-sleep*) occurs

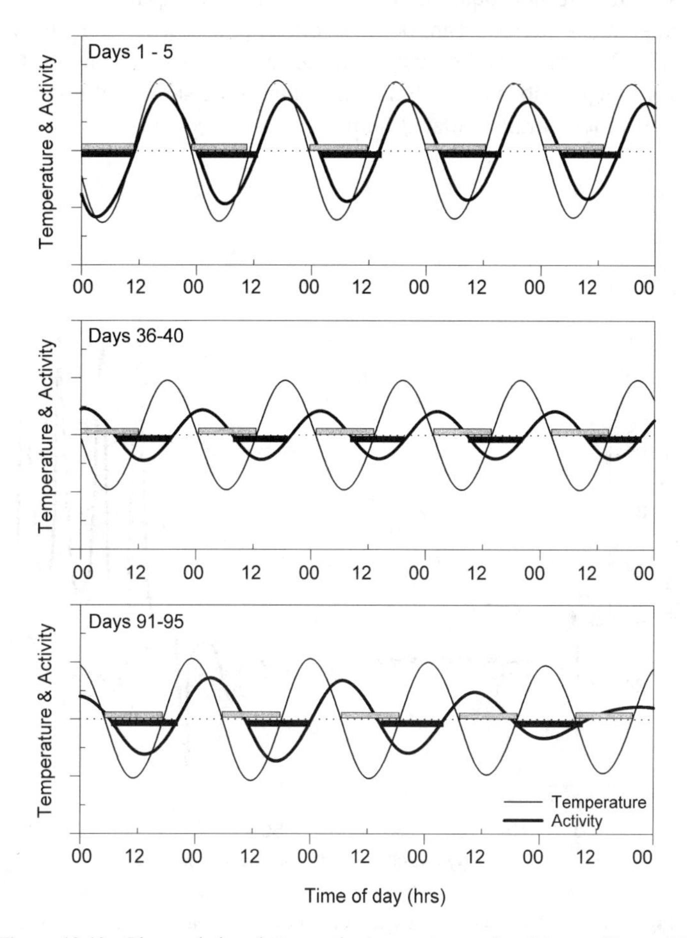


**Figure 10.19** Phase relations between the temperature and activity oscillators following release from 24-hour entrainment by the external zeitgeber. Each panel shows 5 days of simulated behavior. Light waveforms represent temperature, and bold waveforms represent activity. Gray and dark bars represent durations over which temperature and activity, respectively, are below their corresponding mean levels.

approximately 3 hours *after* the point of lowest core body temperature. This gradual delaying of the period of sleep relative to that of lower-than-average body temperature continues until approximately day 35. The day-to-day periods ($\tau_x$ and $\tau_y$) of the two oscillators are shown in Figure 10.20, along with their intrinsic periods ($T_x$ and $T_y$). Note that immediately following release into free run, the activity oscillator abruptly increases its period by about 2 hours, but, almost as abruptly, moves back toward $\tau_x$. From day 36 through day 70, the "drift" in relative phase between $x$ and $y$ ceases to occur; instead, there is a tendency for both oscillators to arrive at a compromise cycle duration and relatively constant phase relationship. This results in the tendency for $\tau_x$ and $\tau_y$ to fluctuate around each other (Figure 10.20). This stage of the free run is referred to as *phase-trapping*, and is illustrated in the middle panel of Figure 10.19. After the 70th day, a new stage begins, known as *internal desynchronization*. Here, each oscillator tends to track its own intrinsic period, as Figure 10.20 quite dramatically illustrates. The oscillations in $\tau_x$ and $\tau_y$ reflect the fact that each oscillator still exerts some influence on the other and they are not totally independent. During internal desynchronization, the sleep and low-temperature periods can become out-of-phase with one another (Figure 10.19, bottom panel). The sleep–wake cycle duration alternates between shorter periods of ~26 hours and very long periods of ~ 35 hours. Again, these simulated results resemble what has been observed in temporally isolated humans.

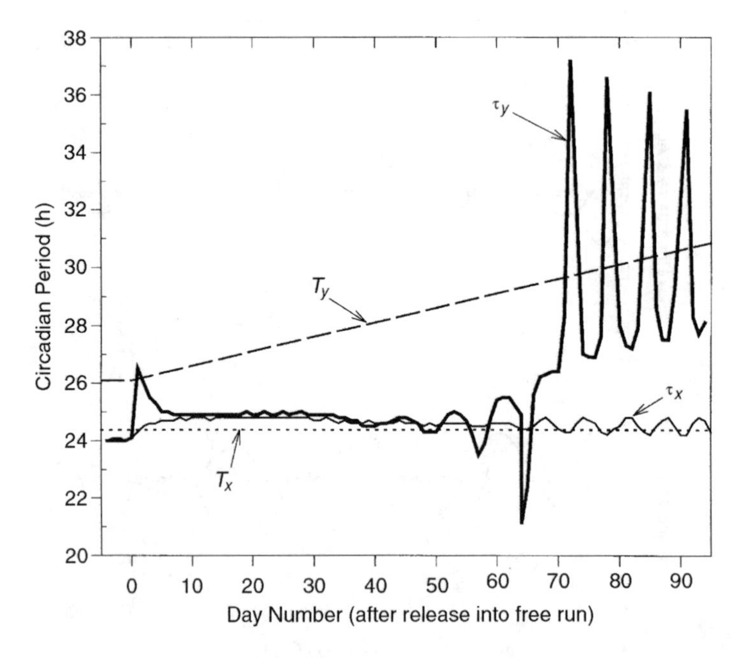


**Figure 10.20** Dynamic changes in the periods of the temperature ($\tau_x$, light tracing) and activity ($\tau_y$, bold tracing) oscillators following release from entrainment by the external 24-hour zeitgeber. The dotted and dashed lines represent the intrinsic periods of the temperature ($T_x$) and activity ($T_y$) oscillators. Both $\tau_x$ and $\tau_y$ assume values between 24 and 25 hours throughout much of the duration of free run. However, after 70 days, there is desynchronization between the two oscillators and $\tau_y$ takes on much larger values.

## 10.4 TIME-VARYING PHYSIOLOGICAL CLOSED-LOOP SYSTEMS: SLEEP APNEA MODEL

In the previous section, we saw how direct input "forcing" from the temperature oscillator affected the dynamics of the sleep–wake (or activity) oscillator, and vice versa. The coupling between related systems is frequently not as direct. For instance, changes in sleep–wake state are known to affect chemoreflex gain, cardiac output, and circulatory delay. These factors are not state variables, but instead comprise the "parameters" in any model of respiratory control. In the previous models of respiratory control that we have discussed (see Sections 3.7, 6.7, and 7.6.2), these parameters were always assumed to take on constant values. A model in which one or more of the key parameters change with time is said to be *time-varying* or *nonstationary*.

In this section, we examine the dynamics that can result from a simple model of obstructive sleep apnea (OSA). A primary mechanism that leads to the obstructive apnea in patients with OSA is the pronounced decrease in tone of the upper airway muscles when sleep sets in. Compounding the effect of this mechanism is the added predisposing factor that the upper airway passage in these patients is already anatomically narrower than in normal individuals. As such, when negative intraluminal pressure is applied to the upper airway during inspiration, the net result is a tendency for the "floppy" airway to collapse, thereby obstructing airflow. There are, of course, many other factors involved, but what we have just described is the basic chain of events that generally occurs during the periodic episodes of sleep apnea. A schematic diagram of the model is shown in Figure 10.21. Figure 10.22a shows the SIMULINK implementation of this model ("`osa.mdl`"). The OSA model is similar to the respiratory control models that we have considered previously, except for three major added features:

1. The first is the addition of the "upper airway conductance" component. In the model, this takes the form of a "gain" that transforms the total respiratory drive into

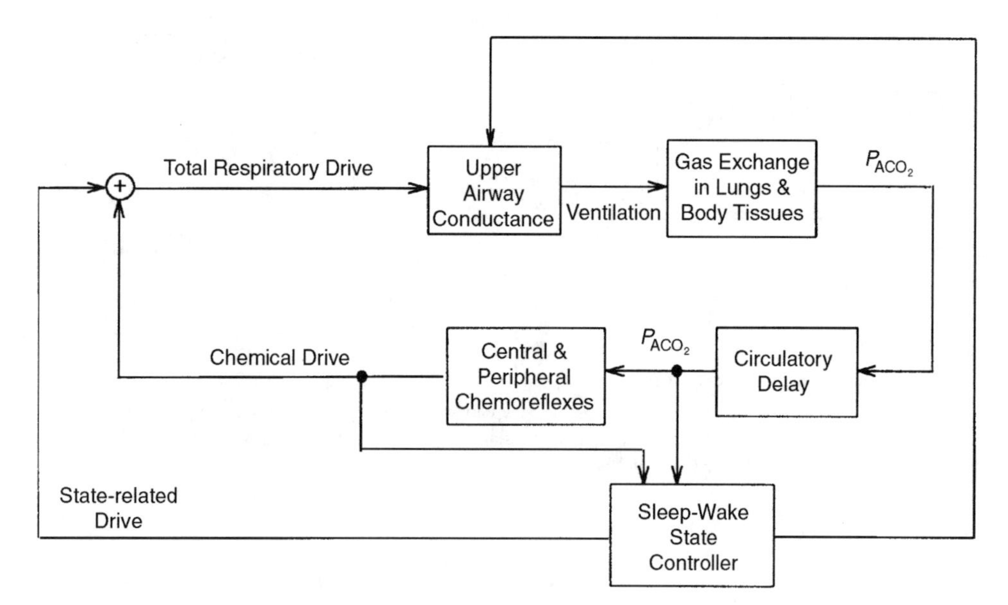


**Figure 10.21** Model of state–chemoreflex interaction in ventilatory control.

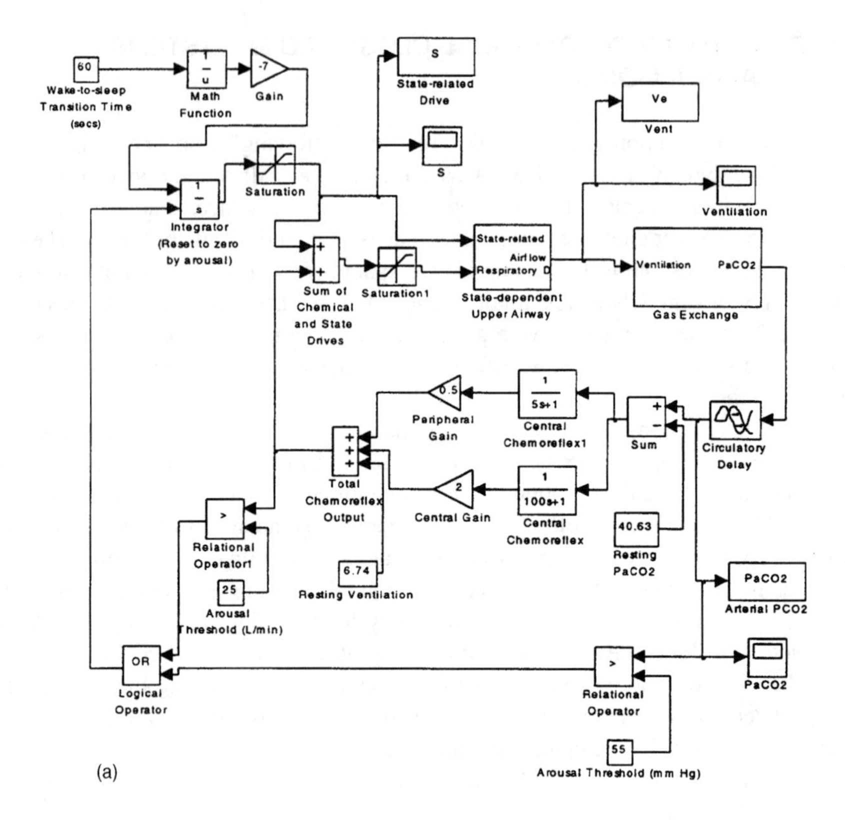


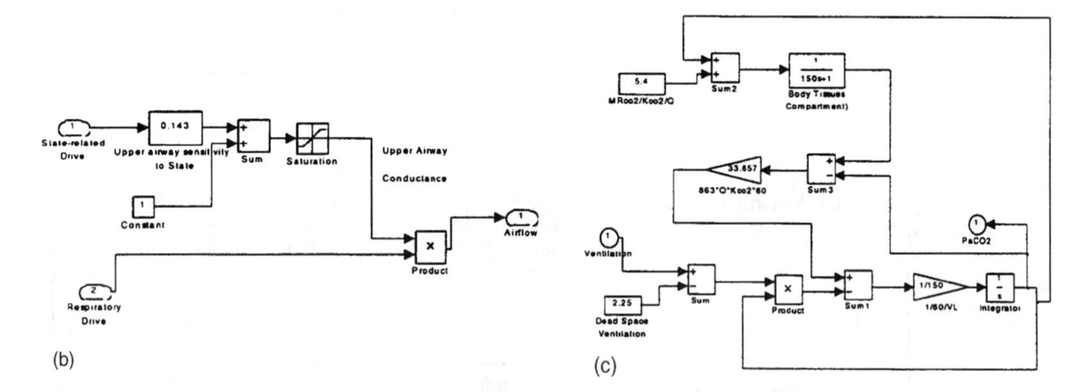


**Figure 10.22** (a) SIMULINK implementation ("`osa.mdl`") of the model of state–chemoreflex interactions in obstructive sleep apnea. (b) Subsystem showing how upper airway conductance is dependent on state. (c) Subsystem that characterizes $CO_2$ exchange in the lungs and body tissues.

ventilation (or airflow). However, this gain is a time-varying parameter. During wakefulness, upper airway conductance is assumed to be equal to unity so that all of the total respiratory drive is converted directly into ventilation. During the transition from wake to sleep, we assume that the upper airway conductance decreases in proportion to the "state-related drive," so that when sleep is attained, conductance becomes zero (i.e., the upper airway is fully obstructed). Thus, the upper airway conductance, as we have defined it, is a normalized quantity that can only assume values between zero and unity. The SIMULINK subsystem block that represents this model component is shown in Figure 10.22b.

2. Aside from the time-varying effect of state changes on upper airway conductance, the model also includes the direct effect of sleep–wake state on respiratory drive. There is much empirical evidence that suggests that a "wakefulness drive" (or "wakefulness stimulus") to breathing that is present during the awake state is withdrawn or inhibited as sleep sets in. This state-related drive is separate from the chemical drive that depends on feedback from the chemoreceptors. Thus, in the model, we have assumed that total respiratory drive consists of the sum of the combined chemoreflex or chemical drive and this state-related drive. During the transition from wake to sleep, we assume a simple linear decrease in this state-related drive, so that it becomes zero when stable sleep has been achieved. As we will demonstrate later, the duration over which this linear decrease (or equivalently, the wake-to-sleep transition) occurs, $\tau$, plays an important role in determining the ventilatory and state dynamics that accompany sleep.
3. The final key component in this model is the "sleep–wake state controller." This controller determines the time-course of the decrease in state-related drive during the transition from wake to sleep. It also receives two forms of feedback from the other parts of the model. First, it monitors the arterial $P_{CO_2}$ ($P_{aCO_2}$). Secondly, it monitors the chemical drive, i.e., the combined output of the central and peripheral chemoreflexes. During sleep or during the transition from wake to sleep, if $P_{aCO_2}$ exceeds 55 mm Hg *or* the chemical drive exceeds 25 L min$^{-1}$, the controller will revert the current state back to wakefulness and restart the transition from wake to sleep. This automatic mechanism simulates, to a first approximation, the arousals that are known to occur when the stimulation of respiratory drive exceeds certain *arousal thresholds*. The arousal mechanism is a potent protective defense against asphyxiation during sleep. However, at the same time, it is the reason why sleep architecture is so severely disrupted in subjects who have OSA.

The equations characterizing the rest of the model are the same as those described in Section 6.7.1, except for the addition of the body tissues compartment in the gas exchange subsystem. $CO_2$ exchange in this compartment is characterized by the differential equation

$$\frac{V_{tis}}{Q}\frac{dP_{vCO_2}}{dt} = (P_{aCO_2} - P_{vCO_2}) + \frac{\dot{V}_{CO_2}}{QK_{CO_2}} \tag{10.7}$$

where $P_{vCO_2}$ is the mixed venous $P_{CO_2}$, $\dot{V}_{CO_2}$ is the metabolic production rate of $CO_2$, $Q$ is the cardiac output and $V_{tis}$ is the effective volume of the body tissues compartment. $K_{CO_2}$ is the slope of the $CO_2$ dissociation curve (approximated as a straight line) for blood. In the simulations, we assume that $V_{tis} = 15$ L, $Q = 0.1$ L s$^{-1}$, $K_{CO_2} = 0.0065$ mm Hg$^{-1}$, and $\dot{V}_{CO_2} = 210$ mL min$^{-1}$. The SIMULINK implementation of the gas exchange subsystem is shown in Figure 10.22c. The values of the other parameters are the same as those employed in Section 6.7.1.

Two simulations with the model showing how sleep onset would affect subsequent ventilatory and state variability are displayed in Figure 10.23. In both cases, the wake-to-sleep transition time, $\tau$, is assumed to be 60 s. The first example (top panel of Figure 10.23) represents a "normal" subject, in which we have assumed upper airway conductance to be unchanged from wake to sleep. (In reality, upper airway resistance increases—or equivalently, upper airway conductance decreases—quite significantly during sleep, even in normals, although not to the point of collapsing the airway.) During the transition from wake to sleep (first 60 s), ventilation ($\dot{V}_E$) decreases and $P_{aCO_2}$ increases as a consequence of the reduction in state-related drive. However, after the transition period, there is some recovery of $P_{aCO_2}$ and $\dot{V}_E$ toward their original equilibrium levels. Finally, the new equilibrium level during sleep is established with $P_{aCO_2}$ about 2 mm Hg higher than during wakefulness and $\dot{V}_E$ about 1 L min$^{-1}$ lower. The main point here is that the wake-to-sleep transition occurs smoothly and without incident. However, in the second simulation representing an OSA subject (lower panel of Figure 10.23), the result is quite different. Here, as the transition from wake to sleep occurs, the rate of decrease in $\dot{V}_E$ takes place more rapidly than in the normal subject. This is due to the fact that, in addition to the decrease in state-related drive, there is also a concomitant decrease in upper airway conductance. $P_{aCO_2}$ also rises at a substantially faster rate. When sleep is established, upper airway conductance becomes zero and, consequently,

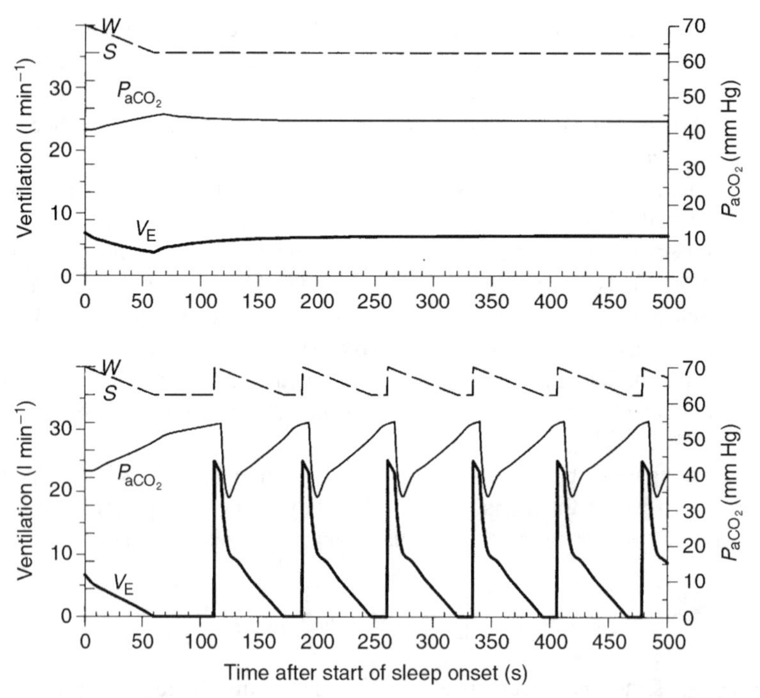


**Figure 10.23** Sample simulations generated by "`osa.mdl`," showing effect of sleep onset on ventilation and $P_{aCO_2}$, as well as subsequent sleep–wake state. In the "normal subject" (top panel), stable levels of ventilation, $P_{aCO_2}$ and sleep are attained. However, in the subject with obstructive sleep apnea (lower panel), there are alternating episodes of upper airway obstruction and arousal-induced hyperpnea.

there is approximately 50 s of obstructive apnea. During this interval, $P_{aCO_2}$ continues to rise until it exceeds the arousal threshold (55 mm Hg). When arousal is triggered, there is an abrupt restoration of the state-related drive as well as a sudden increase in upper airway conductance from zero to unity. This abrupt return to the wake state briefly produces a large increase in $\dot{V}_E$. However, subsequently, a new transition from wake to sleep occurs, leading to obstructive apnea and then another arousal. Hence, a periodic alternation between apnea and hyperpnea, which coincides with the alternation between sleep and wake, occurs. The periodicity of this cyclic behavior is approximately 76 s in duration, which falls in the range of cycle times observed in OSA.

Figure 10.24 shows the effect of altering the wake-to-sleep transition time on the subsequent dynamics of respiration and sleep–wake state. The first example (top panel of Figure 10.24) simulates the case in which $\tau = 20$ s. Although this represents the fast end of the spectrum, it is not entirely unrealistic: EEG measurements in sleep-deprived individuals do show a very rapid change in pattern that reflect fast sleep onset; these changes can occur over time spans as short as a few breaths. The rapid sleep onset leads to a long period of obstructive apnea, during which $P_{aCO_2}$ and chemical drive build up to a point where arousal is triggered. This is followed by a brief hyperpnea and then a subsequent rapid wake-to-sleep transition, and a repetition of the same cycle of events. The period between these apneas (or

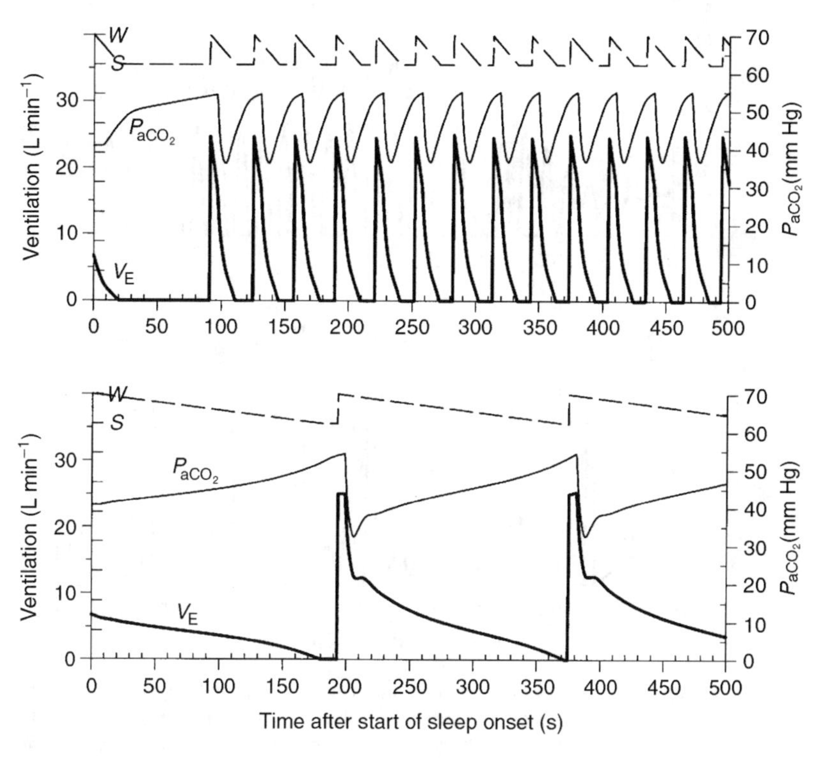


**Figure 10.24** Simulations showing the effect of wake-to-sleep transition time on ventilatory and state stability following sleep onset. Top panel shows results for a rapid wake-to-sleep transition (20 s). Lower panel shows the effects of a slow wake-to-sleep transition (180 s). Although there is periodic ventilation in both cases, the rate of occurrence of arousals is much lower in the case with slow wake-to-sleep transition time.

equivalently, arousals) is 34 s. An alternative means of expressing this result is to convert it into the corresponding *apnea index*, i.e., the number of apneas per hour. In this case, the apnea index is about 106. This is at the high (although not impossible) end of the severity scale for OSA. On the other hand, if $\tau$ is large (i.e., wake-to-sleep transition is very slow), the arousals will be much less frequent and the apneas substantially shorted in duration, as represented in the example in the lower panel of Figure 10.24. Here, $\tau = 180$ s and the corresponding periodicity is about 182 s.

The previous examples assumed central and peripheral chemoreflex sensitivities of 2 and 0.5 L min$^{-1}$ mm Hg$^{-1}$, respectively. Consider what happens when the peripheral chemoreflex gain is increased to 2 L min$^{-1}$ mm Hg$^{-1}$, simulating what would occur if the subject became hypoxic. The top panel of Figure 10.25 shows the result for the simulated "normal" (i.e., upper airway conductance unchanged by sleep onset). Due to the enhanced loop gain, the disturbance produced by the wake-to-sleep transition becomes progressively amplified until periodic breathing results. However, the cycling in $\dot{V}_E$ and $P_{aCO_2}$ is mediated completely by the chemoreflex loops and does not involve the arousal mechanism. As such, a

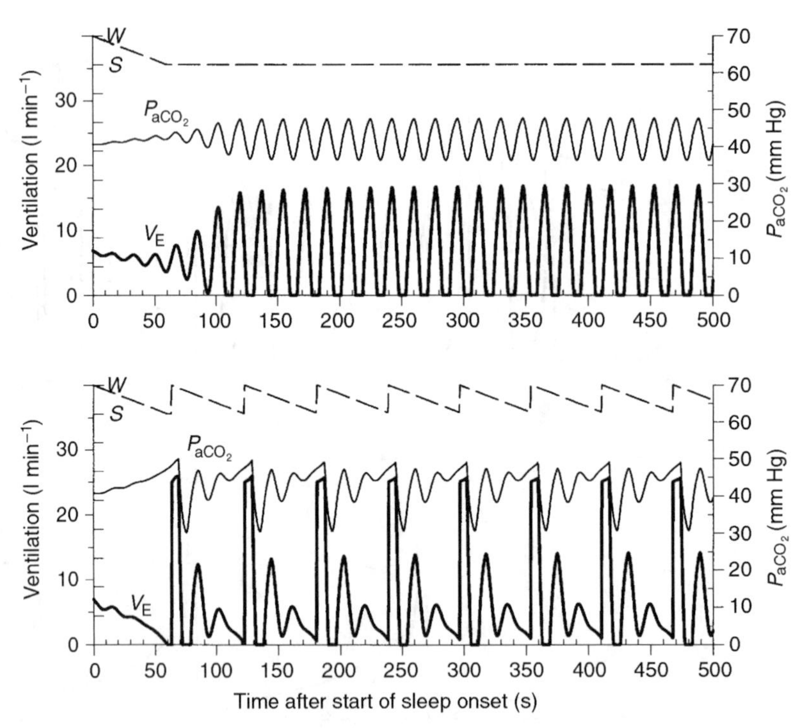


**Figure 10.25** Simulations showing the dynamics of ventilatory and state variability following sleep onset in "subjects" with high chemoreflex gain. The top panel shows the results for a "normal" with no upper airway obstruction: periodic breathing occurs, but the episodes do not elicit periodic arousal. The lower panel shows the corresponding results for the obstructive sleep apnea patient. Periodic arousals that are accompanied by brief durations of strong hyperpnea occur together with chemoreflex-mediated oscillations that do not involve arousal.

stable stage of sleep is attained while the periodic respiration persists. The periodicity in this case is approximately 18 s, similar to the cycle times reported for periodic breathing at altitude. In the corresponding simulation for the OSA subject (lower panel, Figure 10.25), a more complex pattern occurs. Following the initial transition from wake to sleep, there is a brief interval of upper airway obstruction which is punctuated by an arousal. This produces a burst of hyperpnea which drives the $P_{aCO_2}$ down to the point where central apnea (i.e., period of zero respiratory drive) occurs. The central apnea is followed, in turn, by an overshoot in $\dot{V}_E$, a subsequent undershoot and a second (but not so large) overshoot. The key point is that these latter oscillations are mediated by the chemoreflexes and take place while a new wake-to-sleep transition occurs. Thus, in this case, there is a periodic alternation between the large arousal-induced hyperpneas and the chemoreflex-mediated oscillations in ventilation. The arousals are spaced approximately 1 minute apart, while the chemoreflex oscillations have a period of roughly 20 s. This example underscores the kind of interaction that can occur between the arousal and chemoreflex feedback loops as a consequence of the time-varying model parameter representing upper airway conductance.

## 10.5 PROPAGATION OF SYSTEM NOISE IN FEEDBACK LOOPS

We have seen thus far that a variety of factors, such as mixed feedback, time delays, nonlinear coupling, and time-varying properties, can contribute to the complex dynamical behavior of physiological control systems. These factors, however, share a common feature: they are all deterministic characteristics. Random influences are clearly present in physiological systems of all hierarchical levels. We will show through a simple example in this section that one potential source of the spontaneous variability observed in physiological systems may be the responses elicited in these systems by random input perturbations.

We turn once again to the chemoreflex control of respiration model presented in Section 6.7.1. The natural variability in ventilation in wakefulness or sleep is known to be quite substantial, even when the respiratory control system is operating under clearly stable conditions. We propose that some, if not a large part, of this spontaneous variability may be due to the continual perturbation of the chemoreflex dynamics by random noise inputs. Figure 10.26a shows the proposed scheme through which "system noise" might enter the closed-loop structure of the respiratory control system. Since the primary structures that generate the drive to breathe are neural systems, it is reasonable to assume that respiratory drive, which is ultimately converted into ventilation, consists of the chemoreflex-mediated chemical drive plus some random influences that represent neural noise. Another obvious source of "noise" is the gas exchange process itself. Regional inhomogeneities in ventilation and perfusion of the lungs, as well as temporal fluctuations in cardiac output and the circulatory delays, can give rise to noise that contaminates the time-evolution of alveolar $P_{CO_2}$ ($P_{ACO_2}$) and consequently, $P_{aCO_2}$. These considerations have prompted us to select the sites shown in Figure 10.26a as the points in the closed-loop structure at which noise enters the system. The SIMULINK implementation of this model ("`noisycls.mdl`") is displayed in Figure 10.26b. One detail that is of special importance in this example relates to the method by which "breath-by-breath" values of ventilation and $P_{aCO_2}$ are simulated, since the underlying equations assume a continuous-time process. To generate "breaths," the total respiratory output (chemoreflex drive plus "neural noise") is sent through a "`zero-order hold`" block, which samples the continuous respiratory drive waveform every 4 s and holds each sampled value constant for the following 4 s. Then, ventilation, $P_{aCO_2}$, and the noise

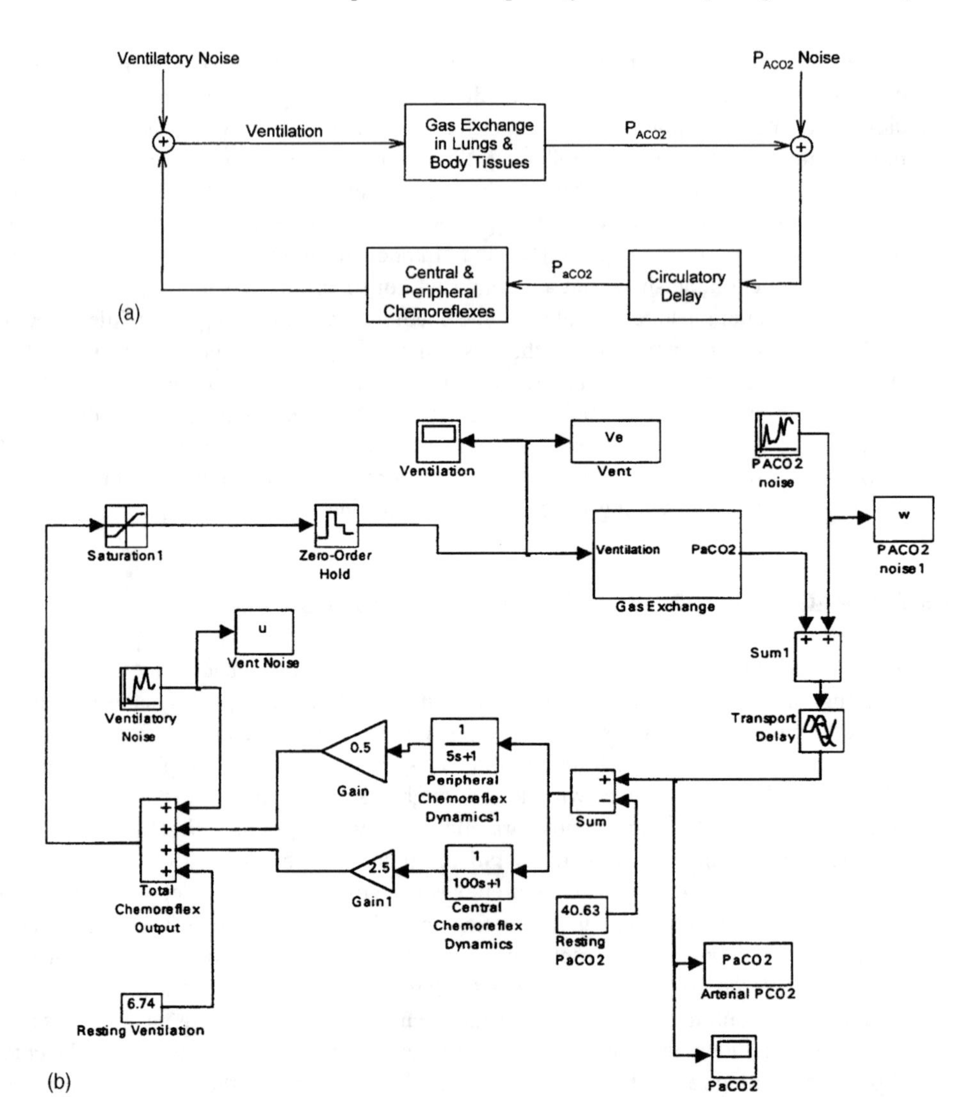


**Figure 10.26** (a) Schematic diagram showing the "sites" at which noise might enter the closed-loop respiratory control system. (b) SIMULINK implementation ("`noisycls.mdl`") of the closed-loop respiratory control model with noise inputs.

processes are sampled at 4 s intervals and saved to the MATLAB workspace for further analysis.

Figure 10.27 shows samples of the spontaneously varying waveforms in ventilation (top panel) and $P_{aCO_2}$ (lower panel) generated by the model. The coefficient of variation (standard deviation divided by mean value) of the simulated ventilation time-course is approximately 12%, which falls in the range commonly observed in resting humans. The peripheral and central chemoreflex gains are assigned values of 0.5 and 2 L min$^{-1}$ mm Hg$^{-1}$, respectively.

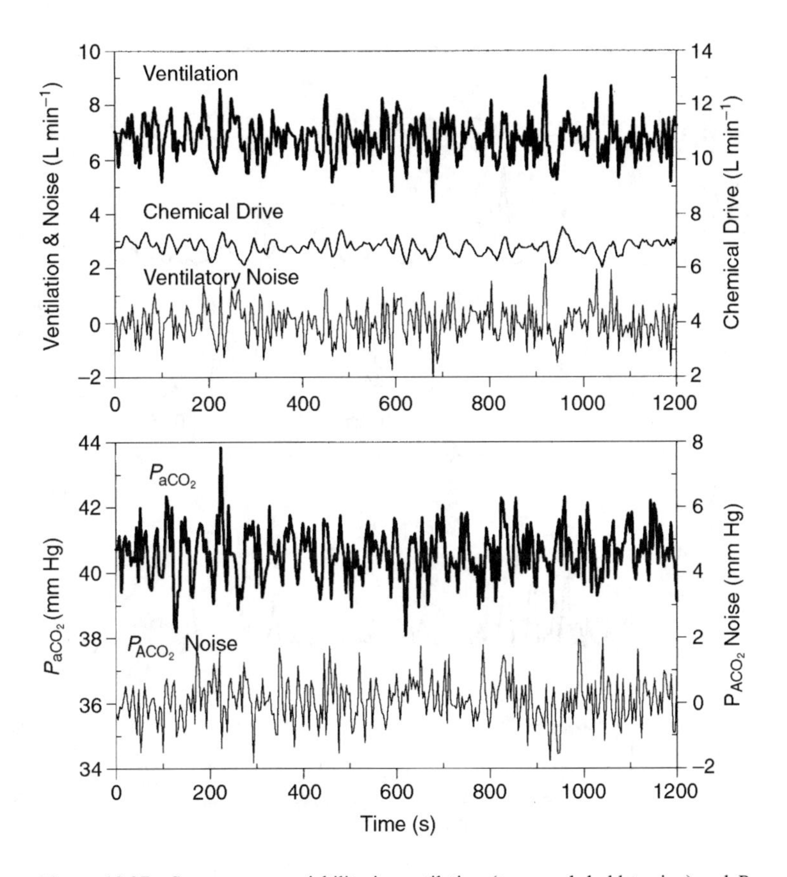


**Figure 10.27** Spontaneous variability in ventilation (top panel, bold tracing) and $P_{aCO_2}$ (lower panel, bold tracing) resulting from the propagation of noise (both panels, light tracings) in the chemoreflex loops of the respiratory control model. "Chemical drive" in the top panel refers to the combined output of the central and peripheral chemoreflexes in response to the "noise" $P_{aCO_2}$.

One may note the strong similarity between the ventilation time-course and the ventilatory noise waveform, since the former is simply the sum of the latter and the combined central and peripheral chemoreflex drives. Ventilation is causally related to current and past values of the ventilatory noise input because of the propagation of the noise around the chemoreflex loops; however, the ventilatory noise is not causally related to past values of ventilation. The waveform labelled "chemical drive" represents the combined response of the chemoreflexes to the variations in $P_{aCO_2}$ elicited in part by the past spontaneous fluctuations in ventilation as well as by the noise entering the system through the gas exchange process ("$P_{ACO_2}$ noise"). The coefficient of variation of the $P_{aCO_2}$ waveform in the lower panel of Figure 10.27 is approximately 2%.

The corresponding spectra of these fluctuations in ventilation and $P_{aCO_2}$ show enhanced power in the 0.01 to 0.05 Hz range (bold tracings in Figure 10.28). These contrast with the much flatter (broad-band) spectra of the noise inputs (light tracings in Figure 10.28). Thus, the propagation of these noise inputs through the feedback loops of the closed-loop system

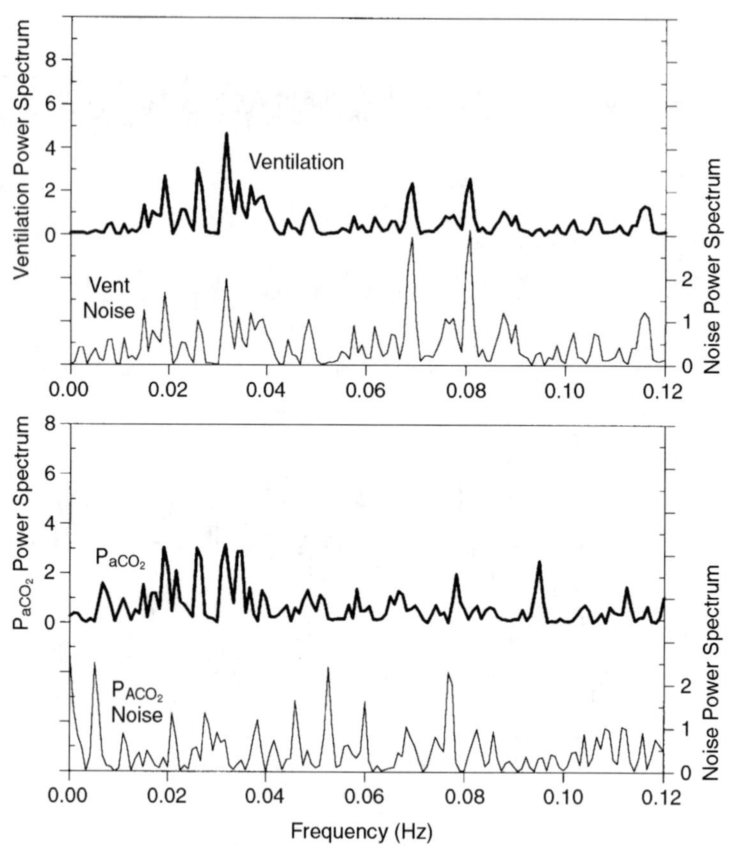


**Figure 10.28** Power spectra of the spontaneous fluctuations in ventilation (top panel, bold tracing) and $P_{aCO2}$ (lower panel, bold tracing) shown in Figure 10.27. Note the increased power in the 0.01 to 0.05 Hz range. This contrasts with the much flatter (broad-band) spectra of the random noise inputs (light tracings).

gives rise to certain oscillations that appear within a bandwidth that is consistent with the dynamic characteristics (i.e., gains, component lags and delays) of the system. In this case, the bandwidth of frequencies 0.01 to 0.05 Hz corresponds to periodicities in the range 20 to 100 s. These are compatible with the oscillation cycle durations that have been observed and reflect the frequency range for "resonance" in the human respiratory control system. When the peripheral chemoreflex gain is increased by 140% to 1.2 L min$^{-1}$ mm Hg$^{-1}$, the resulting simulation clearly shows enhanced oscillatory activity in both ventilation and $P_{aCO_2}$ (Figure 10.29). Furthermore, bursts of oscillations in ventilation and $P_{aCO_2}$ occur somewhat randomly, giving the appearance that the underlying system contains time variations in loop gain although, in fact, the model parameters have been assigned constant values. The increased oscillatory activity is clearly evident in the corresponding power spectra shown in Figure 10.30 (for comparison, see Figure 10.28). However, the spectral composition of the

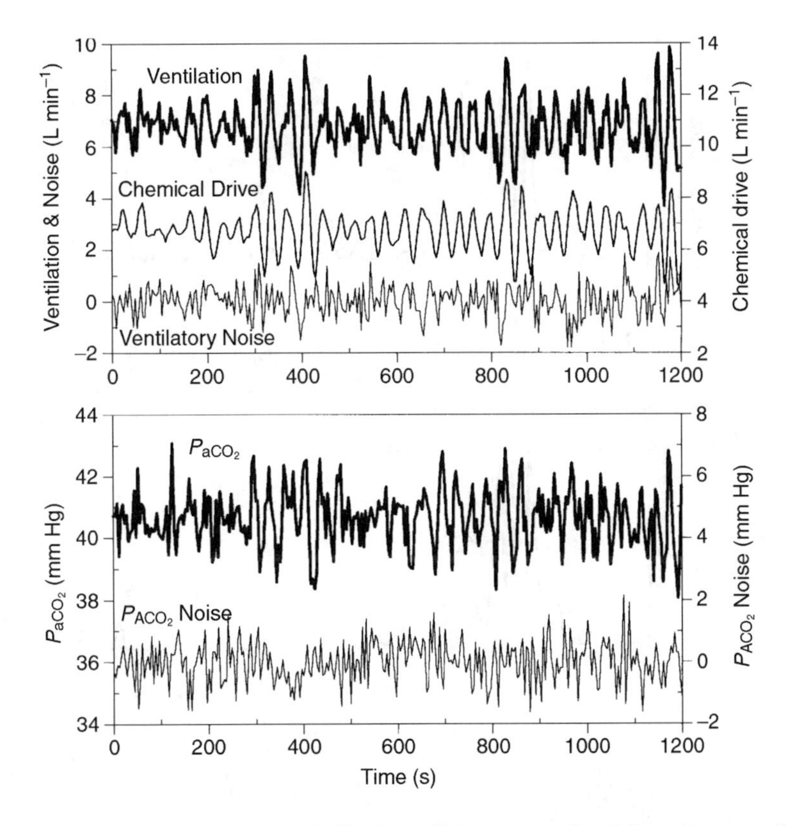


**Figure 10.29** Spontaneous variability in ventilation (top panel) and $P_{aCO2}$ (lower panel) produced by the propagation of random noise through the chemoreflex loops. In this case, peripheral chemoreflex gain has been increased to 1.5 times its nominal value. Notice the much more oscillatory pattern and the appearance of "bursts" of oscillations.

fluctuations in both ventilation and $P_{aCO_2}$ here is concentrated primarily in a narrower band of frequencies that ranges from ~0.02 to 0.04 Hz.

The example presented above demonstrates that the propagation of random influences through the feedback loops of a closed-loop control system can give rise to temporally correlated fluctuations in the system variables. These fluctuations take the form of oscillations that have frequencies consistent with the stability properties of the closed-loop system in question. "Bursts" of oscillations, which represent somewhat lower frequency phenomena, can also appear. In this way, an intrinsically stable closed-loop system can appear to be intermittently unstable as these oscillatory bursts take place.

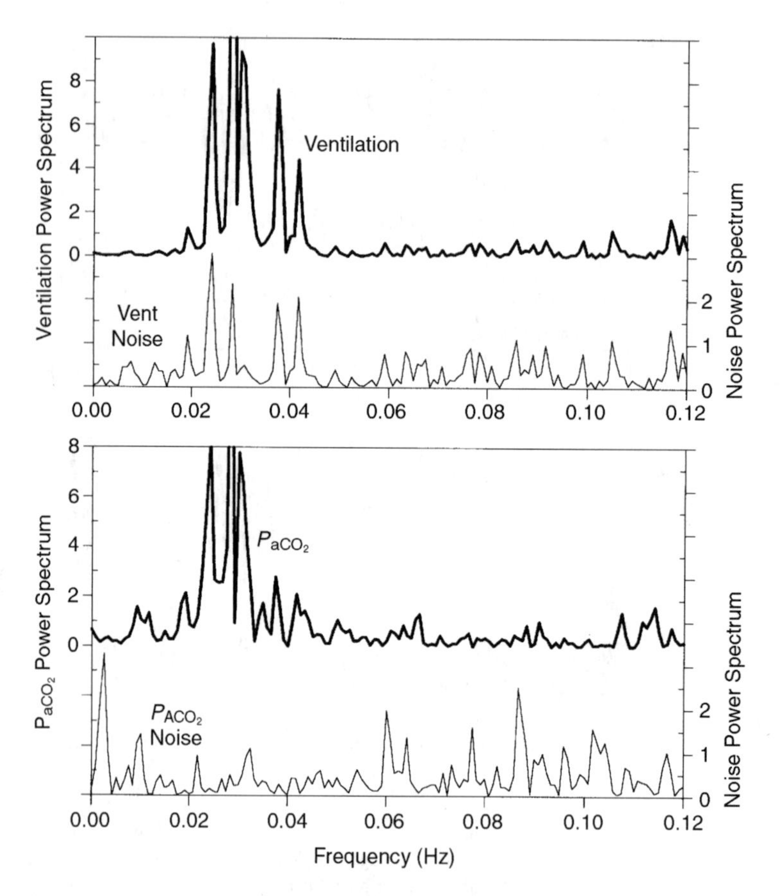


**Figure 10.30** Power spectra of the fluctuations in ventilation (top panel) and $P_{aCO_2}$ (lower panel) when peripheral chemoreflex gain has been increased by 140% of its nominal value. Note the substantially increased power in the 0.02 to 0.04 Hz range (compared to the original condition in Figure 10.28) reflecting the amplification of oscillatory activity visible in Figure 10.29.

## BIBLIOGRAPHY


Bassingthwaighte, J.B., L.S. Liebovitch, and B.J. West. *Fractal Physiology*. Oxford University Press, New York, 1994.

Cavalcanti, S., and E. Belardinelli. Modeling of cardiovascular variability using a differential delay equation. *IEEE Trans. Biomed. Eng.* **43**: 982–989, 1996.

Feder, J. *Fractals*. Plenum Press, New York, 1988.

Feigenbaum, M.J. Universal behavior in nonlinear systems. *Los Alamos Sci.* **1**: 4–27, 1980.

Glass, L., A. Beuter, and D. Larocque. Time delays, oscillations and chaos in physiological control systems. *Math. Biosci.* **90**: 111–125, 1988.

Glass, L., and M.C. Mackey. Pathological conditions resulting from instabilities in physiological control systems. *Ann. N.Y. Acad. Sci.* **316**: 214–235, 1979.

Glass, L., and C.P. Malta. Chaos in multi-looped negative feedback systems. *J. Theor. Biol.* **145**: 217–223, 1990.

Khoo, M.C.K. Modeling the effect of sleep state on respiratory stability. In: *Modeling and Parameter Estimation in Respiratory Control* (edited by M.C.K. Khoo) 1989, pp. 193–204.

Khoo, M.C.K., A. Gottschalk, and A.I. Pack. Sleep-induced periodic breathing and apnea: a theoretical study. *J. Appl. Physiol.* **70**: 2014–2024, 1991.

Kronauer, R.E., C.A. Czeisler, S.F. Pilato, M.C. Moore-Ede, and E.D. Weitzman. Mathematical model of the human circadian system with two interacting oscillators. *Am. J. Physiol.* **242** (*Regulatory Integrative Comp. Physiol.* **11**): R3–R17, 1982.

Mackey, M., and L. Glass. Oscillation and chaos in physiological control systems. *Science* **197**: 287–289, 1977.

May, R.M. Simple mathematical models with very complicated dynamics. *Nature* **261**: 459–467, 1976.

Modarreszadeh, M., and E.N. Bruce. Ventilatory variability induced by spontaneous variations of $P_{aCO_2}$ in humans. *J. Appl. Physiol.* **76**: 2765–2775, 1994.

Moon, F.C. *Chaotic Vibrations: An Introduction for Applied Scientists and Engineers*. Wiley, New York, 1987.

Thompson, J.M.T., and H.B. Stewart. *Nonlinear Dynamics and Chaos*. Wiley, New York, 1986.

West, B.J. *Fractal Physiology and Chaos in Medicine*. World Scientific, Singapore, 1990.

Winfree, A.T. Circadian timing of sleepiness in man and woman. *Am. J. Physiol.* **243** (*Regulatory Integrative Comp. Physiol.* **12**): R193–R204, 1982.

## PROBLEMS

**P10.1.** Develop a MATLAB or SIMULINK program to simulate the "logistic equation" given by

$$x_{n+1} = \alpha(1 - x_n)x_n$$

Use this program to generate time-series for different values of the parameter $\alpha$ that range from 2.9 to 4 in increments of 0.01. Demonstrate that, prior to exhibiting chaotic behavior, this nonlinear dynamical system undergoes several stages of period-doubling. Display the magnitude spectrum of the fast Fourier transforms of each time series to determine how the system dynamical behavior changes in the frequency domain.

**P10.2.** Explore the changes in behavior of the neutrophil regulation model ("`hematop.mdl`") as the shape of the neutrophil production rate function changes with the parameter $n$ increasing from 5 to 20 (see Equation (10.2)). Present phase-plane plots (i.e., $dx/dt$ vs. $x$) of the dynamics for each value of $n$ employed. For your simulations, use the following values for the other parameters: $\beta = 2,\ \theta = 1,\ \gamma = 1$, and $T_d = 2$.

**P10.3.** Determine how the dynamics of the model of cardiovascular variability ("`cvvar2.mdl`") would be affected by conditions that simulate (a) vagal blockade and (b) $\beta$-adrenergic blockade. Simulate vagal blockade by adding a low-pass filter of unit gain and time constant of 10 s to the feedback loop for cardiac period ($T$). To simulate $\beta$-adrenergic blockade, employ a low-pass filter of unit gain and time constant of 0.8 s. In each case, determine how the phase-space plots change as the delay $\tau$ is increased in increments of 0.1 s from 0.5 to 2.5 s. In each case, does the model still exhibit chaotic behavior for certain ranges of $\tau$?

**P10.4.** Use the Kronauer model of circadian oscillators ("`circad.mdl`") to simulate the effect of "jet travel" on the temperature and sleep–wake cycles. First entrain the model to the 24-hour light–dark cycle until a steady state has been attained. Then, "expose" the model to a 24-hour period of light (to simulate a 12-hour flight in continuous daylight from Los Angeles to Tokyo) before it returns to the regular 24-hour light–dark cycle. Determine how this "disturbance" affects the time of sleep onset, the duration of sleep, and the phase relation between the activity and temperature oscillations in the few days that follow.

**P10.5.** Using the model of obstructive sleep apnea ("`osa.mdl`"), determine the dynamics of ventilation and sleep state following sleep onset in a patient whose symptoms have been improved by treatment. Modify the model so that upper airway conductance does not decrease to zero but only falls to one-quarter of its waking value during sleep. Assume a range of wake-to-sleep transiiton times from 20 s to 120 s. What are the cycle durations of the periodic breathing episodes in each case? How would oxygen administration (simulated by eliminating the peripheral chemoreflex gain) affect the ventilation-state dynamics in such a patient?

**P10.6** Modify the model of chemoreflex ventilatory control ("`noisycls.mdl`") to determine how random fluctuations in cardiac output ($Q$) might lead to spontaneous variations in ventilation and $P_{aCO_2}$. Assume a nominal value of 0.1 L $s^{-1}$ for $Q$ and a coefficient of variation of 3% for its random fluctuations. Determine the coefficients of variation for the resulting fluctuations in ventilation and $P_{aCO_2}$, as well as the corresponding spectra.

# I

# Commonly Used Laplace Transform Pairs

| Waveform, $x(t)$ | Laplace Transform, $X(s)$ |
|---|---|
| Unit impulse, $\delta(t)$ | $1$ |
| Unit step, $u(t)$ | $\frac{1}{s}$ |
| $e^{-at}u(t)$ | $\frac{1}{s+a}$ |
| $\frac{t^k}{k!}e^{-at}u(t)$ | $\frac{1}{(s+a)^{k+1}}$ |
| $tu(t)$ | $\frac{1}{s^2}$ |
| $\frac{t^k}{k!}u(t)$ | $\frac{1}{s^{k+1}}$ |
| $\sin(\omega t)u(t)$ | $\frac{\omega}{s^2+\omega^2}$ |
| $\cos(\omega t)u(t)$ | $\frac{s}{s^2+\omega^2}$ |
| $e^{-at}\sin(\omega t)u(t)$ | $\frac{\omega}{(s+a)^2+\omega^2}$ |
| $e^{-at}\cos(\omega t)u(t)$ | $\frac{s+a}{(s+a)^2+\omega^2}$ |
| $e^{-at}\cos(\omega t+\theta)u(t)$ | $\frac{(s+a)\cos(\theta)-\omega\sin(\theta)}{(s+a)^2+\omega^2}$ |
| $t\sin(\omega t)u(t)$ | $\frac{2\omega s}{(s^2+\omega^2)^2}$ |
| $t\cos(\omega t)u(t)$ | $\frac{s^2-\omega^2}{(s^2+\omega^2)^2}$ |

# List of MATLAB and SIMULINK Programs/ Functions

## A MATLAB FUNCTIONS USED

| Function | Purpose/Description |
|---|---|
| abs | Magnitude or absolute value of a complex number |
| angle | Phase angle, in radians, of a complex number |
| bode | Magnitude and phase of frequency response (Bode plot) |
| csd | Cross-spectral power density function of two signals |
| exp | Exponential function |
| fmins | Minimizes a given criterion function with respect to several parameters |
| freqresp | Frequency response of a linear system over specified frequency range |
| gensig | Generates periodic signals (sine, square, pulse) |
| grid | Controls appearance or removal of grid lines in a given plot |
| imag | Imaginary part of a complex number |
| impulse | Simulates impulse response of a given linear, time-invariant system |
| input | Prompt for user input |
| lsim | Simulates response of a given linear system to arbitrary inputs |
| margin | Returns gain margin and phase margin of a given linear system |
| nichols | Produces Nichols chart of given linear system |
| nyquist | Produces Nyquist diagram of given linear system |
| ones | Matrix or vector of ones |
| plot | Produces 2-D plot of given vectors |
| psd | Power spectral density of given signal |
| real | Real part of complex number |
| squeeze | Eliminate singleton dimensions of multidimensional arrays |
| ss | Create state-space formulation of a linear, time-invariant model |
| step | Computes and plots step response of a given linear system |
| subplot | Produces multiple graphs in one screen |
| sum | Sums all elements of a given vector |
| tf | Specify a given linear system in terms of its transfer function |
| tf2ss | Converts transfer function format to state-space description |
| xlabel | Labels the $x$-axis of a plot |
| ylabel | Labels the $y$-axis of a plot |

## B SIMULINK FUNCTION BLOCKS USED

| Function | Description |
|---|---|
| bandlimited white noise | Generates bandlimited white noise of a specified power spectrum |
| chirp signal | Generates sinusoidal signal of continuously increasing frequency |
| constant | Generates a constant value |
| derivative | Computes rate of change of the input signal |
| gain | Multiplies input signal by the specified constant factor |
| hit crossing | Determines whether the input signal attains a given threshold level |
| integrator | Computes time integral of the input signal |
| logical operator | Applies logical operation (e.g., AND or OR) to two or more inputs |
| LTI system | Simulates linear time-invariant system in a variety of formats |
| Matlab fcn | Applies operation specified by MATLAB function to the input |
| memory | Retains value of input from previous integration step |
| mux | Produces a vector output signal from two or more scalar inputs |
| product | Element-by-element product of input vectors |
| pulse generator | Generates rectangular wave of specified parameters |
| relational operator | Implements various relational operations (e.g., <, >, =) |
| repeating sequence | Generates arbitrary periodic signal specified by table of values |
| saturation | Limits output of block to a specified lower and/or upper level |
| scope | Displays time-plot of given scalar or vector input signals |
| slider gain | Allows adjustment of gain during simulation |
| spectrum analyzer | Displays frequency response of a given linear system |
| stop | Stops simulation when the input signal is nonzero |
| subsystem | Encapsulates a network of SIMULINK blocks into one block |
| sum | Produces algebraic sum of all inputs entering the block |
| to file | Saves the input signal into a MATLAB data file (`*.mat`) |
| to workspace | Saves the input signal into a MATLAB array or vector in the workspace |
| transfer fcn | Produces output of a specified transfer function for given input signal |
| transport delay | Simulates the specified (constant) time delay |
| x-y graph | Plots one input signal against another input signal |

## C LIBRARY OF MATLAB SCRIPT FILES

| Function | Description |
|---|---|
| `acs_CO2.m` | Simulates adaptive buffering of spontaneous fluctuations in ventilation |
| `df_resp.m` | Stability analysis of respiratory control model using describing function |
| `fda_llm.m` | Frequency-domain analysis of linearized lung mechanics model |
| `fn_gmm.m` | Computes goodness of fit between glucose–insulin model and data |
| `fn_llm.m` | Computes criterion function for linearized lung mechanics model |
| `fn_rlc.m` | Predicts output of *R–L–C* model for given input |
| `gireg.m` | Computer transfer functions from simulated outputs of "`glucose.mdl`" |
| `gmm_est.m` | Estimates parameters of glucose–insulin minimal model |
| `gpmargin.m` | Computes gain and phase margins of given linear system |
| `nmr_var.m` | Assigns values to parameters of "`msreflex.mdl`" prior to simulation |
| `nyq_resp.m` | Nyquist stability analysis of respiratory control model |
| `optLVEF.m` | Minimization of left ventricular power dissipation |
| `popt_llm.m` | Estimation of lung mechanics model using optimization technique |
| `prbs.m` | Generates pseudorandom binary sequence |
| `pupil.m` | Nyquist stability analysis of pupillary reflex model |

| | |
|---|---|
| rcs_est.m | Estimation of respiratory control model parameters |
| rcssim.m | Simulation of ventilatory control during random inhaled $CO_2$ forcing |
| rsa_tf.m | Computes transfer functions from simulated outputs of "rsa.mdl" |
| rsa_var.m | Assigns values to parameters of RSA model prior to simulation |
| sensanl.m | Performs sensitivity analysis for given model |
| sss_llm.m | Simulation using state-space formulation of lung mechanics model |
| sysid_ls.m | Parameter estimation using least squares technique |
| tra_llm.m | Transient response analysis of linearized lung mechanics model |

## D LIBRARY OF SIMULINK MODEL FILES

| Function | Description |
|---|---|
| bvpmod.mdl | Simulation of Bonhoeffer–van der Pol model |
| circad.mdl | Simulation of Kronauer circadian rhythms model |
| cvvar1.mdl | Simulation of cardiovascular variability (stroke volume constant) |
| cvvar2.mdl | Simulation of cardiovascular variability (stroke volume variable) |
| fdallm2.mdl | Computes frequency response of linearized lung mechanics model |
| glucose.mdl | Simulation of glucose–insulin regulation (Stolwijk and Hardy model) |
| gmm_sim.mdl | Simulation of minimal model of glucose–insulin dynamics (Bergman) |
| hematop.mdl | Simulation of neutrophil density regulation |
| linhhmod.mdl | Simplified and linearized version of Hodgkin–Huxley model |
| msrflx.mdl | Simulation of steady state neuromuscular reflex |
| nmreflex.mdl | Simulation of neuromuscular reflex model |
| noisycls.mdl | Simulation of spontaneous variability in control of ventilation |
| osa.mdl | Simulation of ventilatory instability in obstructive sleep apnea |
| poincare.mdl | Simulation of cardiac dysrhythmias using Poincaré oscillator |
| respm1.mdl | Simulation of patient–ventilator system |
| respm2.mdl | Simulation of patient–ventilator system (alternative model) |
| respss.mdl | Simulation of steady-state respiratory control |
| rsa.mdl | Simulation of respiratory sinus arrhythmia (Saul model) |
| vdpmod.mdl | Simulation of the van der Pol oscillator |

## E DOWNLOADING THE MATLAB SCRIPT AND SIMULINK MODEL FILES

The files listed in Tables C and D above can be downloaded, along with some accompanying data files (*.mat), from the following FTP site of IEEE Press:

ftp : //ftp.ieee.org/uploads/press/khoo

The MATLAB script files are compatible with MATLAB Version 5.1 or higher. The SIMULINK model files are compatible with SIMULINK Version 2.1 or higher.

Comments, corrections of errors and suggestions for future improvement of the book and associated library of MATLAB/SIMULINK programs may be sent to the author at the following email address:

khoo@mizar.usc.edu

# INDEX

## M (*continued*)



## N



## O



## P

## Q



## R



## S



## T



## V



## W

# About the Author

**Michael C. K. Khoo** received the B.Sc.(Eng.) Degree in Mechanical Engineering from Imperial College of Science and Technology, University of London, in 1976, and the S.M. and Ph.D. degrees in Bioengineering from Harvard University, Cambridge, MA, in 1977 and 1981, respectively.

From 1981 to 1983, Dr. Khoo was a research associate at the V.A. Hospital, West Roxbury, MA, and the Brigham and Women's Hospital, Boston, MA. Since 1983, he has been on the faculty of the University of Southern California, Los Angeles, where he is currently a professor of Biomedical Engineering. His current research interests include respiratory and cardiac autonomic control during sleep, biomedical signal processing, and physiological modeling.

Dr. Khoo has been the recipient of a National Institutes of Health Research Career Investigator Award and an American Lung Association Career Investigator Award. He has edited two research volumes: *Bioengineering Approaches to Pulmonary Physiology and Medicine* (New York: Plenum, 1996) and *Modeling and Parameter Estimation in Respiratory Control* (New York: Plenum, 1989), in addition to over 85 journal articles, book chapters and conference papers. He is a member of the Biomedical Engineering Society, IEEE, American Physiological Society, American Heart Association Cardiopulmonary Council, and American Sleep Disorders Association.

# About the Authors